Surface Modification Technology of Biomedical Metals

Surface Modification Technology of Biomedical Metals

Editors

Liqiang Wang
Lechun Xie
Faramarz Djavanroodi

Basel • Beijing • Wuhan • Barcelona • Belgrade • Novi Sad • Cluj • Manchester

Editors

Liqiang Wang
School of Material Science
and Engineering
Shanghai Jiao Tong
University
Shanghai, China

Lechun Xie
Hubei Key Laboratory of
Advanced Technology for
Automotive Components
Wuhan University
of Technology
Wuhan, China

Faramarz Djavanroodi
Department of Mechanical
Engineering, College
of Engineering
Prince Mohammad Bin
Fahd University
Al Khobar, Saudi Arabia

Editorial Office
MDPI
St. Alban-Anlage 66
4052 Basel, Switzerland

This is a reprint of articles from the Special Issue published online in the open access journal *Coatings* (ISSN 2079-6412) (available at: https://www.mdpi.com/journal/coatings/special_issues/biomed_met).

For citation purposes, cite each article independently as indicated on the article page online and as indicated below:

Lastname, A.A.; Lastname, B.B. Article Title. *Journal Name* **Year**, *Volume Number*, Page Range.

ISBN 978-3-0365-9610-5 (Hbk)
ISBN 978-3-0365-9611-2 (PDF)
doi.org/10.3390/books978-3-0365-9611-2

© 2023 by the authors. Articles in this book are Open Access and distributed under the Creative Commons Attribution (CC BY) license. The book as a whole is distributed by MDPI under the terms and conditions of the Creative Commons Attribution-NonCommercial-NoDerivs (CC BY-NC-ND) license.

Contents

Jie Li, Peng Zhou, Shokouh Attarilar and Hongyuan Shi
Innovative Surface Modification Procedures to Achieve Micro/Nano-Graded Ti-Based Biomedical Alloys and Implants
Reprinted from: *Coatings* **2021**, *11*, 647, doi:10.3390/coatings11060647 . . . . **1**

Yingjing Fang, Shokouh Attarilar, Zhi Yang, Guijiang Wei, Yuanfei Fu and Liqiang Wang
Toward Bactericidal Enhancement of Additively Manufactured Titanium Implants
Reprinted from: *Coatings* **2021**, *11*, 668, doi:10.3390/coatings11060668 . . . . **33**

Longfei Shao, Yiheng Du, Kun Dai, Hong Wu, Qingge Wang, Jia Liu, et al.
β-Ti Alloys for Orthopedic and Dental Applications: A Review of Progress on Improvement of Properties through Surface Modification
Reprinted from: *Coatings* **2021**, *11*, 1446, doi:10.3390/coatings11121446 . . . . **59**

Hongfei Sun, Jiuxiao Li, Mingliang Liu, Dongye Yang and Fangjie Li
A Review of Effects of Femtosecond Laser Parameters on Metal Surface Properties
Reprinted from: *Coatings* **2022**, *12*, 1596, doi:10.3390/coatings12101596 . . . . **79**

Zexin Wang, Fei Ye, Liangyu Chen, Weigang Lv, Zhengyi Zhang, Qianhao Zang, et al.
Preparation and Degradation Characteristics of MAO/APS Composite Bio-Coating in Simulated Body Fluid
Reprinted from: *Coatings* **2021**, *11*, 667, doi:10.3390/coatings11060667 . . . . **97**

Boda Liu, Zixin Deng and Defu Liu
Preparation and Properties of Multilayer Ca/P Bio-Ceramic Coating by Laser Cladding
Reprinted from: *Coatings* **2021**, *11*, 891, doi:10.3390/coatings11080891 . . . . **117**

Reghuraj Aruvathottil Rajan, Kaiprappady Kunchu Saju and Ritwik Aravindakshan
Characterization and In Vitro Studies of Low Reflective Magnetite (Fe_3O_4) Thin Film on Stainless Steel 420A Developed by Chemical Method
Reprinted from: *Coatings* **2021**, *11*, 1145, doi:10.3390/coatings11091145 . . . . **135**

Zhaomei Wan, Jiuxiao Li, Dongye Yang and Shuluo Hou
Microstructural and Mechanical Properties Characterization of Graphene Oxide-Reinforced Ti-Matrix Composites
Reprinted from: *Coatings* **2022**, *12*, 120, doi:10.3390/coatings12020120 . . . . **149**

Valentina Vadimovna Chebodaeva, Nikita Andreevich Luginin, Anastasiya Evgenievna Rezvanova, Natalya Valentinovna Svarovskaya, Konstantin Vladimirovich Suliz, Ludmila Yurevna Ivanova, et al.
Formation of Bioresorbable Fe-Cu-Hydroxyapatite Composite by 3D Printing
Reprinted from: *Coatings* **2023**, *13*, 803, doi:10.3390/coatings13040803 . . . . **161**

Diana-Petronela Burduhos-Nergis, Nicanor Cimpoesu, Elena-Luiza Epure, Bogdan Istrate, Dumitru-Doru Burduhos-Nergis and Costica Bejinariu
Ca–Zn Phosphate Conversion Coatings Deposited on Ti6Al4V for Medical Applications
Reprinted from: *Coatings* **2023**, *13*, 1029, doi:10.3390/coatings13061029 . . . . **175**

Shaotian Zhang, Dafu Wei, Xiang Xu and Yong Guan
Transparent, High-Strength, and Antimicrobial Polyvinyl Alcohol/Boric Acid/Poly Hexamethylene Guanidine Hydrochloride Films
Reprinted from: *Coatings* **2023**, *13*, 1115, doi:10.3390/coatings13061115 . . . . **189**

Bingyu Xie and Kai Gao
Research Progress of Surface Treatment Technologies on Titanium Alloys: A Mini Review
Reprinted from: *Coatings* **2023**, *13*, 1486, doi:10.3390/coatings13091486 . . . . **203**

Review

Innovative Surface Modification Procedures to Achieve Micro/Nano-Graded Ti-Based Biomedical Alloys and Implants

Jie Li [1], Peng Zhou [1,*], Shokouh Attarilar [2,*] and Hongyuan Shi [1]

1. School of Aeronautical Materials Engineering, Xi'an Aeronautical Polytechnic Institute, Xi'an 710089, China; 18792771967@163.com (J.L.); shy2008x@163.com (H.S.)
2. State Key Laboratory of Metal Matrix Composites, Shanghai Jiao Tong University, Shanghai 200240, China
* Correspondence: pzhou1975@outlook.com (P.Z.); sh.attarilar@yahoo.com (S.A.)

Abstract: Due to the growing aging population of the world, and as a result of the increasing need for dental implants and prostheses, the use of titanium and its alloys as implant materials has spread rapidly. Although titanium and its alloys are considered the best metallic materials for biomedical applications, the need for innovative technologies is necessary due to the sensitivity of medical applications and to eliminate any potentially harmful reactions, enhancing the implant-to-bone integration and preventing infection. In this regard, the implant's surface as the substrate for any reaction is of crucial importance, and it is accurately addressed in this review paper. For constructing this review paper, an internet search was performed on the web of science with these keywords: surface modification techniques, titanium implant, biomedical applications, surface functionalization, etc. Numerous recent papers about titanium and its alloys were selected and reviewed, except for the section on forthcoming modern implants, in which extended research was performed. This review paper aimed to briefly introduce the necessary surface characteristics for biomedical applications and the numerous surface treatment techniques. Specific emphasis was given to micro/nano-structured topographies, biocompatibility, osteogenesis, and bactericidal effects. Additionally, gradient, multi-scale, and hierarchical surfaces with multifunctional properties were discussed. Finally, special attention was paid to modern implants and forthcoming surface modification strategies such as four-dimensional printing, metamaterials, and metasurfaces. This review paper, including traditional and novel surface modification strategies, will pave the way toward designing the next generation of more efficient implants.

Keywords: titanium; surface modification; surface topographies; multifunctional surfaces; metamaterials; 4D printing

1. Introduction

Titanium (Ti) and its alloys as metallic biomaterials have found lots of application in the biomedical industry, especially for bone implants (dental and orthopedic) [1,2]. This phenomenon is still growing. The utilization of Ti-based biomedical products reached about USD 45.5 billion in sales in 2014, so it is of great importance to make them as efficient as possible. Titanium with respective atomic number and weight of 22 and 47.86 is a transition element located in group IV and period four of Mendeleev's periodic table and, according to its room temperature atomic structure, it can be found as α, near-α, α + β, metastable β, and stable β. The β type Ti alloys are mostly preferred in biomedical applications because of their low elastic modulus (similar to natural bone) and high corrosion resistance [3–7]. In general, the reason behind this ever-increasing utilization of Ti-based materials in the biomedical field is related to their superb properties, including biocompatibility, non-toxic nature, good corrosion resistance, potential to have osteogenic reactions, high specific strength, low Young's modulus, lightweight, and high strength-to-weight ratio compared to steel and other metals [8–11]. As a result of these beneficial properties, from World War

II until now, Ti has found lots of application in the biomedical field, especially in hard tissue replacements such as artificial bones, joints, dental implants, artificial hip joints [12], stents [13], scaffolds [14], and surgical devices.

An implant's surface has a crucial role in its reaction with peripheral live tissue since it is a substrate in which all biological reactions initially occur. As a result, it affects the performance of the implant, as well as its wettability and mechanical and chemical properties. The implant surface highly influences the ambient environment, having a significant effect on osteointegration [15,16], adhesion, and proliferation of osteoblast cells [12]. The implant surface is regarded as an artificial object from live tissue, and the first reaction of the body is to form foreign body giant cells by activated macrophages. Subsequently, the osteoprogenitor cells migrate onto the implant surface and differentiate into osteoblasts that finally produce bone [17]. It is reported that, generally, bone is separated from the implant's surface by a thin layer of non-mineral substances that prevent complete osteointegration [18]. Among other numerous reasons, this issue is one of the main reasons which justifies the need to use surface modification procedures. For example, in Ti implants, manufacturing techniques unfortunately lead to the formation of a layer with a poor condition that contains an oxidized, contaminated, stressed, and non-uniform surface that is not suitable for biomedical applications; hence it is of paramount importance to modify the surface. In addition, the surface must be accurately tailored in order to improve its mechanical bonding with peripheral tissue (bone, blood, muscle, etc.), biocompatibility, corrosion, wear, mechanical properties, roughness, and wettability [19,20]. For instance, Bauer et al. [21] showed the size-selective response of stem cells on anodic TiO_2 and ZrO_2, in which the cell adhesion and spreading were improved for TiO_2 nanotube diameters in the range of 15 to 30 nm, with a significant decrease in diameters higher than 50 nm. Additionally, Park et al. [22] proved the relationship between nanotube diameter and cell fate, and reported that the proliferation, adhesion, and differentiation of stem cells are highly dependent on nanotube diameter. By considering the crucial role of surface condition, up to now, various surface modification techniques have been introduced [23] and practiced, leading to different surface topographies and properties, each of which has its characteristics and applications, and which is intended to be used in a specific location in the body. These methods can be categorized from different aspects such as type of method or resultant topographical size. Figure 1 briefly illustrates the surface modification procedures according to the type of method, but this paper aims to describe them from a structure size point of view.

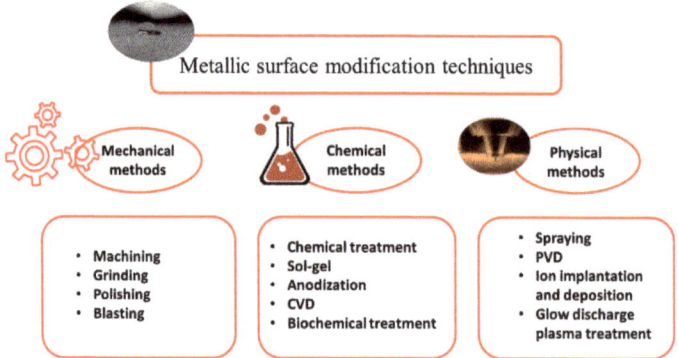

Figure 1. Metallic surface modification classification according to the type of technique.

An implant's surface can be modified through mechanical methods such as machining [24], grinding, polishing, blasting, and attrition [25]. In mechanical methods, the surface is modified by mechanical actions involving physical and attrition treatment, shaping, and removing the surface material. The objective of these methods is to achieve a certain type of roughness and topographies and eliminate the contaminations and oxide layers by removing the surface [26]. Recently, some severe plastic deformation techniques as mechanical methods have also been used to fabricate grain-refined surfaces and enhance the overall performance of biomaterials [27–31]. If the modification method involves chemical reactions, it is classified under chemical surface modification procedures [32]. The chemical techniques include anodic oxidation, sol-gel, chemical vapor deposition (CVD), biochemical modification, acidic, alkaline, and hydrogen peroxide treatment [33,34]. In the case of applying resistance heating, electron beam, laser, or electrical discharge in a vacuum, and of thermal, kinetic, and electrical energies, rather than chemical reactions, the modification method is classified under physical procedures. These physical modification techniques include thermal spraying, physical vapor deposition, ion implantation, and glow discharge plasma treatment.

Previous studies indicate that macro, micro, and nano-structured surface morphologies in implant surfaces can have pivotal effects on biocompatibility, bioactivity, osteogenesis, bone formation, and integration [35,36]. It was proven by many scholars that micro/nano-structured surfaces could improve biocompatibility [37], cell adhesion and proliferation, filament orientation, and even gene expression [38], as well as alkaline phosphatase (ALP) activity, cell mineralization, and osseointegration [39]. Some of these micro/nano-structures, in addition to maintaining biocompatibility and osseointegration, can have bactericidal properties [40]. They can even be used in drug loading applications [41,42], having the potential to be used as multifunctional techniques. It is known that high aspect ratio topographies are bactericidal with high effectiveness, which was numerously proven in Ti material [43,44]. Nanotube formation on Ti can be considered as the multifunctional option, providing biocompatibility, drug loading, and bactericidal effects. López-Pavón et al. [45] loaded the anodically produced nanotubes (on Ti-24Zr-10Nb-2Sn alloy) with gentamicin. The higher length and diameters of nanotubes leads to better a performance of drug loading, with a high percentage of gentamicin release. The micro/nano-scale variations in topology influence the antioxidant characteristic of the implant, as they can affect the oxidative stress microenvironment, enhancing or debilitating osteoinductive ability [46,47]. In this regard, for the first time, Ma et al. [48] reported that nano-structuring of the Ti surface through acid etching and anodic oxidation produces various topologies in which micro/nano-structures have superior oxidative stress resistance toward smooth and small nanotubes. By considering this brief introduction, it is clear that surface topographies are of significant importance in biomedical applications and investigations. In this regard, up-to-date and innovative surface modification methods and procedures are crucial in implant design and industry. These novel technologies will pave the way toward achieving a new generation of Ti implants with superior and multifunctional properties, leading to enhanced quality of human life and even decreased economic burdens on society.

2. Surface Characteristics for Biomedical Applications
2.1. Roughness and Wettability

The surface condition has a significant effect on numerous tissue reactions; it can influence cell proliferation and differentiation [49], protein adsorption [50], osseointegration, etc. In this regard, surface characteristics such as roughness, chemical composition, energy, wettability, biocompatibility, and bactericidal should be thoroughly studied and investigated. Osseointegration, as a direct structural and functional bonding between implant material and bone, is dependent on roughness; the surfaces with higher roughness and waviness can improve osseointegration [51]. It was seen that, in dental implants, rough surfaces have enhanced bone fixation, and they have higher bone-to-implant contact (BIC) value compared to commercially available implants [52], though the rougher surfaces compared to smoother ones have superior bone fixation [53]. Tailoring the surface of the implant with increased micro/submicron-scale roughness with sizes comparable to natural tissues and cells can lead to better osteoblast differentiation and production of local factors [54], with enhanced BIC value in vivo [55] and improved wound healing [49]. Nanorough surfaces with similar features to cell membrane receptors and proteins have a crucial role in improving the performance of the implant and osseointegration. Bone-implant surface interaction at various roughness and topographical conditions is schematically shown in Figure 2. Macroscale features can maintain a suitable mechanical fixation, while micro/submicron features can favorably interact with cells and osteoblasts. In a nanoscale condition, in addition to previous factors, the cell membrane receptors, integrins, and proteins are involved and enhance the overall quality of osseointegration and other biological reactions between bone and implant [56].

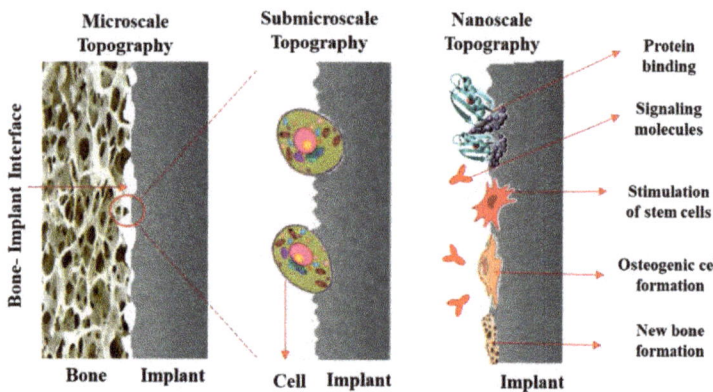

Figure 2. Bone-to-implant surface interaction at various roughness and topographical conditions.

Based on the roughness value, surfaces are classified into four categories: (1) smooth surfaces with Sa less than 0.5 µm can be found in abutments of oral implants; (2) slight roughness with Sa in the range between 0.5 and 1.0 µm in acid-etched samples and Astra Tech implants; (3) moderate roughness with Sa between 1 and 2.0 µm, which comprises nearly all modern implants, such as the Astra Tech, TiOblastTM, Nobel TiUnite, OsseoSpeedTM, Straumann SLA, and Dentsply Cellplus; (4) rough surfaces with a Sa higher than 2 µm, which are plasma-sprayed samples and Dentsply Frialit implants [57]. Since surface roughness is a critical factor modulating osteoblastic function, the optimum value should be determined, and the implant will be tailored according to it. Mustafa et al. [58] investigated the effect of various surface roughness values on the attachment, proliferation, and differentiation of cells on the Ti implant surface. They found out that the proliferation and differentiation of human mandibular bone cells are enhanced by increasing the surface roughness of the titanium implant. According to Rønold et al. [59], an optimal surface roughness ranges between 3.62 and 3.90 µm for bone attachment. It should be mentioned that acid-etched surfaces and Ti coating are the most preferred ways to gain an optimum surface roughness to improve an implant's performance [60].

Wettability, as a tendency of one fluid to spread on or adhere to a solid surface, has a substantial impact on biological interaction between the implant surface and peripheral tissue, and affects protein adsorption phenomenon and cell adhesion [61]. Figure 3 schematically illustrates the wettability behavior on various Ti surfaces; the nanostructures have a more wetted area than the Ti-foil with the smooth surface because of liquid penetration. This better wettability in nanostructures is an advantageous characteristic in biomedical applications.

Figure 3. Wettability in Ti Nanostructures.

2.2. Biocompatibility, Osteogenesis, and Bactericidal Effects

Biocompatibility can be defined as the capability of the material to display a healthy and proper host response and, in this context, the surface condition of materials is highly influential. The biocompatibility of Ti implants is associated with their oxide layer's ability to react with the peripheral environment, water ions, and serum proteins [62]. It was demonstrated that acicular-shaped oxides and topographical features on the surface are more toxic and noncompatible compared to equiaxed ones, and the surface potential can affect the biocompatibility of the implant [63]. In addition, roughness and surface topography can alter cellular behavior, platelet and osteons activation, and protein adsorption, finally changing the material's biocompatibility. Hydrophobicity is another factor that controls biocompatibility, since it determines the type, amount, and binding condition of proteins [64]. In this regard, it is believed that hydrophilic surface features are more compatible and safer compared to the hydrophobic ones, and even lead to better bone integration in Ti implants [65]. The size of surface features can also affect the biocompatibility of implants; for example, in pits with a diameter higher than 25 μm, cells migrate inside and settle there [66,67] or spread them out to fill the pit gap [68]. It was shown that fibronectin has higher adsorption in shallow pits and surface topography can affect the fibronectin distribution [69,70]. Overall, the biocompatibility response of an implant can be governed by its surface condition.

The term osteogenesis refers to the development and formation of new bone tissue by cells named osteoblasts, and osteointegration is known as the direct and through contact between bone and implant. Figure 4 schematically illustrates the osseointegration procedure on metallic implants, which includes three phases [71]. In the first phase after implantation, the surface of the implant is surrounded by blood, biomolecules, proteins, glycoproteins, and lipids, establishing a transitory bioactive film. After about one month, in phase two, the bone tissue is slightly absorbed around the implant, and proliferation and differentiation of osteon cells begin, leading to mineralization. After about three months, the implant surface is covered by many osteoblasts and osteoclasts that gradually mature, leading to progressive osteointegration. Many surface-related variables can impact the osteogenesis and osteointegration conditions of implant material, including wettability, roughness, surface chemistry, energy, etc. [72]. Any failure in proper osteointegration and osteogenesis leads to implant failure and many costs to both patient and the medical system; hence, it is of vital importance to avoid such errors. In this regard, titanium can be considered as a bioinert material with proper bone contact under osteo-permissive conditions; it is better than bio-tolerant materials and, unfortunately, more feeble than bioactive materials, so its surface can be purposefully modified in order to become bioactive. Surface topography and roughness can be accurately tailored to compensate for poor osteogenesis and osteointegration conditions [73]. The optimum roughness is about 1 to 1.5 μm, which maintains suitable bone-to-implant fixation [72]. Lossdörfer et al. [74] showed that, by increasing the surface microroughness, the osteoblast proliferation diminishes while differentiation increases since, in microrough topographical features, osteoblasts secrete some biological factors which improve osteoblast differentiation and diminish osteoclast activity and formation. Surface modification techniques can be incorporated to actively control the bone marrow-derived mesenchymal stem cells' (BMSCs) fate as one of the critical variables in osteogenesis. In this regard, surface technologies such as anodization, micro-arc oxidation, sol-gel, and ion implantation can improve BMSCs' differentiation [75]. Additionally, titania nanotubes (TNTs) are considered as favorable surface modification strategies to improve osteogenesis and osteointegration [76]. Shen et al. [77] reported that TNTs can improve osteogenic self-differentiation and decrease early inflammation of macrophages; the overall cellular response of mesenchymal stem cells (MSCs) and macrophage behavior can be well regulated by proper utilization of TNTs. In another study [78], it was shown that TNTs on Ti surface show the best osteogenesis response in comparison to a micro-scale case (sand blasted-acid etched topography) and a nano-scale case (hybrid sand blasted-acid etched), proving that nano-scale TNTs have the best surface topography for increasing clinical performance.

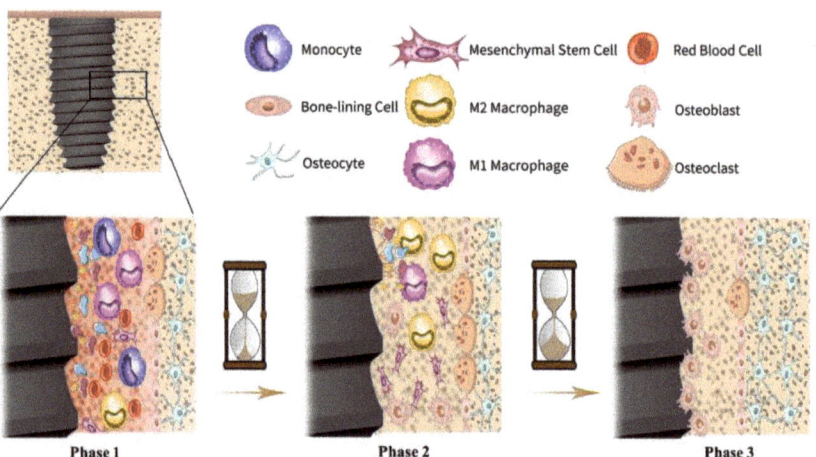

Figure 4. The osseointegration procedure on metallic implants [71]. In phase 1, the surface is surrounded by blood, biomolecules, proteins, glycoproteins, and lipids, establishing a bioactive film. In phase 2, the initial bone tissue absorption is begins simultaneously with osteon cells' proliferation and differentiation, leading to mineralization. In phase 3, the surface is covered by osteoblasts and osteoclasts, causing progressive osteointegration.

By two bacteriostatic and bactericidal strategies. Figure 5 schematically illustrates The bactericidal effect is an action that prevents the growth of bacteria and keeps them in a stationary phase of growth. A bactericide material can also kill bacteria. As is known, implant-related infections are among the most serious problems and issues in implant surgeries, which lead to failure and cause costly subsequent surgeries [79]; hence it is of crucial importance to resolve them. Mainly, two forms of bacteria are responsible for infection-related issues, *Staphylococcus aureus* (*S. aureus*) [80] and *Staphylococcus epidermidis* (*S. epi.*) [81]. Surface modification techniques are highly beneficial in the prevention of biofilm formation and infections, and can be achieved by designing antimicrobial surfaces which are able to avoid any pathogen spread and material deterioration these two mechanisms. These bactericidal surfaces actively prohibit the initial adhesion of living planktonic microbial cells that are created through the killing of bacteria or exorcising the approaching microbial forms. These antibacterial effects can be attained by various methods (physical, chemical, physicochemical, coatings, etc.) and in different scales (macro, micro, nano, atomic, molecular, and textural). In this regard, there are many types of antibacterial macromolecule [82–84], such as inorganic bactericidal metallic elements silver, copper, zinc, etc. [85–87].

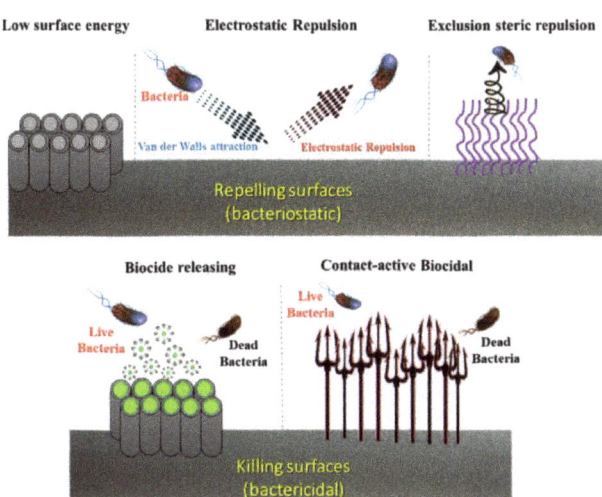

Figure 5. Schematic illustration of two surface modification strategies (bacteriostatic and bactericidal) in designing bactericidal surfaces.

3. Surface Modification Procedures

3.1. Macro-Grade Modification

Macro-scale surface treatments warrant the generation of a rough structure on the surface of titanium implants, and therefore improve their biological performance and osseointegration. Moderate roughness, with the average (Sa) ranging from 0.5 to 2 µm, is suitable for bone ingrowth in dental and orthopedic implants [88]. In this regard, many techniques with the capability to produce macro-rough surfaces have been introduced, such as etching techniques through acid etching [89,90], sandblasting [91,92], three-dimensional printing (3DP) [93], and laser surface texturing [94].

3.1.1. Acid Etching and Sandblasting Techniques

In the surface modification of Ti in biomedical applications, acid etching is of substantial importance, since it can generate fine-rough (1–3 µm), rough (6–10 µm), and macro-rough (10–30 µm) surfaces. Acid-etching can considerably improve early endosseous integration and implant stability; as has been shown, dual acid-etched Ti implants with macro-roughness in the range of 10–30 µm could effectively enhance bone anchorage and early osseointegration in rabbit models [95]. The dual effect of acid etching on cell behavior and mechanical properties was studied by Wang et al. [96], who reported that acid etching in hydrochloric acid could produce very rough surfaces (up to 3.7 mm roughness value), which led to considerable improvements in osteoblast cell adhesion and proliferation. It was seen that roughness is dependent on etching duration and, by increasing etching time, the surface roughness incremented while its yield strength diminishes. As such, a balance between cell response and strength should be considered. Figure 6 shows the immunofluorescence micrographs of osteoblast attachment on Ti samples with various etching durations. The optimal etching time is about 60 min, showing both favorable cell response and strength.

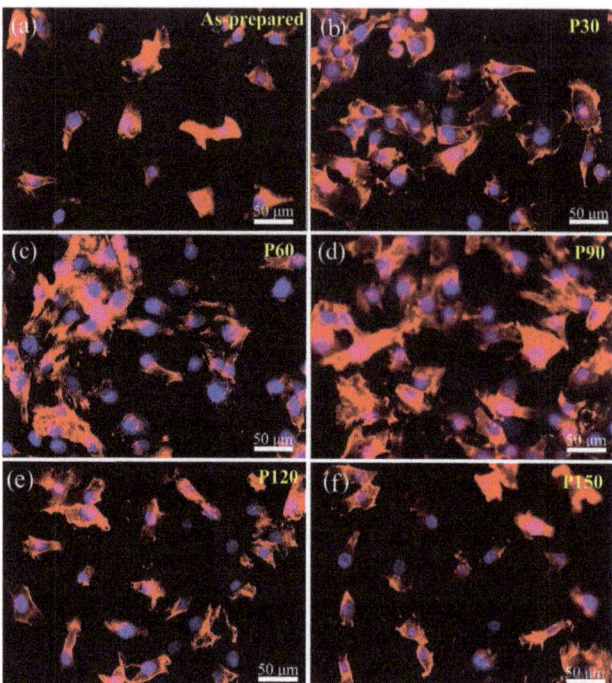

Figure 6. The immunofluorescence micrographs of osteoblast attachment on Ti samples after one-day of incubation. Rhodamine-phalloidin was used to stain the actin filaments and DAPI was utilized to stain cell nuclei. (**a**) Control sample without etching, (**b**–**f**) after etching for 30, 60, 90, 120, and 150 min, respectively [96] (reproduced with permission number: 5066331116663, John Wiley and Sons).

One of the most infamous and oldest modification techniques of Ti implants is sandblasting, in which the pressurized abrasive material is volleyed onto the surface of the implant so that the surface becomes rough and its osseointegration improves. It was shown that acid-etched surfaces have higher roughness magnitudes than sandblasted ones. Sandblasting with alumina powders can be utilized in order to clean and achieve microretentive topography and increase surface area. Sandblasting can also increase the bone anchorage by up to 50% [97]. Sandblasting by hydroxyapatite particles can produce bioactive surfaces, stimulate bone apposition, and facilitate the healing process [98]. Uncontrolled sandblasting may induce unfavorable surface defects and reduce the endurance limit of the implant [99], so the variables in the process should be thoroughly considered and controlled. Li et al. [100] showed that a modified sandblasting treatment can even lead to improved mechanical properties and considerably enhance the shear strength at dental implants. Figure 7 shows a SEM micrograph of a sandblasted Ti sample with aluminum oxide and an untreated sample. While sandblasting is an age-old method, some innovative mediums, post-treatments, and variables are introduced and experienced. In this regard, some complex blasting mediums with a mixture of $Al_2O_3/NaAlSi_3O_8/ZrO_2$-$TiO_2$ were used, and were shown to have improved bonding properties, with obvious "micro-vaccination" regions with enhanced adhesion of porcelain (bond) in titanium–dental porcelain interface [101].

Figure 7. A SEM micrograph of Ti samples: Untreated samples (**a**) 1000× and (**b**) 10,000× magnification; sandblasted samples by aluminum oxide (**c**) 1000× and (**d**) 10,000× magnification [102] (reproduced with permission number: 5071110018393, AIP Publishing).

3.1.2. Three-Dimensional Printing (3DP)

Three-dimensional printing (3DP), also known as additive manufacturing (AM), is a revolutionary procedure affecting all parts of science, industry, and even human life. Due to the very promising benefits of 3DP, its use in medical implants is expected to increase dramatically. These unique advantages include its great potential in the production of porous and complex shapes, even with intricate internal structures. In addition, it has an economic nature and can be utilized in mass production. Other advantages are repeatability, rapid production, and simple design. The most beneficial aspect of 3DP is its great potential in manufacturing patient-specific implants with multifunctional surfaces [103]. Another positive point of 3DP is that it has many diverse techniques, enabling us to use many types of materials, from organic to inorganic compounds, hydrogels, polymers, and metals. Another positive aspect of these numerous 3DP techniques is their ability to use a variety of curing systems according to specific conditions and needs in and intended application, which is a fundamental feature of the biomedical industry [103]. Figure 8 lists the most popular 3DP methods and their respective curing and deposition systems.

The resultant surface features after 3DP are mainly dependent on the deposition technique; for example, the mean roughness value after selective laser melting (SLM) is lower than electron beam melting (EBM), since the laser spot size is smaller than the electron beam. In addition, the SLM technique can produce thinner layers and utilize smaller powder sizes [104]. Many studies concluded that the production of Ti implants through 3DP is highly beneficial in improving cell responses and osseointegration [105,106]. Sirvas et al. [107] produced a 3D printed porous Ti-6Al-4V scaffold (extrusion-based method) with an average roughness of ~5 µm and comparable mechanical properties to natural bone due to a porosity of ~58% and a pore size of ~500 µm. This 3D printed sample shows considerable bone in-growth and in vivo on-growth, and complete filling of pores with the bone after 8 weeks of implantation in the rabbit model; Figure 9 shows the bone in-growth of this 3DP scaffold in the rabbit model [107]. However, the osteogenesis of 3DP manufactured Ti and its alloys are not satisfactory due to biological inertia issues, and should be improved by secondary procedures. In this regard, Wang et al. [108] utilized hydrothermal and alkali-heat treatment to further improve osteogenic differentiation and accelerate osseointegration. They reported a new bone formation and rapid osseointegration in rabbit models due to secondary surface modifications, leading to the generation of micro/nano-topography instead of the initial macro-graded topography in Ti-6Al-4V implants. Additionally, Gulati et al. [109] utilized an anodization procedure to generate dual micro/nano-scale topography on 3DP Ti-6Al-4V orthopedic implants. The results showed an enhanced cellular function, osseointegration, and improved adhesion of osteoblasts.

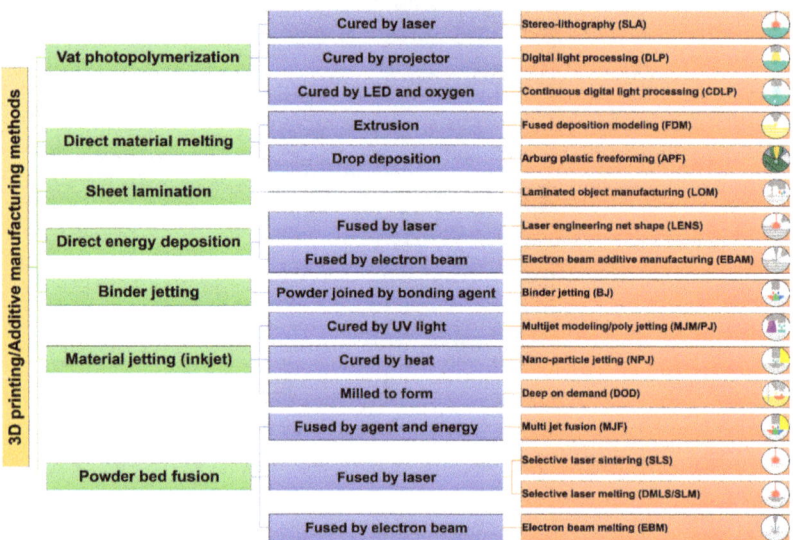

Figure 8. The most important 3DP methods and their corresponding curing and deposition systems [103].

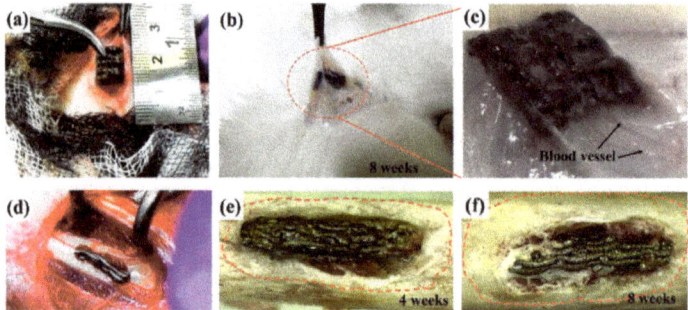

Figure 9. (**a**) Subcutaneous implantation of a 3DPed porous Ti-6Al-4V scaffold in rabbit model; (**b**) scaffold in the rabbit tissue after 8-week implantation; (**c**) magnified view of the circled region in (b), showing the tissue and blood vessel formation around the scaffold; (**d**) scaffold implantation in rabbit's femur and bone in-growth around the scaffold after (**e**) 4 weeks and (**f**) 8 weeks [107] (reproduced with permission number: 5071080659903, Elsevier).

3.1.3. Laser Surface Texturing (LST)

Laser surface processing development over the past 40 years has great potential in terms of surface modification, with numerous promising advantages in the biomaterial field. Laser-based methods are able to modify various surfaces, from macro to nano-scale topographies, and an important aspect of laser utilization is that there is not any need for direct contact, which prevents contaminations. Laser processing can manufacture Ti surfaces with improved tribological, corrosion, and erosion-resistant characteristics [110]. In addition, these techniques are clean, rapid, and easy-automated, with a surface patterning ability and a capability for application on any intricate and complex-shaped samples. In laser-based surface modification techniques, the interaction of an energetic laser beam with a surface induces craters. Diverse topographies and patterns with various dimensions can be generated by controlling the craters' dimensions, depending on laser spot size and mutual interactions of the laser beam, plasma, and surface [111].

Laser surface texturing (LST) is a process that renovates a material's surface properties mainly by modifying its texture and roughness; hence, it can be effectively utilized in implants and biomedical applications. The surface morphology of the Ti sample after the LST method is shown in Figure 10, illustrating the formation of both micro-nano hierarchical structure (MNHS) and the laser-induced periodic surface structures (LIPSS) [112]. The average micropillar dimensions are, respectively, about 20.45, 12.67, and 33.11 μm in length, width, and height. As can be seen in Figure 10, the micropillars' tops are covered with laser-induced nanoparticles, and top caps are surrounded by nanoripples that are organized according to the laser's direction. In general, LST of Ti in biomedical applications is applied under short-pulsed laser conditions [113]. Laser-based methods can effectively produce micro and nano-scales, but macro-grade modifications need much longer processing times compared to micro and nano conditions. The patterning ability in LST methods is crucial, since it can affect cellular response and bonding strength of an implant to tissue; it has also been shown that novel textures by using various patterning plans lead to the formation of diverse topographies, influencing the wettability of the surface, cell integration, and coefficient of friction against bone [94,114]. Pou et al. [113] suggested avoiding high energetic and short pulses in LST to prevent crack formations, and claimed that using an Nd:YAG laser is more appropriate for Ti modification, resulting in more regular shapes and fewer splashes without any variations in the chemistry of the surface. Cunha et al. [115] reported that femtosecond LST processing leads to the formation of nanopillars and laser-induced periodic surface structures (LIPSS) on Ti substrate, which enhances hydrophilicity and surface energy. These topographies are shown to be effective in the reduction of *Staphylococcus aureus* adhesion and biofilm formation on Ti surfaces.

3.2. Micro-Grade Modification

3.2.1. Sandblasting Acid Etching (SLA) Techniques

Sandblasting acid etching (SLA) is a surface treatment leading to improved topographies with enhanced osseointegration due to increased bone-to-implant contact (BIC). SLA is regarded as one of the most investigated and well-known methods for producing micro-rough surfaces. In this method, the sandblasted surface with macro-rough topography is followed by acid etching, which induces microroughness [91]. The SLA process can intensify the osseointegration rate by a combination of grit and acid etching that increases the roughness value on multiple levels. This improved surface with microroughness conditions encourages osteoblasts to proliferate and adhere to the implant surface, resulting in improved implant stability and a treatment time reduction. These SLA-treated implants provide a variety of benefits to patients requiring increased ossification [86]. The in vivo experiments of Buser et al. [87] on miniature pigs showed a considerably higher mean percentage of BIC and bone apposition in a chemical SLA Ti implant. Similar results were reported by Chiang et al. [116] showing the significant beneficial effects of SLA treatment on improving osseointegration, especially at the early stages of bone tissue healing.

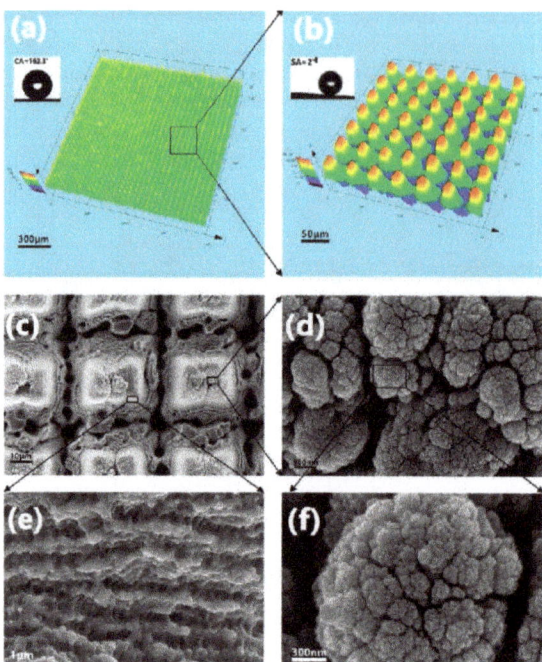

Figure 10. The LST processed Ti samples: (**a**) surface 3D profile, (**b**) magnified 3D profile, (**c–f**) SEM micrographs showing the LIPSS and MNHS topographies with the magnification of rectangular regions [112] (reproduced with permission number: 42, 3936-3939 (2017), Optical Society).

In addition to the traditional SLA method, current modern and complex procedures are proposed with the addition of other schemes to further improve the biological response of Ti implants. For example, Liu et al. [117] utilized a chemical scheme by Cu in addition to SLA treatment, and it was shown that Cu's addition to SLA-produced micro-submicron hybrid structures significantly improves the bactericidal effects toward oral anaerobic types of bacteria (*P. gingivalis* and *S. mutans*), and simultaneously enhances in vitro osteogenic and angiogenic gene expression. In vivo experiments confirm the ability of this combined technique for osseointegration improvement, showing enhanced peri-implant bone formation and favorable bone-binding. Additionally, these Cu-assisted SLA Ti samples, due to Cu-induced antiinfection effects, led to a gain resistance toward bone resorption and improved osseointegration [117]. In this regard, Choi et al. [118] modified the SLA-treated Ti surface with strontium-containing nanostructures through wet chemical treatment, leading to multifunctional effects such as enhanced osteogenic capacity, improved osseointegration, immunoinflammatory macrophage cellular behavior, and early macrophage cell functions. In another study, Kim et al. [119] used Mg ion implantation via a vacuum arc source ion implantation method to further improve human mesenchymal stem cell (hMSC) response in an SLA-treated Ti sample; the results indicated favorable cell adhesion, ALP activity, and calcium accumulation, leading to improved osseointegration. Figure 11 shows these SLA- and Mg-ion-implanted SLA samples with rough and irregular morphology, with an average roughness value of ~2 μm, which is not affected due to Mg-ion implantation.

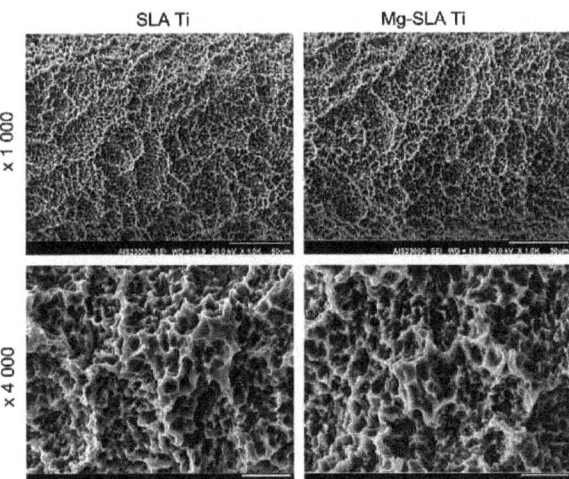

Figure 11. SEM micrograph of SLA-treated Ti samples and Mg-ion implanted SLA-Ti with 1000× and 4000× magnification [119] (reproduced with permission number: 5071100309175, Elsevier).

3.2.2. Other Micro-Grade Methods

One of the well-known methods to modify the surface of Ti implants is grit blasting, in which hard ceramic particles are shot via compressed air at a high velocity through a nozzle. The resultant surface roughness is dependent on the size of the ceramic particles (~100–300 µm), and they should be selected from biocompatible materials such as silicon oxide (SiO_2), aluminum oxide (Al_2O_3), titanium oxide (TiO_2), zirconium oxide (ZrO_2), steel, and calcium phosphate composition [120]. One of the disadvantages of the grit blasting method is the issue of particles remaining on the surface, which are very hard to remove and which may release into the live tissue, leading to allergies and other biologically adverse effects [121]. Studies show that grit blasting can lead to improved BIC values in Ti implants. For example, Ivanoff et al. [122] proved that TiO_2 blasting on Ti micro-implants leads to higher BIC values, while some other studies are syill confirming the promising effects of Al_2O_3 and TiO_2 blasting [123]. The reported clinical studies claimed a high success rate of TiO_2 grit-blasted Ti implants [105].

In addition to macro-grade surface modification, acid etching can also be used for micro-grade modification, in which surface roughening is performed by strong acids such as nitric acid (HNO_3), sulfuric acid (H_2SO_4), hydrofluoric acid (HF), hydrochloric acid (HCl), and other combinatorial acidic solutions. Acid etching procedures are able to produce surfaces with micro-pits with the size range of 0.5–2 µm [124], but acid etching can have some negative effects on the mechanical properties of the implant. Acid etching can stimulate macrophage activity and enhance cell proliferation and the pro-angiogenic response of endothelial cells, leading to increased BIC values, improved osseointegration, and better initial osteoblast anchorage [125,126]. Diomede et al. [127] showed a considerably improved biological response of dual acid-etched Ti samples, with resultant improved cell growth and adhesion, enhanced osteogenic and angiogenic events, as well as a clear osseointegration process. Additionally, Wang et al. [96] proved that acid etching by 20 wt.% HCl induces a remarkable enhancement of osteoblast cell adhesion and proliferation on the porous Ti.

3.3. Nano-Grade Modification

Nano-grade surface modification is of crucial importance, since human tissue morphology includes numerous nanostructures, such as natural bone, which has a hierarchy in terms of the macro-, micro-, and nano-scale structures, with a gradual transition between them [128]. The existence of micro and nanostructures can considerably affect the cell response and the initial bone formation around the implant material [129]; hence, recently, many attentions have been focused on the design of mixed micro/nano topographies with multifunctional properties through modern and innovative techniques [130–134]. These nanofeatures stimulate osteogenic activity and increase the adsorption of proteins, leading to rapid osseointegration and far better performance of implants [135–137]. A comparative study about the effect of micro and nanostructures on osseointegration of Ti implants proves the considerable superiority of nanostructures, with enhanced performance in the pull-out tests, direct bone apposition, and improved osseointegration [123]. Besides, these nanostructures facilitate the functionalization procedures paving the way toward fabricating bactericidal and anti-inflammatory surfaces [138]. Figure 12 shows the schematic of a CeO_2-nanostructured Ti surface in which the formation of CeO_2 nano-rods, -cubes, and -octahedrons can lead to strong antibacterial properties, and in which the nano-octahedrons showed the best anti-inflammatory response [138]. Nowadays, there are many methods for fabricating nano-grade surfaces on Ti implants. In addition to new and modern techniques, other macro and micro-grade methods can also be used in for nanomodification by changing the related variables. In this regard, some methods, such as grit blasting, acid etching, and SLA, were also classified in the nano-gradation group.

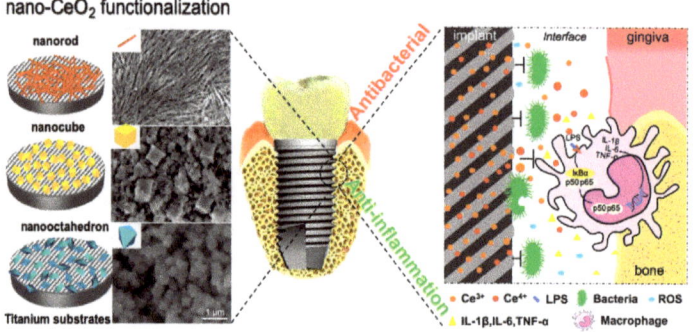

Figure 12. A schematic showing the Ti surface modification by nano-CeO_2 (rod-CeO_2, cube-CeO_2, octa-CeO_2) for antibacterial and anti-inflammatory properties. Readers are referred to [138] (reproduced with permission number: 5071081046141, Elsevier) for more information.

3.3.1. Electrochemical Modification

Recently, much attention has been paid to the fabrication of nanopores and nanotubes with functional properties such as drug loading, being antibacterial, etc. These topographies can be easily attained through electrochemical methods such as anodization. Figure 13 shows the generation of various topographies, with related mechanisms on the metallic substrate by the anodization method. Briefly, anodization is an electrochemical technique in which an ordered oxide film is grown on a metallic sample connected to the anode of an electrochemical cell [139]. Charging the double electric layer at the metal–electrolyte interface generates the anodic oxide film. Subsequently, the dissolution of oxide film by the electric field leads to the formation of soluble salt consisting of an anion and metal cation in the electrolyte solution. Finally, the electrochemical reactions (oxidation and reduction), conjointly with field-driven ion diffusion, generate an oxide layer on the anode's surface [140]. The dimension of nano-topographies is easily controlled by changing the applied potential, electrical current power, anodization time, electrolyte composition, and temperature [141]. The formation of these anodized-produced nano-topographies have many advantages, including their promising potential in delivering many types of drugs [142], such as bioactive molecules [143], and growth factors [144] to enhance the cellular response, affecting the contact surface area and leading to enhanced wettability, inducing mechano-bactericidal effects [145], improving implant to bone bonding, etc.

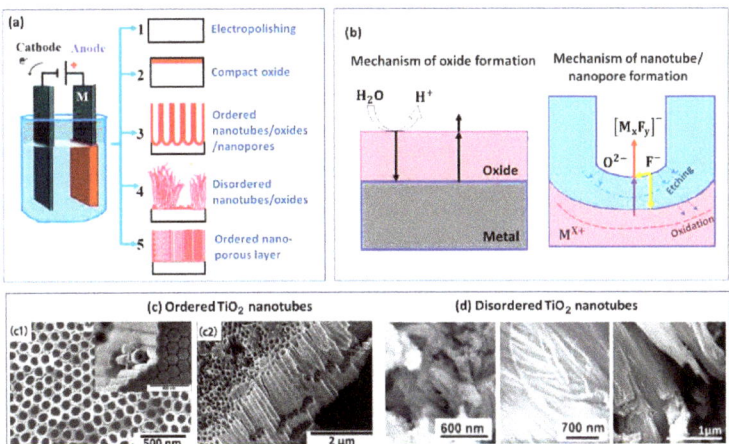

Figure 13. (**a**) The fabrication of various topographies on metallic substrates by the anodization method; (**b**) the mechanism of oxide, nanotube, and nanopore formation by anodization; (**c1**) and (**c2**) ordered TiO$_2$ nanotubes [146]; (**d**) disordered TiO$_2$ nanotubes [147] (reproduced with permission number: 5071091497388, Elsevier).

Together with the type of fabricated nanostructure by anodization, their dimension is important. The nanotube length can influence biocompatibility and its diameter can affect cell adhesion and proliferation [148,149]. Many studies have shown that favorable osteoconductive responses experimented on in the range of 30–120 nm can be achieved in a ~70 nm diameter, and optimal differentiation and proliferation can be achieved in ~80 nm [129]. Su et al. [149] claimed that in vitro and in vivo experiments prove the positive effect of TiO$_2$ nanotube formation on osseointegration, osteogenic activity, cell differentiation, proliferation, mineralization, and anti-microbial properties. Besides, this electrochemical-based surface treatment can be performed on pre-treated macro and microporous Ti implant surfaces, fabricating favorable textures for nanoscale cellular interactions [149]. In addition, mechanical pull-out and histological tests in rabbit models indicate that about nine-fold improved bone-bonding was achieved in TiO$_2$ nanotube-modified Ti surfaces [149]. Lee et al. [150] produced TiO$_2$ nanotube arrays through a two-step anodic oxidation procedure on Ti dental implants in a solution containing ethylene glycol with 0.5 wt.% NH4F. The in vivo studies on rabbit models show further improved osseointegration results compared to machined, sandblasted, large-grit and acid-etched surfaces; these nanotubes can also be used as reservoirs for recombinant human bone morphogenetic protein-2 (rhBMP-2). Hu et al. [151] indicated far better interfacial adhesion and osseointegration in anodic-produced TiO$_2$ nanotubes on ultra-fine-grained titanium; a grain refinement strategy of high-pressure torsion processing was used in order to enhance the weak bonding of anodic TiO$_2$ nanotubes to the substrate.

3.3.2. Plasma Spraying

Plasma spraying is known as a reliable nano-scale coating technique that can also be used to roughen a surface. Figure 14 shows the schematic of this process including an electrical power source, water-based cooling system, gas flow control, powder injector, cathode, anode, and insulators, along with the affected variables [23]. In this method, a direct current arc plasma gun is utilized to spray the melted powder material onto the Ti substrate; it can also roughen and increase the surface area of the implant material [152]. It was reported that the plasma spraying technique can facilitate and accelerate the formation of a bone-to-implant interface [153].

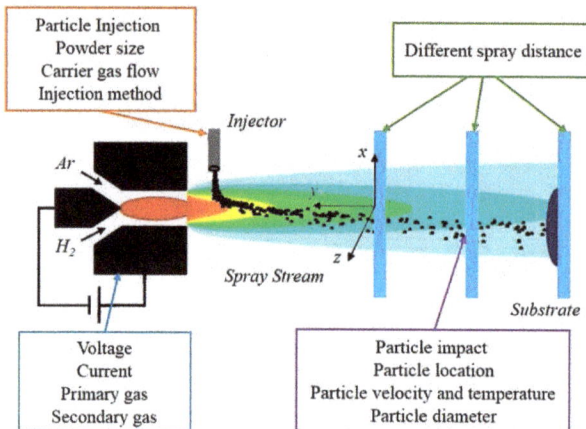

Figure 14. A schematic of a plasma spraying technique along with the affecting factors [23] (reproduced with permission number: 5071091227684, Elsevier).

Until now, numerous studies have been focused on plasma sprayed coatings on Ti implant surfaces, showing their great potential to improve the properties and functionality of Ti implants; meanwhile, hydroxyapatite (HA) and calcium phosphate coatings are of special interest for enhancing the bioactivity of these surfaces [154–156]. Coating of composite materials can also be achieved by the utilization of plasma spraying, Li et al. [157] prepared a nano-TiO_2/Ag coating to enhance bioactivity and bactericidal properties. Functionally graded (HA)/Ti-6Al-4V composite coatings were produced by plasma spraying. This three-layered structure, without any distinct interfaces among layers, had improved and graded mechanical properties comparable to the natural bone, with enhanced tensile adhesion strength and toughness [158]. Additionally, a biomimetic nano-porous composite (50HA–50TiO_2) was produced by plasma spraying for orthopedic applications on Ti6Al4V alloy, showing the natural bone-like nano-porous morphology favorable for effective bone bonding with an implant's surface [159]. Hameed et al. [160] used axial suspension plasma spraying (SPS) and atmospheric plasma spraying (APS) procedures to induce HA coating and tailor a Ti6Al4V surface for orthopedic implant applications. In this study, various coatings using different processing variables were produced, including two HA coatings by APS (P1 and P2) and four coatings with SPS HA (S1, S2, S3, and S4). The results indicated the optimal properties of S3 (1.3 times increased adhesion strength and 9.5 times higher corrosion resistance compared to P1). After S3, the best results confirmed in the P1 sample, both of these samples have favorable biocompatibility [160]. Figure 15 shows a brief illustration of the research. For the preparation of the S3 sample, the following parameters were used: 150 (L/min) gas flow rate, 6.10 (kJ/mol) enthalpy, 220 A arc current, 70 mm stand-off distance, and 36 μm coating thickness [160]. Briefly, the SPS method indicated better hMSCs cell viability and corrosion performance.

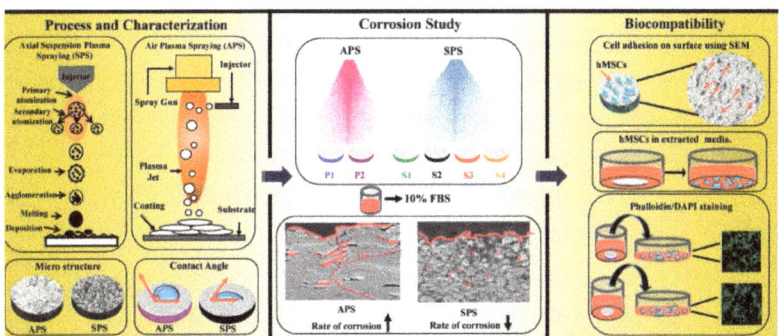

Figure 15. A schematic of SPS and APS methods, leading to different microstructures and wettability; corrosion studies in 10% fetal bovine serum (FBS) with 0.1% sodium azide, and in vitro biocompatibility experiments by hMSCs [160] (reproduced with permission number: 5071090717980, Elsevier).

3.3.3. Pulsed Laser Deposition

The pulsed laser deposition (PLD) technique is one of the physical vapor deposition (PVD) derivatives in which a high-power laser beam with a narrow frequency bandwidth is utilized to vaporize the substrate. PLD techniques are generally used to produce a variety of nanotubes, nanopores, nano-powders, and quantum dots. The most advantageous aspect of PLD is that it can be applied to almost any target material [161,162]. In addition, PLD has the potential to further tailor nanostructure morphology by changing the process parameters such as wavelength, laser energy, gas pressure, etc. Many in vitro and in vivo investigations show the beneficial aspects of PLD utilization in the production of ceramic, titania, calcium phosphate-based, and silicon oxide coatings for biomedical applications [163,164], indicating the superior osseointegration properties of CaP functionalized metallic implants via the PLD technique. PLD techniques can produce HA coatings (as a popular surface strategy for Ti implants), including doped species. [165]. Overall, PLD is one of the most favorable techniques for fabricating bioactive coatings on metallic implants. PLD can promote implant cytocompatibility [166], corrosion resistance [167], antibacterial effect [168], drug-eluting characteristics [169], osteogenesis [170], and mechanical properties [171]. Even some organic animal-based material can be used in the PLD method; for example, Duta et al. [172] produced the ovine and bovine-derived hydroxyapatite thin films on a Ti substrate with a rougher and more adhesive nature. Despite the numerous benefits of PLD, it also has some limitations and shortcomings. PLD may lead to compositional changes and irreversible destruction of chemical bonds and initial material structures. This issue usually occurs in complex and delicate biomolecules, drugs, and biopolymers. Finally, these compositional changes and destructions affect the quality of the deposited film [173].

4. Multifunctional Biomimetic Surfaces and Their Applications

Nature as the best teacher has numerous examples of multifunctional hierarchical micro/nanostructures. These natural structures are generated to actively adapt to extremely harsh environmental conditions. They also provide other required properties such as self-cleaning, bactericidal effects, and even fabricating fascinating colors to attract the attention of possible mates; some of these examples can be found in studies by Vijayan et al. [174] and Hu et al. [175]. In this regard, tailoring a multifunctional biomimetic surface on Ti material, as one of the most used materials in biomedical applications, is of high importance, and has attracted much attention from academic society, with a sharp increase in the number of studies conducted in this area; most of these studies on Ti focused on the fabrication of bioactive, bactericidal, and drug loading structures. Figure 16 lists the strategies for achieving bioactive and antibacterial properties on Ti surfaces through various types of surface modification [176].

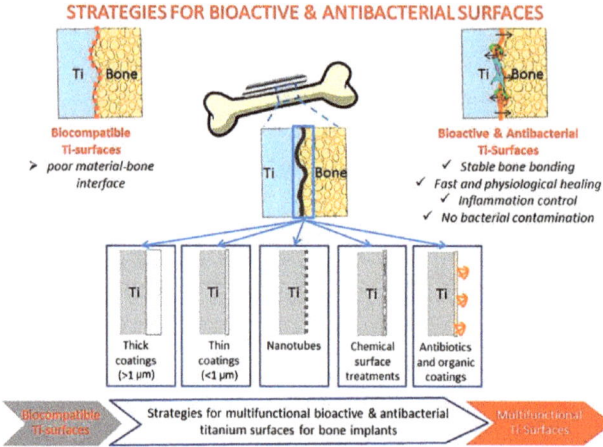

Figure 16. The strategies for achieving bioactive and antibacterial properties on Ti surfaces through various types of surface modifications [176] (reproduced with permission number: 5071090149493, Elsevier).

Spriano et al. [177] produced two types of multifunctional Ti surface. The first one, with an inorganic modification aiming to induce in vivo apatite precipitation with complex micro/nano-roughness and modified chemistry (full of hydroxyls groups), showed enhanced wettability and protein adsorption. The second surface was additionally processed with alkaline phosphatase (ALP) grafting. The results showed increased cell proliferation rates and higher osteoblast differentiation, with more filopodia in these nanotextures compared to traditional polished and grit blasted samples [177]. Biomimetic growth of nanostructured TiO_2 can be performed through plasma treatment. Liu et al. [178] reported the fabrication of bioactive nanostructured TiO_2 surfaces with grain sizes less than 50 nm using plasma spraying following hydrogen plasma immersion ion implantation (PIII), leading to bone-like apatite generation. It was seen that a hydrogenated surface increases the negatively charged functional groups, and a refined microstructure improves the surface adsorption, leading to the facilitation of apatite formation and bioactivity [178]. One of the well-known methods for the loading and releasing of various drugs, as well as anti-inflammation and antibacterial compounds, in Ti implants is the production of TiO_2 nanotubes through electrochemical methods, which could be beneficial in orthopedic implants [179]. A coating strategy can also be used to functionalize the Ti surface by the utilization of antimicrobial peptides and amphiphilic oligopeptides. These peptides show multifunctional behaviors including a reduction in pro-inflammatory cytokines, along with the increment of anti-inflammatory cytokines, down-regulation of macrophage activation, prevention of the adhesion of bacteria, and increases in osteoblast viability [180–182]. Achieving simultaneous bioactive and antibacterial surfaces on Ti can be achieved by the utilization of either inorganic (metallic ions, nanoparticles, and their oxides, e.g., Ag, Cu, Zn, and Ce) or organic (antibiotics) antibacterial agents. It seems that the inorganic scheme is better than the organic ones due to its capability to respond to polymicrobial infections without having an issue with resistant bacterial strains, which is among the main issues in antibiotics [176,183]. In addition to the so-called studies, Table 1 briefly introduces some of the investigations about multifunctional Ti material production using various techniques, some of which are not categorized in surface modification techniques. The advantages of biomimetic multifunctional tailoring of Ti surfaces are vast and, because of that, make it a crucial research topic with the potential to revolutionize biomedical engineering. As such, the future of this field is dependent on developing new methods and improving the current ones. This paper provides a brief review about these methods and aims to incite future investigations.

Table 1. The multifunctional properties in Ti and Ti6Al4V alloy, techniques, and surface types.

Substrate	Technique	Surface Type or Coating	Multifunctional Properties	Ref.
C.P. Ti	Layer-by-layer self-assembly	Phospholipid-based multifunctional coating with phospholipids-based polymers, type I collagen (Col-I), and Arg-Glu-Asp-Val (REDV) peptide	Inhibit platelet adhesion, smooth muscle cells, and endothelial cells proliferation	[184]
C.P. Ti	Plasma electrolytic processing (PEP)	Ag substituted hydroxyapatite/TiO_2 composite	Corrosion-resistant, bioactive, antibacterial	[185]
C.P. Ti	Electrodeposition	Cu-substituted carbonated hydroxyapatite coating	Antibacterial function against *Escherichia coli*, corrosion-resistant, favorable osteoblast function	[60]
C.P. Ti	Aqueous precipitation (electrochemical)	Ag-doped β-$Ca_3(PO_4)_2$/chitosan hybrid composite coatings	Antibacterial, biocompatible, corrosion-resistant	[186]
C.P. Ti	Micro arc oxidation	Zn-incorporated TiO_2 porous coating	Antibacterial, corrosion resistant	[187]
C.P. Ti	Micro arc oxidation	Cu NP-incorporated TiO_2 porous coating	Antibacterial activity against *Staphylococcus aureus*, the enhanced cellular activity of osteoblasts and endothelial cells	[188]
C.P. Ti	Micro arc oxidation + dopamine dip coating + $AgNO_3$ reduction	Hierarchical coating by Ag NP deposition on micro-nano-porous TiO_2	Anticorrosion, antibacterial properties against *Staphylococcus aureus*, optimal osteoblast cell function	[189]
C.P. Ti	Plasma electrolytic oxidation	TiO_2 + ZnO NP in phosphate-based electrolyte	Anticorrosion, antibacterial effect against both Gram-positive and Gram-negative bacteria	[190]
C.P. Ti	Electrostatic spraying	Ag-incorporated hydroxyapatite coating	Antibacterial activity against *Escherichia coli*, optimal osteoblast cell function	[191]
C.P. Ti	Anodic oxidation (TiO_2) and electrodeposition (Ca-P)	TiO_2 + calcium Phosphate, (Ca-P) coating	Antibacterial function against *Staphylococcus aureus*, anticorrosion	[192]
C.P. Ti	Anodic oxidation and electrodeposition	Ag-Mn-doped double-layer hydroxyapatite coating	Super-hydrophilic, corrosion-resistant, improved osteoblast cell function	[193]
C.P. Ti	Electrochemical and heat treatment	Ag-hydroxyapatite composite coatings	Antibacterial function against *Escherichia coli*	[194]
C.P. Ti	Hydrothermal method	Ag- and Sr-substituted hydroxyapatite coating on dopamine functionalized titanium	Antibacterial function against *Escherichia coli* and *Staphylococcus aureus*, reduction of Ag cytotoxicity	[195]
Ti6Al4V	Micro arc oxidation	Multi-layer HA/TiO_2 coatings containing Ag	Enhancement of bioactivity, antibacterial effect	[196]
Ti6Al4V	Hybrid approach of magnetron sputtering and micro-arc oxidation	Zn-doped ZrO_2/TiO_2 porous coatings	Antibacterial property against *Staphylococcus aureus*, corrosion-resistant, cytocompatibility	[197]
Ti6Al4V	Electrodeposition	Zn-halloysite nanotubes /Sr^{2+}, Sm^{2+} substituted hydroxyapatite bilayer coating	Corrosion-resistant, bioactive, favorable antibacterial function	[198]

5. Forthcoming Modern Implants

The development process in medical applications is extremely fast, especially in relation to medical implants due to the growing number of the world's aging population. In this regard, many modern materials, procedures, and technologies have been proposed and studied. In addition, in the case of implant surface technologies, new concepts have been introduced which cannot be categorized in traditional classifications. This section aims to present and discuss these modern technologies; some of them have not yet been studied directly on titanium, but they may inspire future studies and pave the way for new advances in the field.

Modern hydrophobic surfaces are highly interested in the dental and ophthalmological communities. Hydrophobic surfaces can prevent dental enamel erosion [199] and have many applications in the production of eye lenses [200]. A superhydrophobic surface can be produced by various methods and shows promising properties. Ma et al. [201] produced a fluorine-based superhydrophobic surface via electrochemical etching with anticorrosion and anti-abrasion characteristics. Additionally, Bains et al. [202] fabricated a hierarchical hydrophobic surface with long-term antibacterial properties, minimizing wetting through biological secretions and inhibiting corrosion. They developed this kind of superhydrophobicity by the production of benzimidazolium ionic liquids ILs-1(a–d)-based metal hybrid nanocomposites by the utilization of different metals, such as silver, gold, and copper. In addition, self-healing hydrophobic coatings with high transparency, excellent stability, and favorable adhesion were introduced [203], which can be further improved for use in the implant industry. A study by Tang et al. [204] proved the effectiveness of TiO_2 nanotube-based superhydrophobic surfaces, with more than 150° contact angle in the prevention of bacterial contamination. The superhydrophobic surfaces of Ti show considerable self-cleaning, prevention of bacterial adhesion, and enhanced anticoagulant characteristics; these surfaces can be produced by various methods such as laser ablation, electrochemical processes, high-speed micro-milling, electrodeposition, anodic oxidation, and fluoroalkylsilane modifications [205]. Recently, new approaches in bone regeneration of implants have also been introduced and practiced. Digital and visualized guided bone regeneration (GBR) is among these new technologies, with promising benefits in precision and controllability of bone augmentation procedures [206]. Yin et al. [207] used a novel dental implant design to preserve the alveolar ridge height by mechanical memory in which, through 3D printing technology, a micron-sized pore-channel structure was fabricated, with a more bony ingrowth, thus assuring the required horizontal mechanical force and mimicking natural teeth force. In this design, the pore-channel considerably assists stem cell differentiation and tissue morphogenesis.

5.1. Four-Dimensional (4D) Printing

The production of shapeshifting materials using four-dimensional (4D) printing can be a revolutionary approach in biomedical applications. These 4D printed materials have the potential to reconfigure themselves upon demand when exposed to changes in temperature, electric current, stress, etc. The future of 4D printing can address complex medical issues and further improve and facilitate patient-specific designs. Through 4D printing technology, a 3D physical object can be fabricated by adding smart material layer by layer via computer-operated computer-aided design (CAD) data [208]. The smart material function adds the fourth dimension, with the capability to transform over time, in which the printed products become sensitive to parameters such as temperature, humidity, time, electricity, magnetism, etc. [209]. The main applications for 4D printing in medicine were reported in dentistry [210], prosthetics [211], and implants [212]. In addition, 4D printing technology can be used in drug delivery systems, scaffolds [213], and stents [214]. The 4D printed drug delivery systems can be stimulated by external parameters to release drugs, and 4D-printed containers [215,216] are popular examples, with the other example being expandable gastroretentive devices [217].

5.2. Metasurfaces and Metamaterials

Metamaterials are modern artificial composite structures with exceptional material properties and applications displaying exotic physical properties that surpass or complement the usual properties seen in nature. Metamaterial, as a new frontier in science and technology, is a multi-disciplinary field involving physics, material science, engineering, and chemistry. Metamaterials (MM) as synthetic composite structures use a conventional type of material such as metals and plastics, but they act entirely differently to bulk materials, presenting exotic and exceptional properties. Metamaterials' structures are composed of microscopic patterns that can interact with light, electromagnetic, elastic, acoustic, and thermal properties in unconventional ways.

Metasurfaces are thin-films which include individual elements that have been introduced to defeat the obstacles that metamaterials are confronted with. The unique benefits of metasurfaces and metamaterials can be utilized in biomedical applications. For instance, the stress shielding issue mostly seen in orthopedic bone implants can be solved by lattice and shell-type architecture in bone scaffolds, in which the varied topology of nodal connections has great potential to control the relative rigidity of the metamaterial [218]. Three-dimensional printing technology, along with multi-objective genetic algorithm (GA) optimization with the finite element (FE) simulation, can be used to produce an optimum force-displacement response in designing printable tunable stiffness metamaterial for bone healing [219]. Mechanical 3D metamaterial with a porous structure is among the best materials for bone implants, with a graded Poisson's ratio distribution to optimize stress and micromotion distributions. In this regard, Ghavidelnia et al. [220] analytically designed an auxetic 3D re-entrant structure with tailored elastic modulus and Poisson's ratio, and which has great potential to solve the stress shielding problem. Kolken et al. [221] produced a novel meta-implant by 3D printing technology. Their designed non-auxetic meta-biomaterials with a deformable porous outer layer were intended to be used in acetabular revision surgeries. During implantation, the outer layer plastically deforms into the defects, enhancing the initial stability and stimulating the surrounding bone ingrowth. This space-filling behavior with 3D printed lattice (including six-unit cell) is highly beneficial for improving the mechanical performance of implants and enhancing bone-mimicking characteristics. Figure 17 shows this space-filling meta-implant.

Figure 17. The meta-implant design has the potential to deform and fill up defects, restoring the physiological loading condition. The implants were compressed in bone-mimicking molds to evaluate their deformability [221].

6. Conclusions

In recent years, titanium implant modification methods have shifted from improving mechanical strength and reducing yield strength toward multifunctional designs. Multifunctional surface designs can improve biocompatibility and osseointegration. They can also lead to an optimal cellular response, wettability, roughness, and even drug loading and antibacterial properties. In order to achieve all these crucial requirements in biomedical applications, developing modern and optimum techniques is vital. Besides, the existing methods must be improved and modified accordingly. Recent investigations in strength, roughness, and wettability issues have shifted toward hierarchical structures, mimicking the periphery tissue (specifically bone, in the case of titanium), along with activating the surface with organic, inorganic, and biochemical materials. Many techniques based on mechanical, physical, chemical, and biochemical methods were introduced and explained in the literature. Besides, the combination of these methods is highly interesting for the community, since each of them has its advantages and disadvantages. Three-dimensional printing technology emerged in the last decade and, despite its infancy, has opened new horizons for developing modern implants with patient-specific properties. Based on the literature review, it seems that the importance of micro/nano-gradation in the surface modification of Ti and its alloys, along with providing multifunctional properties, is more than the utilized technique; hence, in this paper, special attention was paid to tailoring micro/nano-grade surface structures, which is well suited to cellular and biosystems' dimensions and can actively mimic their functions. In this regard, firstly, basic requirements such as roughness, wettability, biocompatibility, osteogenesis, and bactericidal properties were discussed. Then, macro-, micro-, and nano-grade surface modification techniques were thoroughly explained, including acid-etching, sandblasting, 3D printing, and laser surface texturing in the macro-grade group, SLA techniques in the micro-grade group, and electrochemical surface treatment, plasma spraying, and PLD in the nano-grade group. Subsequently, the multifunctional biomimetic surfaces were discussed. Finally, the forthcoming modern implants, with attention paid to 4D printing and novel metasurfaces and metamaterials, were explained. It seems that these revolutionary techniques have a very promising future in the medical implant industry. Overall, it was suggested that improving the exciting traditional techniques, further modifying novel methods such as metasurfaces, or designing new ones by combining the so-called methods, along with developing modern procedures, will be highly beneficial. By collecting the required information from the literature, encompassing both traditional and modern procedures, this review paper can pave the way toward tailoring modern and more efficient processes in the surface treatment of Ti and its alloys.

Author Contributions: J.L. and P.Z. designed this study. P.Z. and H.S. analyzed the data and helped in data collection. S.A. and J.L. participated in its design and wrote the paper. All authors have read and agreed to the published version of the manuscript.

Funding: This work was supported by Youth Innovation Team of Shaanxi Universities of Education Department of Shaanxi Provincial Government (Grant No. 2019-73), Education Department of Shaanxi Provincial Government (Grant No. 20JK0700), Shaanxi Key Research and Development Program of Shaanxi Science and Technology Department (Grant No. 2020GY-316).

Institutional Review Board Statement: Not applicable.

Informed Consent Statement: Not applicable.

Data Availability Statement: The data presented in this study are available in references.

Conflicts of Interest: The authors declare no conflict of interest.

References

1. Elias, C.N.; Lima, J.H.C.; Valiev, R.; Meyers, M.A. Biomedical applications of titanium and its alloys. *JOM* **2008**, *60*, 46–49. [CrossRef]
2. Niinomi, M.; Boehlert, C.J. Titanium Alloys for Biomedical Applications. In *Advances in Metallic Biomaterials*; Springer: Berlin/Heidelberg, Germany, 2015; pp. 179–213, ISBN 978-3-662-46836-4.
3. Chen, L.-Y.; Cui, Y.-W.; Zhang, L.-C. Recent development in beta titanium alloys for biomedical applications. *Metals* **2020**, *10*, 1139. [CrossRef]
4. Niinomi, M.; Liu, Y.; Nakai, M.; Liu, H.; Li, H. Biomedical titanium alloys with Young's moduli close to that of cortical bone. *Regen. Biomater.* **2016**, *3*, 173–185. [CrossRef] [PubMed]
5. Gode, C.; Attarilar, S.; Eghbali, B.; Ebrahimi, M. Electrochemical behavior of equal channel angular pressed titanium for biomedical application. *AIP Conf. Proc.* **2015**, *1653*, 20041. [CrossRef]
6. Wang, L.; Lu, W.; Qin, J.; Zhang, F.; Zhang, D. Effect of precipitation phase on microstructure and superelasticity of cold-rolled beta titanium alloy during heat treatment. *Mater. Des.* **2009**, *30*, 3873–3878. [CrossRef]

7. Wang, L.; Xie, L.; Lv, Y.; Zhang, L.; Chen, L.; Meng, Q.; Qu, J.; Zhang, D.; Lu, W. Microstructure evolution and superelastic behavior in Ti-35Nb-2Ta-3Zr alloy processed by friction stir processing. *Acta Mater.* **2017**, *131*, 499–510. [CrossRef]
8. Zhang, L.; Chen, L. A review on biomedical titanium alloys: Recent progress and prospect. *Adv. Eng. Mater.* **2019**, *21*, 1801215. [CrossRef]
9. Wang, L.; Xie, L.; Zhang, L.; Chen, L.; Ding, Z.; Lv, Y.; Zhang, W.; Lu, W.; Zhang, D. Microstructure evolution and superelasticity of layer-like NiTiNb porous metal prepared by eutectic reaction. *Acta Mater.* **2018**, *143*, 214–226. [CrossRef]
10. Wang, L.; Wang, C.; Lu, W.; Zhang, D. Superelasticity of NiTi–Nb metallurgical bonding via nanoindentation observation. *Mater. Lett.* **2015**, *161*, 255–258. [CrossRef]
11. Attarilar, S.; Yang, J.; Ebrahimi, M.; Wang, Q.; Liu, J.; Tang, Y.; Yang, J. The toxicity phenomenon and the related occurrence in metal and metal oxide nanoparticles: A brief review from the biomedical perspective. *Front. Bioeng. Biotechnol.* **2020**, *8*, 822. [CrossRef]
12. Liu, X.; Chu, P.K.; Ding, C. Surface modification of titanium, titanium alloys, and related materials for biomedical applications. *Mater. Sci. Eng. R Rep.* **2004**, *47*, 49–121. [CrossRef]
13. Kirby, R.; Heard, S.; Miller, P.; Eardley, I.; Holmes, S.; Vale, J.; Bryan, J.; Liu, S. Use of the ASI titanium stent in the management of bladder outflow obstruction due to benign prostatic hyperplasia. *J. Urol.* **1992**, *148*, 1195–1197. [CrossRef]
14. Khodaei, M.; Valanezhad, A.; Watanabe, I.; Yousefi, R. Surface and mechanical properties of modified porous titanium scaffold. *Surf. Coat. Technol.* **2017**, *315*, 61–66. [CrossRef]
15. Kang, H.-G.; Jeong, Y.-S.; Huh, Y.-H.; Park, C.-J.; Cho, L.-R. Impact of surface chemistry modifications on speed and strength of osseointegration. *Int. J. Oral Maxillofac. Implants* **2018**, *33*, 780–787. [CrossRef]
16. Huanhuan, J.; PengJie, H.; Sheng, X.; Binchen, W.; Li, S. The effect of strontium-loaded rough titanium surface on early osseointegration. *J. Biomater. Appl.* **2017**, *32*, 561–569. [CrossRef]
17. Ratner, B.D. A Perspective on Titanium Biocompatibility. In *Titanium in Medicine*; Springer: Berlin/Heidelberg, Germany, 2001; pp. 1–12.
18. Thomsen, P.; Larsson, C.; Ericson, L.E.; Sennerby, L.; Lausmaa, J.; Kasemo, B. Structure of the interface between rabbit cortical bone and implants of gold, zirconium and titanium. *J. Mater. Sci. Mater. Electron.* **1997**, *8*, 653–665. [CrossRef]
19. Rupp, F.; Liang, L.; Geis-Gerstorfer, J.; Scheideler, L.; Hüttig, F. Surface characteristics of dental implants: A review. *Dent. Mater.* **2018**, *34*, 40–57. [CrossRef] [PubMed]
20. Swain, S.; Rautray, T.R. Effect of Surface Roughness on Titanium Medical Implants. In *Nanostructured Materials and Their Applications*; Swain, B.P., Ed.; Springer: Singapore, 2021; pp. 55–80, ISBN 978-981-15-8307-0.
21. Bauer, S.; Park, J.; Faltenbacher, J.; Berger, S.; Von Der Mark, K.; Schmuki, P. Size selective behavior of mesenchymal stem cells on ZrO_2 and TiO_2 nanotube arrays. *Integr. Biol.* **2009**, *1*, 525–532. [CrossRef]
22. Park, J.; Bauer, S.; von der Mark, K.; Schmuki, P. Nanosize and vitality: TiO_2 nanotube diameter directs cell fate. *Nano Lett.* **2007**, *7*, 1686–1691. [CrossRef] [PubMed]
23. Xue, T.; Attarilar, S.; Liu, S.; Liu, J.; Song, X.; Li, L.; Zhao, B.; Tang, Y. Surface modification techniques of titanium and its alloys to functionally optimize their biomedical properties: Thematic review. *Front. Bioeng. Biotechnol.* **2020**, *8*, 1–19. [CrossRef] [PubMed]
24. Prakash, C.; Kansal, H.K.; Pabla, B.; Puri, S.; Aggarwal, A. Electric discharge machining—A potential choice for surface modification of metallic implants for orthopedic applications: A review. *Proc. Inst. Mech. Eng. Part B J. Eng. Manuf.* **2016**, *230*, 331–353. [CrossRef]
25. Kim, Y.-W. Surface modification of Ti dental implants by grit-blasting and micro-arc oxidation. *Mater. Manuf. Process.* **2010**, *25*, 307–310. [CrossRef]
26. Brunette, D.M.; Tengvall, P.; Textor, M.; Thomsen, P. *Titanium in Medicine: Material Science, Surface Science, Engineering, Biological Responses and Medical Applications*; Springer: Berlin/Heidelberg, Germany; New York, NY, USA, 2001; ISBN 3540669361.
27. Wang, L.; Qu, J.; Chen, L.; Meng, Q.; Zhang, L.; Qin, J.; Zhang, D.; Lu, W. Investigation of deformation mechanisms in β-type Ti-35Nb-2Ta-3Zr alloy via FSP leading to surface strengthening. *Met. Mater. Trans. A* **2015**, *46*, 4813–4818. [CrossRef]
28. Xie, L.; Wang, L.; Jiang, C.; Lu, W. The variations of microstructures and hardness of titanium matrix composite (TiB+TiC)/Ti–6Al–4V after shot peening. *Surf. Coat. Technol.* **2014**, *244*, 69–77. [CrossRef]
29. Guo, X.; Lu, W.; Wang, L.; Qin, J. A research on the creep properties of titanium matrix composites rolled with different deformation degrees. *Mater. Des.* **2014**, *63*, 50–55. [CrossRef]
30. Attarilar, S.; Salehi, M.-T.; Djavanroodi, F. Microhardness evolution of pure titanium deformed by equal channel angular extrusion. *Met. Res. Technol.* **2019**, *116*, 408. [CrossRef]
31. Attarilar, S.; Djavanroodi, F.; Irfan, O.; Al-Mufadi, F.; Ebrahimi, M.; Wang, Q. Strain uniformity footprint on mechanical performance and erosion-corrosion behavior of equal channel angular pressed pure titanium. *Results Phys.* **2020**, *17*, 103141. [CrossRef]
32. Zareidoost, A.; Yousefpour, M.; Ghaseme, B.; Amanzadeh, A. The relationship of surface roughness and cell response of chemical surface modification of titanium. *J. Mater. Sci. Mater. Med.* **2012**, *23*, 1479–1488. [CrossRef]
33. Wei, Q.; Wang, L.; Fu, Y.; Qin, J.; Lu, W.; Zhang, D. Influence of oxygen content on microstructure and mechanical properties of Ti–Nb–Ta–Zr alloy. *Mater. Des.* **2011**, *32*, 2934–2939. [CrossRef]
34. Li, J.; Wang, L.; Qin, J.; Chen, Y.; Lu, W.; Zhang, D. The effect of heat treatment on thermal stability of Ti matrix composite. *J. Alloys Compd.* **2011**, *509*, 52–56. [CrossRef]

35. Rangel, A.L.R.; Falentin-daudré, C.; Natália, B.; Pimentel, S.; Eduardo, C.; Migonney, V.; Alves, A.P.R. Nanostructured titanium alloy surfaces for enhanced osteoblast response: A combination of morphology and chemistry. *Surf. Coat. Technol.* **2020**, *383*, 125226. [CrossRef]
36. Tian, L.; Tang, N.; Ngai, T.; Wu, C.; Ruan, Y.; Huang, L.; Qin, L. Hybrid fracture fixation systems developed for orthopaedic applications: A general review. *J. Orthop. Transl.* **2019**, *16*, 1–13. [CrossRef]
37. Wang, T.; Wan, Y.; Liu, Z. Effects of superimposed micro/nano-structured titanium alloy surface on cellular behaviors in vitro. *Adv. Eng. Mater.* **2016**, *18*, 1259–1266. [CrossRef]
38. Lai, M.; Cai, K.; Hu, Y.; Yang, X.; Liu, Q. Regulation of the behaviors of mesenchymal stem cells by surface nanostructured titanium. *Colloids Surf. B Biointerfaces* **2012**, *97*, 211–220. [CrossRef]
39. Moon, S.-K.; Kwon, J.-S.; Uhm, S.-H.; Lee, E.-J.; Gu, H.-J.; Eom, T.-G.; Kim, K.-N. Biological evaluation of micro–nano patterned implant formed by anodic oxidation. *Curr. Appl. Phys.* **2014**, *14*, S183–S187. [CrossRef]
40. Sirdeshmukh, N.; Dongre, G. Laser micro & nano surface texturing for enhancing osseointegration and antimicrobial effect of biomaterials: A review. *Mater. Today Proc.* **2021**, *44*, 2348–2355. [CrossRef]
41. Ständert, V.; Borcherding, K.; Bormann, N.; Schmidmaier, G.; Grunwald, I.; Wildemann, B. Antibiotic-loaded amphora-shaped pores on a titanium implant surface enhance osseointegration and prevent infections. *Bioact. Mater.* **2021**, *6*, 2331–2345. [CrossRef] [PubMed]
42. Zhang, C.; Zhang, T.; Geng, T.; Wang, X.; Lin, K.; Wang, P. Dental implants loaded with bioactive agents promote osseointegration in osteoporosis: A review. *Front. Bioeng. Biotechnol.* **2021**, *9*. [CrossRef] [PubMed]
43. Tsimbouri, P.M.; Fisher, L.; Holloway, N.; Sjostrom, T.; Nobbs, A.H.; Meek, R.M.D.; Su, B.; Dalby, M.J. Osteogenic and bactericidal surfaces from hydrothermal titania nanowires on titanium substrates. *Sci. Rep.* **2016**, *6*, 36857. [CrossRef] [PubMed]
44. Cai, Y.; Bing, W.; Xu, X.; Zhang, Y.; Chen, Z.; Gu, Z. Topographical nanostructures for physical sterilization. *Drug Deliv. Transl. Res.* **2021**, 1–14. [CrossRef]
45. López-Pavón, L.; Dagnino-Acosta, D.; López-Cuéllar, E.; Meléndez-Anzures, F.; Zárate-Triviño, D.; Barrón-González, M.; Moreno-Cortez, I.; Kim, H.Y.; Miyazaki, S. Synthesis of nanotubular oxide on Ti–24Zr–10Nb–2Sn as a drug-releasing system to prevent the growth of *Staphylococcus aureus*. *Chem. Pap.* **2021**, *75*, 2441–2450. [CrossRef]
46. Bonifacio, M.A.; Cerqueni, G.; Cometa, S.; Licini, C.; Sabbatini, L.; Mattioli-Belmonte, M.; De Giglio, E. Insights into arbutin effects on bone cells: Towards the development of antioxidant titanium implants. *Antioxidants* **2020**, *9*, 579. [CrossRef] [PubMed]
47. Ueno, T.; Ikeda, T.; Tsukimura, N.; Ishijima, M.; Minamikawa, H.; Sugita, Y.; Yamada, M.; Wakabayashi, N.; Ogawa, T. Novel antioxidant capability of titanium induced by UV light treatment. *Biomaterials* **2016**, *108*, 177–186. [CrossRef]
48. Ma, P.; Yu, Y.; Yie, K.H.R.; Fang, K.; Zhou, Z.; Pan, X.; Deng, Z.; Shen, X.; Liu, J. Effects of titanium with different micro/nano structures on the ability of osteoblasts to resist oxidative stress. *Mater. Sci. Eng. C* **2021**, *123*, 111969. [CrossRef] [PubMed]
49. Gittens, R.A.; McLachlan, T.; Olivares-Navarrete, R.; Cai, Y.; Berner, S.; Tannenbaum, R.; Schwartz, Z.; Sandhage, K.H.; Boyan, B.D. The effects of combined micron-/submicron-scale surface roughness and nanoscale features on cell proliferation and differentiation. *Biomaterials* **2011**, *32*, 3395–3403. [CrossRef] [PubMed]
50. Wang, Z.; Yan, Y.; Qiao, L. Protein adsorption on implant metals with various deformed surfaces. *Colloids Surf. B Biointerfaces* **2017**, *156*, 62–70. [CrossRef] [PubMed]
51. Guglielmotti, M.B.; Olmedo, D.G.; Cabrini, R.L. Research on implants and osseointegration. *Periodontology 2000* **2019**, *79*, 178–189. [CrossRef] [PubMed]
52. Grassi, S.; Piattelli, A.; De Figueiredo, L.C.; Feres, M.; De Melo, L.; Iezzi, G.; Alba, R.C., Jr.; Shibli, J.A. Histologic evaluation of early human bone response to different implant surfaces. *J. Periodontol.* **2006**, *77*, 1736–1743. [CrossRef]
53. Wennerberg, A.; Albrektsson, T.; Johansson, C.; Andersson, B. Experimental study of turned and grit-blasted screw-shaped implants with special emphasis on effects of blasting material and surface topography. *Biomaterials* **1996**, *17*, 15–22. [CrossRef]
54. Raines, A.L.; Olivares-Navarrete, R.; Wieland, M.; Cochran, D.L.; Schwartz, Z.; Boyan, B.D. Regulation of angiogenesis during osseointegration by titanium surface microstructure and energy. *Biomaterials* **2010**, *31*, 4909–4917. [CrossRef]
55. Cochran, D.L.; Schenk, R.K.; Lussi, A.; Higginbottom, F.L.; Buser, D. Bone response to unloaded and loaded titanium implants with a sandblasted and acid-etched surface: A histometric study in the canine mandible. *J. Biomed. Mater. Res.* **1998**, *40*, 1–11. [CrossRef]
56. Gittens, R.A.; Olivares-Navarrete, R.; Schwartz, Z.; Boyan, B.D. Implant osseointegration and the role of microroughness and nanostructures: Lessons for spine implants. *Acta Biomater.* **2014**, *10*, 3363–3371. [CrossRef] [PubMed]
57. Wennerberg, A.; Albrektsson, T. Implant surfaces beyond micron roughness. experimental and clinical knowledge of surface topography and surface chemistry. *Int. Dent. SA* **2004**, *8*, 14–18.
58. Mustafa, K.; Wroblewski, J.; Lopez, B.S.; Wennerberg, A.; Hultenby, K.; Arvidson, K. Determining optimal surface roughness of TiO_2 blasted titanium implant material for attachment, proliferation and differentiation of cells derived from human mandibular alveolar bone. *Clin. Oral Implants Res.* **2001**, *12*, 515–525. [CrossRef] [PubMed]
59. Rønold, H.; Lyngstadaas, S.; Ellingsen, J. Analysing the optimal value for titanium implant roughness in bone attachment using a tensile test. *Biomaterials* **2003**, *24*, 4559–4564. [CrossRef]
60. Jemat, A.; Ghazali, M.J.; Razali, M.; Otsuka, Y. Surface modifications and their effects on titanium dental implants. *BioMed Res. Int.* **2015**, *2015*, 1–11. [CrossRef] [PubMed]

61. Akkas, T.; Citak, C.; Sirkecioglu, A.; Güner, F.S. Which is more effective for protein adsorption: Surface roughness, surface wettability or swelling? Case study of polyurethane films prepared from castor oil and poly(ethylene glycol). *Polym. Int.* **2013**, *62*, 1202–1209. [CrossRef]
62. De Jonge, L.T.; Leeuwenburgh, S.; Wolke, J.G.C.; Jansen, J.A. Organic-inorganic surface modifications for titanium implant surfaces. *Pharm. Res.* **2008**, *25*, 2357–2369. [CrossRef]
63. Adabi, M.; Naghibzadeh, M.; Adabi, M.; Zarrinfard, M.A.; Esnaashari, S.S.; Seifalian, A.M.; Faridi-Majidi, R.; Aiyelabegan, H.T.; Ghanbari, H. Biocompatibility and nanostructured materials: Applications in nanomedicine. *Artif. Cells Nanomed. Biotechnol.* **2017**, *45*, 833–842. [CrossRef]
64. Cedervall, T.; Lynch, I.; Foy, M.; Berggård, T.; Donnelly, S.C.; Cagney, G.; Linse, S.; Dawson, K.A. Detailed identification of plasma proteins adsorbed on copolymer nanoparticles. *Angew. Chem. Int. Ed.* **2007**, *46*, 5754–5756. [CrossRef]
65. Vasak, C.; Busenlechner, D.; Schwarze, U.Y.; Leitner, H.F.; Guzon, F.M.; Hefti, T.; Schlottig, F.; Gruber, R. Early bone apposition to hydrophilic and hydrophobic titanium implant surfaces: A histologic and histomorphometric study in minipigs. *Clin. Oral Implants Res.* **2013**, *25*, 1378–1385. [CrossRef]
66. Berry, C.C.; Campbell, G.; Spadiccino, A.; Robertson, M.; Curtis, A.S. The influence of microscale topography on fibroblast attachment and motility. *Biomaterials* **2004**, *25*, 5781–5788. [CrossRef] [PubMed]
67. Mata, A.; Hsu, L.; Capito, R.; Aparicio, C.; Henrikson, K.; Stupp, S.I. Micropatterning of bioactive self-assembling gels. *Soft Matter* **2009**, *5*, 1228–1236. [CrossRef]
68. Dalby, M.J.; McCloy, D.; Robertson, M.; Wilkinson, C.D.; Oreffo, R.O. Osteoprogenitor response to defined topographies with nanoscale depths. *Biomaterials* **2006**, *27*, 1306–1315. [CrossRef] [PubMed]
69. Pérez-Garnes, M.; González-García, C.; Moratal, D.; Rico, P.; Salmerón-Sánchez, M. Fibronectin distribution on demixed nanoscale topographies. *Int. J. Artif. Organs* **2011**, *34*, 54–63. [CrossRef]
70. González-García, C.; Sousa, S.R.; Moratal, D.; Rico, P.; Salmerón-Sánchez, M. Effect of nanoscale topography on fibronectin adsorption, focal adhesion size and matrix organisation. *Colloids Surf. B Biointerfaces* **2010**, *77*, 181–190. [CrossRef] [PubMed]
71. Zhou, A.; Yu, H.; Liu, J.; Zheng, J.; Jia, Y.; Wu, B.; Xiang, L. Role of hippo-YAP signaling in osseointegration by regulating osteogenesis, angiogenesis, and osteoimmunology. *Front. Cell Dev. Biol.* **2020**, *8*, 8. [CrossRef]
72. Damiati, L.; Eales, M.G.; Nobbs, A.H.; Su, B.; Tsimbouri, P.M.; Salmeron-Sanchez, M.; Dalby, M.J. Impact of surface topography and coating on osteogenesis and bacterial attachment on titanium implants. *J. Tissue Eng.* **2018**, *9*. [CrossRef] [PubMed]
73. Le Guéhennec, L.; Soueidan, A.; Layrolle, P.; Amouriq, Y. Surface treatments of titanium dental implants for rapid osseointegration. *Dent. Mater.* **2007**, *23*, 844–854. [CrossRef]
74. Lossdorfer, S.; Schwartz, Z.; Wang, L.; Lohmann, C.; Turner, J.D.; Wieland, M.; Cochran, D.L.; Boyan, B.D. Microrough implant surface topographies increase osteogenesis by reducing osteoclast formation and activity. *J. Biomed. Mater. Res. Part A* **2004**, *70*, 361–369. [CrossRef]
75. Huo, S.-C.; Yue, B. Approaches to promoting bone marrow mesenchymal stem cell osteogenesis on orthopedic implant surface. *World J. Stem Cells* **2020**, *12*, 545–561. [CrossRef]
76. Gulati, K.; Maher, S.; Findlay, D.M.; Losic, D. Titania nanotubes for orchestrating osteogenesis at the bone–implant interface. *Nanomedicine* **2016**, *11*, 1847–1864. [CrossRef]
77. Shen, X.; Yu, Y.; Ma, P.; Luo, Z.; Hu, Y.; Li, M.; He, Y.; Zhang, Y.; Peng, Z.; Song, G.; et al. Titania nanotubes promote osteogenesis via mediating crosstalk between macrophages and MSCs under oxidative stress. *Colloids Surf. B Biointerfaces* **2019**, *180*, 39–48. [CrossRef]
78. Huang, J.; Zhang, X.; Yan, W.; Chen, Z.; Shuai, X.; Wang, A.; Wang, Y. Nanotubular topography enhances the bioactivity of titanium implants. *Nanomed. Nanotechnol. Biol. Med.* **2017**, *13*, 1913–1923. [CrossRef]
79. Broggini, N.; McManus, L.M.; Hermann, J.S.; Medina, R.U.; Schenk, R.K.; Buser, D.; Cochran, D.L. Peri-implant inflammation defined by the implant-abutment interface. *J. Dent. Res.* **2006**, *85*, 473–478. [CrossRef]
80. Teterycz, D.; Ferry, T.; Lew, D.; Stern, R.; Assal, M.; Hoffmeyer, P.; Bernard, L.; Uçkay, I. Outcome of orthopedic implant infections due to different staphylococci. *Int. J. Infect. Dis.* **2010**, *14*, e913–e918. [CrossRef]
81. Gomes, F.I.; Teixeira, P.; Oliveira, R. Mini-review: *Staphylococcus epidermidis* as the most frequent cause of nosocomial infections: Old and new fighting strategies. *Biofouling* **2014**, *30*, 131–141. [CrossRef] [PubMed]
82. Uppu, D.S.S.M.; Konai, M.M.; Sarkar, P.; Samaddar, S.; Fenserseifer, I.C.M.; Farias-Junior, C.; Krishnamoorthy, P.; Shome, B.R.; Franco, O.L.; Haldar, J. Membrane-active macromolecules kill antibiotic-tolerant bacteria and potentiate antibiotics towards Gram-negative bacteria. *PLoS ONE* **2017**, *12*, e0183263. [CrossRef] [PubMed]
83. Lakshmipraba, J.; Prabhu, R.N.; Sivasankar, V. Polymer Macromolecules to Polymeric Nanostructures: Efficient Antibacterial Candidates. In *Nanostructures for Antimicrobial and Antibiofilm Applications*; Springer: Cham, Switzerland, 2020; pp. 209–232. [CrossRef]
84. Li, S.; Dong, S.; Xu, W.; Tu, S.; Yan, L.; Zhao, C.; Ding, J.; Chen, X. Antibacterial hydrogels. *Adv. Sci.* **2018**, *5*, 1700527. [CrossRef] [PubMed]
85. Vimbela, G.V.; Ngo, S.M.; Fraze, C.; Yang, L.; Stout, D.A. Antibacterial properties and toxicity from metallic nanomaterials. *Int. J. Nanomed.* **2017**, *12*, 3941–3965. [CrossRef]

86. Gupta, N.; Santhiya, D.; Murugavel, S.; Kumar, A.; Aditya, A.; Ganguli, M.; Gupta, S. Effects of transition metal ion dopants (Ag, Cu and Fe) on the structural, mechanical and antibacterial properties of bioactive glass. *Colloids Surf. A Physicochem. Eng. Asp.* **2018**, *538*, 393–403. [CrossRef]
87. Goudouri, O.-M.; Kontonasaki, E.; Lohbauer, U.; Boccaccini, A.R. Antibacterial properties of metal and metalloid ions in chronic periodontitis and peri-implantitis therapy. *Acta Biomater.* **2014**, *10*, 3795–3810. [CrossRef] [PubMed]
88. Wennerberg, A.; Albrektsson, T. Effects of titanium surface topography on bone integration: A systematic review. *Clin. Oral Implants Res.* **2009**, *20*, 172–184. [CrossRef] [PubMed]
89. Szmukler-Moncler, S.; Testori, T.; Bernard, J.P. Etched implants: A comparative surface analysis of four implant systems. *J. Biomed. Mater. Res.* **2004**, *69*, 46–57. [CrossRef]
90. Li, D.; Ferguson, S.J.; Beutler, T.; Cochran, D.L.; Sittig, C.; Hirt, H.P.; Buser, D. Biomechanical comparison of the sandblasted and acid-etched and the machined and acid-etched titanium surface for dental implants. *J. Biomed. Mater. Res.* **2002**, *60*, 325–332. [CrossRef]
91. Roehling, S.K.; Meng, B.; Cochran, D.L. Sandblasted and Acid-Etched Implant Surfaces with or without High Surface Free Energy: Experimental and Clinical Background BT. In *Implant Surfaces and Their Biological and Clinical Impact*; Wennerberg, A., Albrektsson, T., Jimbo, R., Eds.; Springer: Berlin/Heidelberg, Germany, 2015; pp. 93–136, ISBN 978-3-662-45379-7.
92. Yang, G.-L.; He, F.-M.; Yang, X.-F.; Wang, X.-X.; Zhao, S.-F. Bone responses to titanium implants surface-roughened by sandblasted and double etched treatments in a rabbit model. *Oral Sur. Oral Med. Oral Pathol. Oral Radiol. Endodontol.* **2008**, *106*, 516–524. [CrossRef]
93. Ni, J.; Ling, H.; Zhang, S.; Wang, Z.; Peng, Z.; Benyshek, C.; Zan, R.; Miri, A.K.; Li, Z.; Zhang, X.; et al. Three-dimensional printing of metals for biomedical applications. *Mater. Today Bio* **2019**, *3*, 100024. [CrossRef] [PubMed]
94. Tiainen, L.; Abreu, P.; Buciumeanu, M.; Silva, F.; Gasik, M.; Guerrero, R.S.; Carvalho, O. Novel laser surface texturing for improved primary stability of titanium implants. *J. Mech. Behav. Biomed. Mater.* **2019**, *98*, 26–39. [CrossRef]
95. Klokkevold, P.R.; Johnson, P.; Dadgostari, S.; Davies, J.E.; Caputo, A.; Nishimura, R.D. Early endosseous integration enhanced by dual acid etching of titanium: A torque removal study in the rabbit. *Clin. Oral Implants Res.* **2001**, *12*, 350–357. [CrossRef] [PubMed]
96. Wang, D.; He, G.; Tian, Y.; Ren, N.; Liu, W.; Zhang, X. Dual effects of acid etching on cell responses and mechanical properties of porous titanium with controllable open-porous structure. *J. Biomed. Mater. Res. Part B Appl. Biomater.* **2020**, *108*, 2386–2395. [CrossRef]
97. Szmukler-Moncler, S.; Perrin, D.; Ahossi, V.; Magnin, G.; Bernard, J.P. Biological properties of acid etched titanium implants: Effect of sandblasting on bone anchorage. *J. Biomed. Mater. Res.* **2004**, *68*, 149–159. [CrossRef] [PubMed]
98. Yabutsuka, T.; Mizuno, H.; Takai, S. Fabrication of bioactive titanium and its alloys by combination of doubled sandblasting process and alkaline simulated body fluid treatment. *J. Ceram. Soc. Jpn.* **2019**, *127*, 669–677. [CrossRef]
99. Baleani, M.; Viceconti, M.; Toni, A. The effect of sandblasting treatment on endurance properties of titanium alloy hip prostheses. *Artif. Organs* **2000**, *24*, 296–299. [CrossRef] [PubMed]
100. Li, D.-H.; Liu, B.-L.; Zou, J.-C.; Xu, K.-W. Improvement of osseointegration of titanium dental implants by a modified sandblasting surface treatment. *Implant Dent.* **1999**, *8*, 289–294. [CrossRef]
101. Lubas, M. The impact of an innovative sandblasting medium on the titanium-dental porcelain joint. *Ceram. Mater.* **2019**, *71*, 276–285.
102. Yuda, A.W.; Supriadi, S.; Saragih, A.S. Surface modification of Ti-alloy based bone implant by sandblasting. In Proceedings of the 4th Biomedical Engineering's Recent Progress in Biomaterials, Drugs Development, Health, and Medical Devices, Padang, Indonesia, 22–24 July 2019; p. 020015.
103. Attarilar, S.; Ebrahimi, M.; Djavanroodi, F.; Fu, Y.; Wang, L.; Yang, J. 3D Printing technologies in metallic implants: A thematic review on the techniques and procedures. *Int. J. Bioprint.* **2020**, *7*. [CrossRef]
104. Fousová, M.; Vojtěch, D.; Doubrava, K.; Daniel, M.; Lin, C.-F. Influence of inherent surface and internal defects on mechanical properties of additively manufactured Ti6Al4V alloy: Comparison between selective laser melting and electron beam melting. *Materials* **2018**, *11*, 537. [CrossRef]
105. Wang, Q.; Zhou, P.; Liu, S.; Attarilar, S.; Ma, R.L.-W.; Zhong, Y.; Wang, L. Multi-scale surface treatments of titanium implants for rapid osseointegration: A review. *Nanomaterials* **2020**, *10*, 1244. [CrossRef]
106. Kunrath, M.F. Customized dental implants: Manufacturing processes, topography, osseointegration and future perspectives of 3D fabricated implants. *Bioprinting* **2020**, *20*, e00107. [CrossRef]
107. Srivas, P.K.; Kapat, K.; Dadhich, P.; Pal, P.; Dutta, J.; Datta, P.; Dhara, S. Osseointegration assessment of extrusion printed Ti6Al4V scaffold towards accelerated skeletal defect healing via tissue in-growth. *Bioprinting* **2017**, *6*, 8–17. [CrossRef]
108. Wang, H.; Liu, J.; Wang, C.; Shen, S.G.; Wang, X.; Lin, K. The synergistic effect of 3D-printed microscale roughness surface and nanoscale feature on enhancing osteogenic differentiation and rapid osseointegration. *J. Mater. Sci. Technol.* **2021**, *63*, 18–26. [CrossRef]
109. Gulati, K.; Prideaux, M.; Kogawa, M.; Lima-Marques, L.; Atkins, G.J.; Findlay, D.M.; Losic, D. Anodized 3D-printed titanium implants with dual micro- and nano-scale topography promote interaction with human osteoblasts and osteocyte-like cells. *J. Tissue Eng. Regen. Med.* **2017**, *11*, 3313–3325. [CrossRef]

110. Baker, T. Laser surface modification of titanium alloys. In *Surface Engineering of Light Alloys*; Elsevier: Amsterdam, The Netherlands, 2010; pp. 398–443.
111. Kurella, A.; Dahotre, N.B. Review paper: Surface modification for bioimplants: The role of laser surface engineering. *J. Biomater. Appl.* 2005, *20*, 5–50. [CrossRef]
112. He, A.; Yang, H.; Xue, W.; Sun, K.; Cao, Y. Tunable coffee-ring effect on a superhydrophobic surface. *Opt. Lett.* 2017, *42*, 3936–3939. [CrossRef] [PubMed]
113. Pou, P.; Riveiro, A.; Val, J.; Comesaña, R.; Penide, J.; Arias-González, F.; Sotoa, R.; Lusquiños, F.; Pou, J. Laser surface texturing of Titanium for bioengineering applications. *Procedia Manuf.* 2017, *13*, 694–701. [CrossRef]
114. Mirhosseini, N.; Crouse, P.; Schmidt, M.; Li, L.; Garrod, D. Laser surface micro-texturing of Ti–6Al–4V substrates for improved cell integration. *Appl. Surf. Sci.* 2007, *253*, 7738–7743. [CrossRef]
115. Cunha, A.; Elie, A.-M.; Plawinski, L.; Serro, A.P.; Rego, A.M.B.D.; Almeida, A.; Urdaci, M.C.; Durrieu, M.-C.; Vilar, R. Femtosecond laser surface texturing of titanium as a method to reduce the adhesion of *Staphylococcus aureus* and biofilm formation. *Appl. Surf. Sci.* 2016, *360*, 485–493. [CrossRef]
116. Chiang, H.-J.; Hsu, H.-J.; Peng, P.-W.; Wu, C.-Z.; Ou, K.-L.; Cheng, H.-Y.; Walinski, C.J.; Sugiatno, E. Early bone response to machined, sandblasting acid etching (SLA) and novel surface-functionalization (SLAffinity) titanium implants: Characterization, biomechanical analysis and histological evaluation in pigs. *J. Biomed. Mater. Res. Part A* 2016, *104*, 397–405. [CrossRef]
117. Liu, R.; Tang, Y.; Liu, H.; Zeng, L.; Ma, Z.; Li, J.; Zhao, Y.; Ren, L.; Yang, K. Effects of combined chemical design (Cu addition) and topographical modification (SLA) of Ti-Cu/SLA for promoting osteogenic, angiogenic and antibacterial activities. *J. Mater. Sci. Technol.* 2020, *47*, 202–215. [CrossRef]
118. Choi, S.-M.; Park, J.-W. Multifunctional effects of a modification of SLA titanium implant surface with strontium-containing nanostructures on immunoinflammatory and osteogenic cell function. *J. Biomed. Mater. Res. Part A* 2018, *106*, 3009–3020. [CrossRef] [PubMed]
119. Kim, B.-S.; Kim, J.S.; Park, Y.M.; Choi, B.-Y.; Lee, J. Mg ion implantation on SLA-treated titanium surface and its effects on the behavior of mesenchymal stem cell. *Mater. Sci. Eng. C* 2013, *33*, 1554–1560. [CrossRef]
120. Bobzin, K.; Ote, M.; Linke, T.F.; Sommer, J.; Liao, X. Influence of process parameter on grit blasting as a pretreatment process for thermal spraying. *J. Therm. Spray Technol.* 2016, *25*, 3–11. [CrossRef]
121. Galvan, J.C.; Saldaña, L.; Multigner, M.; Calzado-Martín, A.; Larrea, M.T.; Serra, C.; Vilaboa, N.; González-Carrasco, J.L. Grit blasting of medical stainless steel: Implications on its corrosion behavior, ion release and biocompatibility. *J. Mater. Sci. Mater. Electron.* 2012, *23*, 657–666. [CrossRef]
122. Ivanoff, C.-J.; Widmark, G.; Hallgren, C.; Sennerby, L.; Wennerberg, A. Histologic evaluation of the bone integration of TiO_2 blasted and turned titanium microimplants in humans. *Clin. Oral Implants Res.* 2001, *12*, 128–134. [CrossRef]
123. Salou, L.; Hoornaert, A.; Louarn, G.; Layrolle, P. Enhanced osseointegration of titanium implants with nanostructured surfaces: An experimental study in rabbits. *Acta Biomater.* 2015, *11*, 494–502. [CrossRef]
124. Kurup, A.; Dhatrak, P.; Khasnis, N. Surface modification techniques of titanium and titanium alloys for biomedical dental applications: A review. *Mater. Today Proc.* 2021, *39*, 84–90. [CrossRef]
125. Smeets, R.; Stadlinger, B.; Schwarz, F.; Beck-Broichsitter, B.; Jung, O.; Precht, C.; Kloss, F.; Gröbe, A.; Heiland, M.; Ebker, T. Impact of dental implant surface modifications on osseointegration. *BioMed Res. Int.* 2016, *2016*, 1–16. [CrossRef] [PubMed]
126. Saghiri, M.; Asatourian, A.; Garcia-Godoy, F.; Sheibani, N. The role of angiogenesis in implant dentistry part I: Review of titanium alloys, surface characteristics and treatments. *Med. Oral Patol. Oral Cir. Bucal* 2016, *21*, e514–e525. [CrossRef] [PubMed]
127. Diomede, F.; Marconi, G.D.; Cavalcanti, M.F.X.B.; Pizzicannella, J.; Pierdomenico, S.D.; Fonticoli, L.; Piattelli, A.; Trubiani, O. VEGF/VEGF-R/RUNX2 Upregulation in human periodontal ligament stem cells seeded on dual acid etched titanium disk. *Materials* 2020, *13*, 706. [CrossRef] [PubMed]
128. Liu, Y.; Luo, D.; Wang, T. Hierarchical structures of bone and bioinspired bone tissue engineering. *Small* 2016, *12*, 4611–4632. [CrossRef]
129. Souza, J.C.; Sordi, M.B.; Kanazawa, M.; Ravindran, S.; Henriques, B.; Silva, F.S.; Aparicio, C.; Cooper, L.F. Nano-scale modification of titanium implant surfaces to enhance osseointegration. *Acta Biomater.* 2019, *94*, 112–131. [CrossRef]
130. Coelho, P.G.; Jimbo, R.; Tovar, N.; Bonfante, E.A. Osseointegration: Hierarchical designing encompassing the macrometer, micrometer, and nanometer length scales. *Dent. Mater.* 2015, *31*, 37–52. [CrossRef] [PubMed]
131. Attarilar, S.; Salehi, M.T.; Al-Fadhalah, K.J.; Djavanroodi, F.; Mozafari, M. Functionally graded titanium implants: Characteristic enhancement induced by combined severe plastic deformation. *PLoS ONE* 2019, *14*, e0221491. [CrossRef] [PubMed]
132. Attarilar, S.; Djavanroodi, F.; Ebrahimi, M.; Al-Fadhalah, K.J.; Wang, L.; Mozafari, M. Hierarchical microstructure tailoring of pure titanium for enhancing cellular response at tissue-implant interface. *J. Biomed. Nanotechnol.* 2021, *17*, 115–130. [CrossRef] [PubMed]
133. Meirelles, L.; Uzumaki, E.T.; Lima, J.H.C.; Lambert, C.S.; Muller, C.A.; Albrektsson, T.; Wennerberg, A. A novel technique for tailored surface modification of dental implants—A step wise approach based on plasma immersion ion implantation. *Clin. Oral Implants Res.* 2011, *24*, 461–467. [CrossRef] [PubMed]
134. Liu, J.; Liu, J.; Attarilar, S.; Wang, C.; Tamaddon, M.; Yang, C.; Xie, K.; Yao, J.; Wang, L.; Liu, C.; et al. Nano-modified titanium implant materials: A way toward improved antibacterial properties. *Front. Bioeng. Biotechnol.* 2020, *8*, 576969. [CrossRef] [PubMed]

135. Xu, R.; Hu, X.; Yu, X.; Wan, S.; Wu, F.; Ouyang, J.; Deng, F. Micro-/nano-topography of selective laser melting titanium enhances adhesion and proliferation and regulates adhesion-related gene expressions of human gingival fibroblasts and human gingival epithelial cells. *Int. J. Nanomed.* **2018**, *13*, 5045–5057. [CrossRef]
136. Mendonca, G.; Mendonça, D.B.; Aragão, F.J.; Cooper, L.F. Advancing dental implant surface technology—From micron-to nanotopography. *Biomaterials* **2008**, *29*, 3822–3835. [CrossRef] [PubMed]
137. Yoo, D.; Tovar, N.; Jimbo, R.; Marin, C.; Anchieta, R.B.; Machado, S.; Montclare, J.; Guastaldi, F.P.S.; Janal, M.N.; Coelho, P.G. Increased osseointegration effect of bone morphogenetic protein 2 on dental implants: An in vivo study. *J. Biomed. Mater. Res. Part A* **2014**, *102*, 1921–1927. [CrossRef] [PubMed]
138. Li, X.; Qi, M.; Sun, X.; Weir, M.D.; Tay, F.R.; Oates, T.W.; Dong, B.; Zhou, Y.; Wang, L.; Xu, H.H. Surface treatments on titanium implants via nanostructured ceria for antibacterial and anti-inflammatory capabilities. *Acta Biomater.* **2019**, *94*, 627–643. [CrossRef] [PubMed]
139. İzmir, M.; Ercan, B. Anodization of titanium alloys for orthopedic applications. *Front. Chem. Sci. Eng.* **2019**, *13*, 28–45. [CrossRef]
140. Yao, C.; Lu, J.; Webster, T.J. Titanium and cobalt–chromium alloys for hips and knees. In *Biomaterials for Artificial Organs*; Elsevier: Amsterdam, The Netherlands, 2011; pp. 34–55.
141. Minagar, S.; Berndt, C.C.; Wang, J.; Ivanova, E.; Wen, C. A review of the application of anodization for the fabrication of nanotubes on metal implant surfaces. *Acta Biomater.* **2012**, *8*, 2875–2888. [CrossRef] [PubMed]
142. Wang, Q.; Huang, J.-Y.; Li, H.-Q.; Zhao, A.Z.-J.; Wang, Y.; Zhang, K.-Q.; Sun, H.-T.; Lai, Y.-K. Recent advances on smart TiO_2 nanotube platforms for sustainable drug delivery applications. *Int. J. Nanomed.* **2016**, *12*, 151–165. [CrossRef] [PubMed]
143. Peng, L.; Mendelsohn, A.D.; LaTempa, T.J.; Yoriya, S.; Grimes, C.A.; Desai, T.A. Long-term small molecule and protein elution from TiO_2 nanotubes. *Nano Lett.* **2009**, *9*, 1932–1936. [CrossRef] [PubMed]
144. Setyawati, M.I.; Sevencan, C.; Bay, B.H.; Xie, J.; Zhang, Y.; Demokritou, P.; Leong, D.T. Nano-TiO_2 drives epithelial-mesenchymal transition in intestinal epithelial cancer cells. *Small* **2018**, *14*, e1800922. [CrossRef]
145. Linklater, D.P.; Baulin, V.A.; Juodkazis, S.; Crawford, R.J.; Stoodley, P.; Ivanova, E.P. Mechano-bactericidal actions of nanostructured surfaces. *Nat. Rev. Genet.* **2021**, *19*, 8–22. [CrossRef]
146. Smith, Y.R.; Ray, R.S.; Carlson, K.; Sarma, B.; Misra, M. Self-ordered titanium dioxide nanotube arrays: Anodic synthesis and their photo/electro-catalytic applications. *Materials* **2013**, *6*, 2892–2957. [CrossRef] [PubMed]
147. Fu, Y.; Mo, A. A review on the electrochemically self-organized titania nanotube arrays: Synthesis, modifications, and biomedical applications. *Nanoscale Res. Lett.* **2018**, *13*, 1–21. [CrossRef]
148. Wang, N.; Li, H.; Lü, W.; Li, J.; Wang, J.; Zhang, Z.; Liu, Y. Effects of TiO_2 nanotubes with different diameters on gene expression and osseointegration of implants in minipigs. *Biomaterials* **2011**, *32*, 6900–6911. [CrossRef]
149. Su, E.P.; Justin, D.F.; Pratt, C.R.; Sarin, V.K.; Nguyen, V.S.; Oh, S.; Jin, S. Effects of titanium nanotubes on the osseointegration, cell differentiation, mineralisation and antibacterial properties of orthopaedic implant surfaces. *Bone Jt. J.* **2018**, *100-B*, 9–16. [CrossRef]
150. Choi, W.-Y.; Lee, J.-K.; Choi, D.-S.; Jang, I. Improved osseointegration of dental titanium implants by TiO_2 nanotube arrays with recombinant human bone morphogenetic protein-2: A pilot in vivo study. *Int. J. Nanomed.* **2015**, *10*, 1145–1154. [CrossRef] [PubMed]
151. Hu, N.; Wu, Y.; Xie, L.; Yusuf, S.M.; Gao, N.; Starink, M.J.; Tong, L.; Chu, P.K.; Wang, H. Enhanced interfacial adhesion and osseointegration of anodic TiO_2 nanotube arrays on ultra-fine-grained titanium and underlying mechanisms. *Acta Biomater.* **2020**, *106*, 360–375. [CrossRef] [PubMed]
152. Sargin, F.; Erdogan, G.; Kanbur, K.; Turk, A. Investigation of in vitro behavior of plasma sprayed Ti, TiO_2 and HA coatings on PEEK. *Surf. Coat. Technol.* **2021**, *411*, 126965. [CrossRef]
153. Lee, B.-H.; Kim, J.K.; Kim, Y.D.; Choi, K.; Lee, K.H. In vivo behavior and mechanical stability of surface-modified titanium implants by plasma spray coating and chemical treatments. *J. Biomed. Mater. Res.* **2004**, *69*, 279–285. [CrossRef]
154. Singh, H.; Kumar, R.; Prakash, C.; Singh, S. HA-based coating by plasma spray techniques on titanium alloy for orthopedic applications. *Mater. Today Proc.* **2021**. [CrossRef]
155. Roy, M.; Balla, V.K.; Bandyopadhyay, A.; Bose, S. Compositionally graded hydroxyapatite/tricalcium phosphate coating on Ti by laser and induction plasma. *Acta Biomater.* **2011**, *7*, 866–873. [CrossRef] [PubMed]
156. Van Oirschot, B.A.J.A.; Bronkhorst, E.; van den Beucken, J.J.J.P.; Meijer, G.J.; Jansen, J.A.; Junker, R. A systematic review on the long-term success of calcium phosphate plasma-spray-coated dental implants. *Odontology* **2016**, *104*, 347–356. [CrossRef]
157. Li, B.; Liu, X.; Meng, F.; Chang, J.; Ding, C. Preparation and antibacterial properties of plasma sprayed nano-titania/silver coatings. *Mater. Chem. Phys.* **2009**, *118*, 99–104. [CrossRef]
158. Khor, K.; Gu, Y.; Quek, C.; Cheang, P. Plasma spraying of functionally graded hydroxyapatite/Ti–6Al–4V coatings. *Surf. Coat. Technol.* **2003**, *168*, 195–201. [CrossRef]
159. Singh, H.; Rana, P.K.; Singh, J.; Singh, S.; Prakash, C.; Królczyk, G. Plasma Spray Deposition of HA–TiO_2 Composite Coating on Ti–6Al–4V Alloy for Orthopedic Applications. In *Advances in Materials Processing*; Singh, S., Prakash, C., Ramakrishna, S., Królczyk, G., Eds.; Springer: Singapore, 2020; pp. 13–20, ISBN 978-981-15-4748-5.
160. Hameed, P.; Gopal, V.; Bjorklund, S.; Ganvir, A.; Sen, D.; Markocsan, N.; Manivasagam, G. Axial suspension plasma spraying: An ultimate technique to tailor Ti6Al4V surface with HAp for orthopaedic applications. *Colloids Surf. B Biointerfaces* **2019**, *173*, 806–815. [CrossRef]

161. Dietsch, R.; Holz, T.; Mai, H.; Panzner, M.; Völlmar, S. Pulsed laser deposition (PLD)—An advanced state for technical applications. *Opt. Quantum Electron.* **1995**, *27*, 1385–1396. [CrossRef]
162. Prasanna, S.R.V.S.; Balaji, K.; Pandey, S.; Rana, S. Metal Oxide Based Nanomaterials and Their Polymer Nanocomposites. In *Nanomaterials and Polymer Nanocomposites*; Elsevier: Amsterdam, The Netherlands, 2019; pp. 123–144.
163. Duta, L. In vivo assessment of synthetic and biological-derived calcium phosphate-based coatings fabricated by pulsed laser deposition: A review. *Coatings* **2021**, *11*, 99. [CrossRef]
164. Galindo-Valdés, J.; Cortés-Hernández, D.; Ortiz-Cuellar, J.; De la O-Baquera, E.; Escobedo-Bocardo, J.; Acevedo-Dávila, J. Laser deposition of bioactive coatings by in situ synthesis of pseudowollastonite on Ti6Al4V alloy. *Opt. Laser Technol.* **2021**, *134*, 106586. [CrossRef]
165. Cao, J.; Lian, R.; Jiang, X.; Rogachev, A.V. In vitro degradation assessment of calcium fluoride-doped hydroxyapatite coating prepared by pulsed laser deposition. *Surf. Coat. Technol.* **2021**, *416*, 127177. [CrossRef]
166. Cao, J.; Lian, R.; Jiang, X. Magnesium and fluoride doped hydroxyapatite coatings grown by pulsed laser deposition for promoting titanium implant cytocompatibility. *Appl. Surf. Sci.* **2020**, *515*, 146069. [CrossRef]
167. Zaveri, N.; Mahapatra, M.; Deceuster, A.; Peng, Y.; Li, L.; Zhou, A. Corrosion resistance of pulsed laser-treated Ti–6Al–4V implant in simulated biofluids. *Electrochim. Acta* **2008**, *53*, 5022–5032. [CrossRef]
168. Vishwakarma, V.; Josephine, J.; George, R.P.; Krishnan, R.; Dash, S.; Kamruddin, M.; Kalavathi, S.; Manoharan, N.; Tyagi, A.K.; Dayal, R.K. Antibacterial copper–nickel bilayers and multilayer coatings by pulsed laser deposition on titanium. *Biofouling* **2009**, *25*, 705–710. [CrossRef]
169. Rajesh, P.; Mohan, N.; Yokogawa, Y.; Varma, H. Pulsed laser deposition of hydroxyapatite on nanostructured titanium towards drug eluting implants. *Mater. Sci. Eng. C* **2013**, *33*, 2899–2904. [CrossRef]
170. Chen, L.; Komasa, S.; Hashimoto, Y.; Hontsu, S.; Okazaki, J. In vitro and in vivo osteogenic activity of titanium implants coated by pulsed laser deposition with a thin film of fluoridated hydroxyapatite. *Int. J. Mol. Sci.* **2018**, *19*, 1127. [CrossRef] [PubMed]
171. Pelletier, H.; Nelea, V.; Mille, P.; Muller, D. Mechanical properties of pulsed laser-deposited hydroxyapatite thin film implanted at high energy with N+ and Ar+ ions. Part I: Nanoindentation with spherical tipped indenter. *Nucl. Instrum. Methods Phys. Res. Sect. B Beam Interact. Mater. At.* **2004**, *216*, 269–274. [CrossRef]
172. Duta, L.M.; Oktar, F.; Stan, G.; Popescu-Pelin, G.; Serban, N.; Luculescu, C.; Mihailescu, I. Novel doped hydroxyapatite thin films obtained by pulsed laser deposition. *Appl. Surf. Sci.* **2013**, *265*, 41–49. [CrossRef]
173. Mihailescu, I.N.; Ristoscu, C.; Bigi, A.; Mayer, I. Advanced Biomimetic Implants Based on Nanostructured Coatings Synthe-sized by Pulsed Laser Technologies. In *Laser-Surface Interactions for New Materials Production*; Springer: New York, NY, USA, 2010; pp. 235–260.
174. Vijayan, P.P.; Puglia, D. Biomimetic multifunctional materials: A review. *Emergent Mater.* **2019**, *2*, 391–415. [CrossRef]
175. Han, Z.; Mu, Z.; Yin, W.; Li, W.; Niu, S.; Zhang, J.; Ren, L. Biomimetic multifunctional surfaces inspired from animals. *Adv. Colloid Interface Sci.* **2016**, *234*, 27–50. [CrossRef]
176. Spriano, S.; Yamaguchi, S.; Baino, F.; Ferraris, S. A critical review of multifunctional titanium surfaces: New frontiers for improving osseointegration and host response, avoiding bacteria contamination. *Acta Biomater.* **2018**, *79*, 1–22. [CrossRef] [PubMed]
177. Spriano, S.; Ferraris, S.; Pan, G.; Cassinelli, C.; Vernè, E. Multifunctional titanium: Surface modification process and biological response. *J. Mech. Med. Biol.* **2015**, *15*, 1540001. [CrossRef]
178. Liu, X.; Zhao, X.; Fu, R.K.; Ho, J.P.; Ding, C.; Chu, P.K. Plasma-treated nanostructured TiO_2 surface supporting biomimetic growth of apatite. *Biomaterials* **2005**, *26*, 6143–6150. [CrossRef] [PubMed]
179. Gulati, K.; Ramakrishnan, S.; Aw, M.S.; Atkins, G.J.; Findlay, D.M.; Losic, D. Biocompatible polymer coating of titania nanotube arrays for improved drug elution and osteoblast adhesion. *Acta Biomater.* **2012**, *8*, 449–456. [CrossRef] [PubMed]
180. Xu, D.; Yang, W.; Hu, Y.; Luo, Z.; Li, J.; Hou, Y.; Liu, Y.; Cai, K. Surface functionalization of titanium substrates with cecropin B to improve their cytocompatibility and reduce inflammation responses. *Colloids Surf. B Biointerfaces* **2013**, *110*, 225–235. [CrossRef]
181. Zhou, L.; Lin, Z.; Ding, J.; Huang, W.; Chen, J.; Wu, D. Inflammatory and biocompatibility evaluation of antimicrobial peptide GL13K immobilized onto titanium by silanization. *Colloids Surf. B Biointerfaces* **2017**, *160*, 581–588. [CrossRef]
182. Nie, B.; Long, T.; Li, H.; Wang, X.; Yue, B. A comparative analysis of antibacterial properties and inflammatory responses for the KR-12 peptide on titanium and PEGylated titanium surfaces. *RSC Adv.* **2017**, *7*, 34321–34330. [CrossRef]
183. Wang, L.; Xie, L.; Shen, P.; Fan, Q.; Wang, W.; Wang, K.; Lu, W.; Hua, L.; Zhang, L.-C. Surface microstructure and mechanical properties of Ti-6Al-4V/Ag nanocomposite prepared by FSP. *Mater. Charact.* **2019**, *153*, 175–183. [CrossRef]
184. Li, P.; Li, X.; Cai, W.; Chen, H.; Chen, H.; Wang, R.; Zhao, Y.; Wang, J.; Huang, N. Phospholipid-based multifunctional coating via layer-by-layer self-assembly for biomedical applications. *Mater. Sci. Eng. C* **2020**, *116*, 111237. [CrossRef]
185. Venkateswarlu, K.; Rameshbabu, N.; Bose, A.C.; Muthupandi, V.; Subramanian, S.; MubarakAli, D.; Thajuddin, N. Fabrication of corrosion resistant, bioactive and antibacterial silver substituted hydroxyapatite/titania composite coating on Cp Ti. *Ceram. Int.* **2012**, *38*, 731–740. [CrossRef]
186. Singh, R.K.; Awasthi, S.; Dhayalan, A.; Ferreira, J.; Kannan, S. Deposition, structure, physical and invitro characteristics of Ag-doped β-Ca3(PO4)2/chitosan hybrid composite coatings on Titanium metal. *Mater. Sci. Eng. C* **2016**, *62*, 692–701. [CrossRef] [PubMed]

187. Zhang, X.; Wang, H.; Li, J.; He, X.; Hang, R.; Huang, X.; Tian, L.; Tang, B. Corrosion behavior of Zn-incorporated antibacterial TiO$_2$ porous coating on titanium. *Ceram. Int.* **2016**, *42*, 17095–17100. [CrossRef]
188. Zhang, X.; Li, J.; Wang, X.; Wang, Y.; Hang, R.; Huang, X.; Tang, B.; Chu, P.K. Effects of copper nanoparticles in porous TiO$_2$ coatings on bacterial resistance and cytocompatibility of osteoblasts and endothelial cells. *Mater. Sci. Eng. C* **2018**, *82*, 110–120. [CrossRef] [PubMed]
189. Jia, Z.; Xiu, P.; Li, M.; Xu, X.; Shi, Y.; Cheng, Y.; Wei, S.; Zheng, Y.; Xi, T.; Cai, H.; et al. Bioinspired anchoring AgNPs onto micro-nanoporous TiO$_2$ orthopedic coatings: Trap-killing of bacteria, surface-regulated osteoblast functions and host responses. *Biomaterials* **2016**, *75*, 203–222. [CrossRef]
190. Roknian, M.; Fattah-Alhosseini, A.; Gashti, S.O.; Keshavarz, M.K. Study of the effect of ZnO nanoparticles addition to PEO coatings on pure titanium substrate: Microstructural analysis, antibacterial effect and corrosion behavior of coatings in Ringer's physiological solution. *J. Alloys Compd.* **2018**, *740*, 330–345. [CrossRef]
191. Gokcekaya, O.; Webster, T.J.; Ueda, K.; Narushima, T.; Ergun, C. In vitro performance of Ag-incorporated hydroxyapatite and its adhesive porous coatings deposited by electrostatic spraying. *Mater. Sci. Eng. C* **2017**, *77*, 556–564. [CrossRef]
192. El-Rab, S.M.G.; Fadl-Allah, S.A.; Montser, A. Improvement in antibacterial properties of Ti by electrodeposition of biomimetic Ca-P apatite coat on anodized titania. *Appl. Surf. Sci.* **2012**, *261*, 1–7. [CrossRef]
193. Huang, Y.; Wang, W.; Zhang, X.; Liu, X.; Xu, Z.; Han, S.; Su, Z.; Liu, H.; Gao, Y.; Yang, H. A prospective material for orthopedic applications: Ti substrates coated with a composite coating of a titania-nanotubes layer and a silver-manganese-doped hydroxyapatite layer. *Ceram. Int.* **2018**, *44*, 5528–5542. [CrossRef]
194. Zhang, X.; Chaimayo, W.; Yang, C.; Yao, J.; Miller, B.L.; Yates, M.Z. Silver-hydroxyapatite composite coatings with enhanced antimicrobial activities through heat treatment. *Surf. Coat. Technol.* **2017**, *325*, 39–45. [CrossRef]
195. Geng, Z.; Cui, Z.; Li, Z.; Zhu, S.; Liang, Y.; Liu, Y.; Li, X.; He, X.; Yu, X.; Wang, R.; et al. Strontium incorporation to optimize the antibacterial and biological characteristics of silver-substituted hydroxyapatite coating. *Mater. Sci. Eng. C* **2016**, *58*, 467–477. [CrossRef] [PubMed]
196. Muhaffel, F.; Cempura, G.; Menekse, M.; Czyrska-Filemonowicz, A.; Karaguler, N.; Cimenoglu, H. Characteristics of multi-layer coatings synthesized on Ti6Al4V alloy by micro-arc oxidation in silver nitrate added electrolytes. *Surf. Coat. Technol.* **2016**, *307*, 308–315. [CrossRef]
197. Wang, R.; He, X.; Gao, Y.; Zhang, X.; Yao, X.; Tang, B. Antimicrobial property, cytocompatibility and corrosion resistance of Zn-doped ZrO$_2$/TiO$_2$ coatings on Ti6Al4V implants. *Mater. Sci. Eng. C* **2017**, *75*, 7–15. [CrossRef]
198. Chozhanathmisra, M.; Ramya, S.; Kavitha, L.; Gopi, D. Development of zinc-halloysite nanotube/minerals substituted hydroxyapatite bilayer coatings on titanium alloy for orthopedic applications. *Colloids Surf. A Physicochem. Eng. Asp.* **2016**, *511*, 357–365. [CrossRef]
199. Patiño-Herrera, R.; González-Alatorre, G.; Estrada-Baltazar, A.; Escoto-Chavéz, S.; Pérez, E. Hydrophobic coatings for prevention of dental enamel erosion. *Surf. Coat. Technol.* **2015**, *275*, 148–154. [CrossRef]
200. Tetz, M.; Jorgensen, M.R. New hydrophobic IOL materials and understanding the science of glistenings. *Curr. Eye Res.* **2015**, *40*, 969–981. [CrossRef] [PubMed]
201. Ma, N.; Chen, Y.; Zhao, S.; Li, J.; Shan, B.; Sun, J. Preparation of super-hydrophobic surface on Al–Mg alloy substrate by electrochemical etching. *Surf. Eng.* **2019**, *35*, 394–402. [CrossRef]
202. Bains, D.; Singh, G.; Kaur, N.; Singh, N. Development of an ionic liquid metal-based nanocomposite-loaded hierarchical hydrophobic surface to the aluminum substrate for antibacterial properties. *ACS Appl. Bio Mater.* **2020**, *3*, 4962–4973. [CrossRef]
203. Zhang, D.; Yuan, T.; Wei, G.; Wang, H.; Gao, L.; Lin, T. Preparation of self-healing hydrophobic coating on AA6061 alloy surface and its anti-corrosion property. *J. Alloys Compd.* **2019**, *774*, 495–501. [CrossRef]
204. Tang, P.; Zhang, W.; Wang, Y.; Zhang, B.; Wang, H.; Lin, C.; Zhang, L. Effect of superhydrophobic surface of titanium on *Staphylococcus aureus* adhesion. *J. Nanomater.* **2011**, *2011*, 1–8. [CrossRef]
205. Zhang, X.; Wan, Y.; Ren, B.; Wang, H.; Yu, M.; Liu, A.; Liu, Z. Preparation of superhydrophobic surface on titanium alloy via micro-milling, anodic oxidation and fluorination. *Micromachines* **2020**, *11*, 316. [CrossRef] [PubMed]
206. Li, S.; Zhang, T.; Zhou, M.; Zhang, X.; Gao, Y.; Cai, X. A novel digital and visualized guided bone regeneration procedure and digital precise bone augmentation: A case series. *Clin. Implant Dent. Relat. Res.* **2021**, *23*, 19–30. [CrossRef] [PubMed]
207. Yin, S.; Zhang, W.; Tang, Y.; Yang, G.; Wu, X.; Lin, S.; Liu, X.; Cao, H.; Jiang, X. Preservation of alveolar ridge height through mechanical memory: A novel dental implant design. *Bioact. Mater.* **2021**, *6*, 75–83. [CrossRef] [PubMed]
208. Momeni, F.; Hassani, S.M.; Liu, X.; Ni, J. A review of 4D printing. *Mater. Des.* **2017**, *122*, 42–79. [CrossRef]
209. Javaid, M.; Haleem, A. 4D printing applications in medical field: A brief review. *Clin. Epidemiol. Glob. Health* **2019**, *7*, 317–321. [CrossRef]
210. Khorsandi, D.; Fahimipour, A.; Abasian, P.; Saber, S.S.; Seyedi, M.; Ghanavati, S.; Ahmad, A.; De Stephanis, A.A.; Taghavinezhaddilami, F.; Leonova, A.; et al. 3D and 4D printing in dentistry and maxillofacial surgery: Printing techniques, materials, and applications. *Acta Biomater.* **2021**, *122*, 26–49. [CrossRef] [PubMed]
211. Zarek, M.; Mansour, N.; Shapira, S.; Cohn, D. 4D Printing of shape memory-based personalized endoluminal medical devices. *Macromol. Rapid Commun.* **2017**, *38*, 1600628. [CrossRef]
212. Banks, J. Adding value in additive manufacturing: Researchers in the United Kingdom and Europe look to 3D printing for customization. *IEEE Pulse* **2013**, *4*, 22–26. [CrossRef]

213. Zhang, C.; Cai, D.; Liao, P.; Su, J.-W.; Deng, H.; Vardhanabhuti, B.; Ulery, B.D.; Chen, S.-Y.; Lin, J. 4D Printing of shape-memory polymeric scaffolds for adaptive biomedical implantation. *Acta Biomater.* **2021**, *122*, 101–110. [CrossRef] [PubMed]
214. Lin, C.; Zhang, L.; Liu, Y.; Liu, L.; Leng, J. 4D printing of personalized shape memory polymer vascular stents with negative Poisson's ratio structure: A preliminary study. *Sci. China Ser. E Technol. Sci.* **2020**, *63*, 578–588. [CrossRef]
215. Azam, A.; Laflin, K.E.; Jamal, M.; Fernandes, R.; Gracias, D.H. Self-folding micropatterned polymeric containers. *Biomed. Microdevices* **2011**, *13*, 51–58. [CrossRef]
216. Lukin, I.; Musquiz, S.; Erezuma, I.; Al-Tel, T.H.; Golafshan, N.; Dolatshahi-Pirouz, A.; Orive, G. Can 4D bioprinting revolutionize drug development? *Expert Opin. Drug Discov.* **2019**, *14*, 953–956. [CrossRef] [PubMed]
217. Melocchi, A.; Uboldi, M.; Inverardi, N.; Vangosa, F.B.; Baldi, F.; Pandini, S.; Scalet, G.; Auricchio, F.; Cerea, M.; Foppoli, A.; et al. Expandable drug delivery system for gastric retention based on shape memory polymers: Development via 4D printing and extrusion. *Int. J. Pharm.* **2019**, *571*, 118700. [CrossRef] [PubMed]
218. Evdokimov, P.V.; Putlayev, V.I.; Orlov, N.K.; Tikhonov, A.A.; Tikhonova, S.A.; Garshev, A.V.; Milkin, P.A.; Klimashina, E.S.; Zuev, D.M.; Filippov, Y.Y.; et al. Adaptable metamaterials based on biodegradable composites for bone tissue regeneration. *Inorg. Mater. Appl. Res.* **2021**, *12*, 404–415. [CrossRef]
219. Hashemi, M.S.; McCrary, A.; Kraus, K.H.; Sheidaei, A. A novel design of printable tunable stiffness metamaterial for bone healing. *J. Mech. Behav. Biomed. Mater.* **2021**, *116*, 104345. [CrossRef]
220. Ghavidelnia, N.; Bodaghi, M.; Hedayati, R. Femur auxetic meta-implants with tuned micromotion distribution. *Materials* **2020**, *14*, 114. [CrossRef] [PubMed]
221. Kolken, H.; de Jonge, C.; van der Sloten, T.; Garcia, A.F.; Pouran, B.; Willemsen, K.; Weinans, H.; Zadpoor, A. Additively manufactured space-filling meta-implants. *Acta Biomater.* **2021**, *125*, 345–357. [CrossRef]

Review

Toward Bactericidal Enhancement of Additively Manufactured Titanium Implants

Yingjing Fang [1], Shokouh Attarilar [2], Zhi Yang [1], Guijiang Wei [2,3], Yuanfei Fu [1,*] and Liqiang Wang [2,*]

[1] Shanghai Key Laboratory of Stomatology, National Center for Stomatology, National Clinical Research Center for Oral Diseases, Department of Prosthodontics, Shanghai Ninth People's Hospital, Shanghai Jiao Tong University School of Medicine; College of Stomatology, Shanghai Jiao Tong University, Shanghai 200011, China; fangyj1998@mail.sjtu.edu.cn (Y.F.); yangzhi412@alumni.sjtu.edu.cn (Z.Y.)

[2] State Key Laboratory of Metal Matrix Composites, School of Material Science and Engineering, Shanghai Jiao Tong University, Shanghai 200240, China; at.shokooh@gmail.com (S.A.); sjtuwgj@mail.sjtu.edu.cn (G.W.)

[3] Department of Medical Laboratory, Affiliated Hospital of Youjiang Medical University for Nationalities, Baise 533000, China

* Correspondence: fuyf1421@sh9hospital.org.cn (Y.F.); wang_liqiang@sjtu.edu.cn (L.W.)

Citation: Fang, Y.; Attarilar, S.; Yang, Z.; Wei, G.; Fu, Y.; Wang, L. Toward Bactericidal Enhancement of Additively Manufactured Titanium Implants. *Coatings* 2021, *11*, 668. https://doi.org/10.3390/coatings11060668

Academic Editor: Grzegorz Dercz

Received: 8 April 2021
Accepted: 27 May 2021
Published: 31 May 2021

Publisher's Note: MDPI stays neutral with regard to jurisdictional claims in published maps and institutional affiliations.

Copyright: © 2021 by the authors. Licensee MDPI, Basel, Switzerland. This article is an open access article distributed under the terms and conditions of the Creative Commons Attribution (CC BY) license (https://creativecommons.org/licenses/by/4.0/).

Abstract: Implant-associated infections (IAIs) are among the most intractable and costly complications in implant surgery. They can lead to surgery failure, a high economic burden, and a decrease in patient quality of life. This manuscript is devoted to introducing current antimicrobial strategies for additively manufactured (AM) titanium (Ti) implants and fostering a better understanding in order to pave the way for potential modern high-throughput technologies. Most bactericidal strategies rely on implant structure design and surface modification. By means of rational structural design, the performance of AM Ti implants can be improved by maintaining a favorable balance between the mechanical, osteogenic, and antibacterial properties. This subject becomes even more important when working with complex geometries; therefore, it is necessary to select appropriate surface modification techniques, including both topological and chemical modification. Antibacterial active metal and antibiotic coatings are among the most commonly used chemical modifications in AM Ti implants. These surface modifications can successfully inhibit bacterial adhesion and biofilm formation, and bacterial apoptosis, leading to improved antibacterial properties. As a result of certain issues such as drug resistance and cytotoxicity, the development of novel and alternative antimicrobial strategies is urgently required. In this regard, the present review paper provides insights into the enhancement of bactericidal properties in AM Ti implants.

Keywords: additive manufacturing; porous titanium; implant coatings; antibacterial agent

1. Introduction

Bone infection is one the most serious and destructive risks associated with bone implant surgeries. According to the results of the International Consensus Meeting on Musculoskeletal Infection in 2018, the infection incidences for all orthopedic subspecialties ranged from 0.1% to 30%, and the cost of each patient ranged from USD 17,000 to 150,000 [1]. *Staphylococcus aureus* is considered to be the most common pathogen isolated from implant-associated osteomyelitis [2], and an increasing number of cases (more than 50%) are caused by refractory methicillin-resistant *S. aureus* (MRSA) strains [3]. Implants act as carriers for bacterial growth, increasing the bacterial virulence on the surface [4]. Bacteria adhere to the implant surface, before cell proliferation and biofilm formation [5]. A biofilm is a type of microbially derived fixation community that is characterized by cells that are irreversibly adhered to a substrate or interface between them, embedded in a matrix made up of their own extracellular polymeric substances [6]. Bacterial biofilms are resistant to antimicrobial treatments and evade host defenses by providing a physical barrier [7]. As a result of restricted blood flow, it is difficult to deliver antibiotics to the area around the implant, and due to recurrence and drug resistance, the infection is often difficult to treat and will become chronic [8]. As a result, in these cases, the removal of the implant is sometimes necessary.

In recent years, various novel approaches have been proposed to prevent and treat implant-associated infections (IAIs). In order to endow titanium (Ti) implant surfaces with effective antimicrobial properties, two models were proposed: Passive antimicrobial mechanisms, which prevent the initial bacterial attachment, and active antimicrobial mechanisms, which release antimicrobial agents to kill adherent or planktonic bacteria. In addition, the local innate immune response should be enhanced, in order to stimulate immune cells to kill the bacteria [9]. Passive antimicrobial surfaces do not release antimicrobial agents; by changing the physicochemical properties of the surface, such as the roughness, hydrophilicity, or nanotopological structure, or by covalently immobilizing active molecules on the surface, they inhibit surface bacterial adhesion or kill bacteria by contact [10]. Active antimicrobial surfaces are loaded with antimicrobial agents through a variety of surface modification processes. They facilitate the controlled local release of antimicrobial agents that kill bacteria on the surface of the implant and around the tissue. Antibiotics [11], antibacterial active metals (such as silver (Ag) [12–14], copper (Cu) [15], zinc (Zn) [16]), and metal oxides [17] are widely applied in the study of implant antimicrobial surfaces. In a meta-analysis of 23 studies, Tsikopoulos et al. showed that a combination of active and passive antibacterial surfaces reduced the risk of IAIs [18].

Additive manufacturing (AM), often referred to as "3D printing", enables the fabrication of scaffolds with high geometric complexity [19]. AM technology enables the fabrication of patient-tailored and structurally optimized porous implants through the precise design of external and internal structures [20–22]. Metal AM is composed of electron beam melting (EBM) [23,24], selective laser sintering (SLS) [25,26], selective laser melting (SLM) [27], laser engineered net shaping (LENS) [28,29], direct metal laser sintering (DMLS) [30,31], and laser aided AM methods [32]. Figure 1 shows several widely applied metal AM processes.

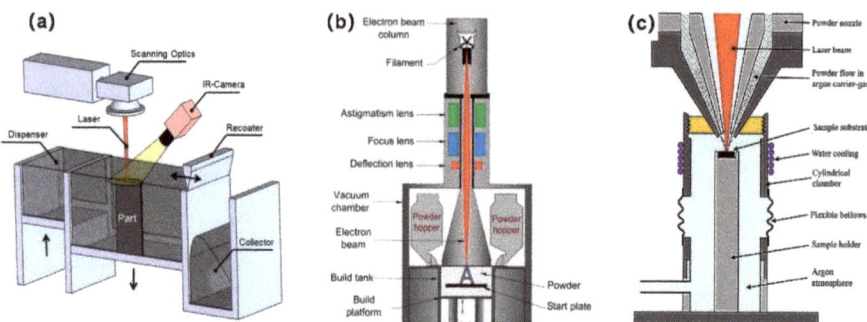

Figure 1. Schematic illustrations of the widely applied metal additive manufacturing (AM) processes: (**a**) Selective laser melting (SLM). Reprinted with permission from ref. [33]; (**b**) electron beam melting (EBM). Reprinted with permission from ref. [34]. Copyright 2017 Elsevier; (**c**) laser metal deposition (LMD). Reprinted with permission from ref. [35]. Copyright 2019 Elsevier.

At present, most implants in clinical use are bulk solid Ti with an elastic modulus much higher than that of bone, resulting in stress shielding [36]. The mechanical mismatch between the implant and the surrounding natural bone results in bone resorption [37]. AM technology enables the fabrication of patient-tailored and structurally optimized porous implants through the precise design of external and internal structures. Porous Ti implants have a suitable porosity and topological structure. This is achieved by mimicking the trabecular structure of natural bone tissue so that their mechanical properties match with the surrounding tissue, which can effectively transfer loads, reduce the stress shielding effect, and promote osseointegration [38–40]. In addition, trabecular-like porous implants provide an open interconnecting space for cell growth and the transport of nutrient and metabolic waste. Osteoblasts and mesenchymal cells migrate and proliferate within the pores, which is accompanied by vascularization, facilitating bone ingrowth [41,42]. Although, at present, there is no consensus criteria, most studies show that improving porosity and pore size promotes bone tissue ingrowth into pores and the osseointegration of the implants on the basis of ensuring mechanical properties [41,43,44].

For metallic AM implants, a large number of studies focus on conferring antimicrobial properties through indirect means. For instance, the additional surface area of a porous or specially shaped AM implant is used to enhance the antimicrobial properties of surface biofunctionalized implants for protection against IAIs [45]. The same biofunctionalization treatments in in vitro tests showed that porous implants with an increased surface area release a significantly higher amount of antimicrobial agent and have an increased inhibition zone size as compared to solid implants with the same dimensions [46]. The difference is that for AM polymer implants, which are manufactured by fused deposition modeling (FDM) or stereolithography (SLA), antimicrobial agents such as antibiotics [47,48], quaternary ammonium salts [49,50], chitosan [51], and antimicrobial metal particles [52] are usually added to the raw material during the manufacturing stage to facilitate the generation of bactericidal surfaces. At present, there are very few studies on the direct preparation of antimicrobial Ti implants using AM. The following are various starting points from which to approach this: (1) Improving the AM processing parameters to obtain optimized surface physical properties and reduce bacterial adhesion [53]; (2) developing nano-AM technology to prepare nanoantibacterial structures on the implant surface [54]; (3) adding antimicrobial metal elements and exploring the appropriate proportion of elements to prepare the antimicrobial alloy using AM [55].

From another point of view, a porous structure is more conducive to bacterial growth than implants with smooth surfaces, and the increased surface area may also increase the number of bacteria that survive the disinfection process or cling to the surface before surgery [45]. There is, however, no clear evidence proving that porous implants have a higher incidence of IAIs than solid implants [56]. For example, trabecular metal implants are known as implants that have a porous structure. A great number of clinical trials compare the prognosis of trabecular metal implants and nontrabecular metal implants. Different clinical centers reported different revision rates for IAIs of the two types of implants, but the results were not statistically significant ($p > 0.05$) [57–59]. It may be inferred that the increased surface area of AM porous implants has more effect on the antibacterial properties of the coating than the residual bacteria on the surface before surgery.

This review summarizes the various recently proposed strategies that improve the antibacterial properties of AM Ti implants. Firstly, the effect of AM technology on the antimicrobial properties of implants is discussed, and optimization schemes are proposed. Then, from the perspective of antimicrobial drug loading, drug release is explored through drug filling and implant surface bio-functionalization. Finally, the typical methods of antimicrobial functionalization of AM implants are listed, such as antibiotic coatings and antimicrobial active metal coatings.

2. The Effect of AM Technology on Antimicrobial Properties

An in vitro study found that Ti discs produced using the AM DMLS method could change the distribution of microbial species in subgingival biofilm and decrease the total counts of *Porphyromonas gingvalis*, which is the most widespread pathogen found in peri-implantitis [60]. Currently, AM processing parameters are being optimized and new AM technologies are being developed to improve antimicrobial performance.

Research shows that AM processing parameters were able to change the surface parameters and roughness of a Ti scaffold, such as the inclination angle, laser power, and beam diameter [61]. The inclination angle is a key design parameter, and it has been shown to be selective for the attachment of bacteria and tissue cells [53,62,63]. In the manufacturing stage, by reducing the SLM build inclination angle, a lower biofilm-covered surface morphology can be constructed without changing the surface chemical properties of the scaffolds, which is characterized by a reduction in partially melted metal particles, leading to a decrease in the roughness and hydrophobicity (Figure 2) [53]. Villapún et al. [62] established a mathematical model to optimize the orientation of customized SLM implants, which can accurately predict and optimize the surface roughness of scaffolds. A case study focusing on the customized implant proved the feasibility of this method. The optimization of the inclination angle facilitates the rapid fabrication and functionalization of implants in a single-step process, without postprocessing or with only local processing. Ginestra et al. [64] found that the inclination angle has a limited effect on surface topography, highlighting the influence of different AM processing techniques on surface properties. It is not a simple task to explore the influence of a single parameter on antibacterial properties, and further in vitro and in vivo studies are required to this end.

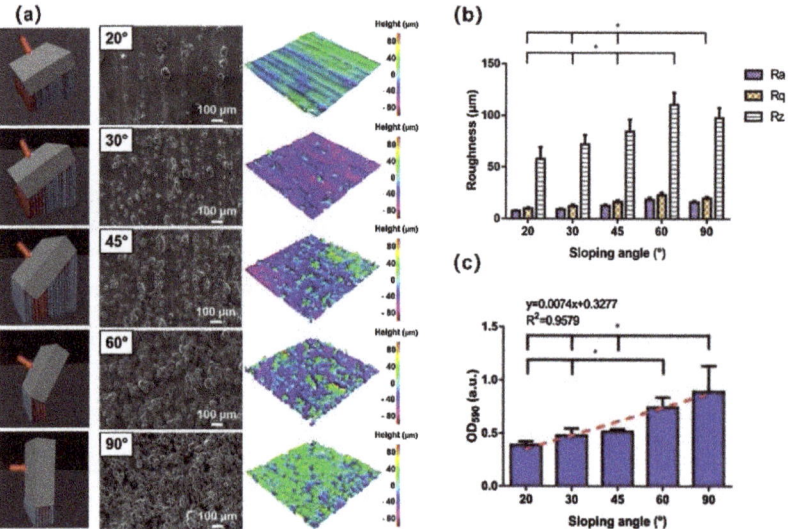

Figure 2. The SLM build inclination angle affects the surface roughness, which in turn affects bacterial adhesion. (**a**) Three-dimensional model of the build inclination angle, SEM micrograph images, and topographic scan of as-built Ti6Al4V samples with sloping angles from 20° to 90°; (**b**) measured arithmetic mean height (R_a), root mean squared height (R_q), and maximum height of profile (R_z) as average values of 20 scans, where * signifies p-value < 0.05 for R_a. (**c**) After 24 h of colonization, the biomass of *S. epidermidis* was quantified using crystal violet staining. Reprinted with permission from ref. [62]. Copyright 2020 Elsevier.

In the manufacturing process, the surface of the Ti implant inevitably retains some residual powders or partially melted particles. According to the report, it should be noted that this can inhibit the osteogenic activity of human bone marrow mesenchymal stem cells and enhance bacterial adhesion [65]. Therefore, AM scaffold surfaces are not suitable for direct use. Appropriate post-treatment care must be carried out before use [66]. The most common post-treatment methods include ultrasonic cleaning, sandblasting, chemical polishing, mechanical polishing, etc. The ability of ultrasonic cleaning to remove residual powder between the trabeculae on the surface is poor. In contrast, chemical polishing has a superior ability to remove powder residue and correlates with a decreased number of staphylococcal cells on the surface [67]. Junka et al. [68] speculated that fluoride and nitrogen on the surface of scaffolds could inhibit biofilm formation after chemical polishing. Sand blasting is a standard process for improving the surface finish in SLM production. However, it is worth noting that sandblasting may increase the risk of bacterial adhesion on the implant surface if the sandblasting medium is not adequately removed [66,69]. Interestingly, Szymczyk-Ziolkowska et al. [70] found that different microorganisms have a species-specific ability to form biofilms on different types of implant modification. For patients with a history of *S. aureus*, *Pseudomonas aeruginosa*, or *Candida albicans* infection, the use of as-built, sandblasted, or acid-etched alloys are not recommended, respectively.

Metal AM technology is common and is processed on the microscale. For example, the surface arithmetic mean roughness (R_a) of SLM is 5–20 μm and the EBM is 20–50 μm [71]. Several AM technologies with a nanoscale resolution have been developed, such as electron beam induced deposition (EBID) [72] and two-photon polymerization (TPP) [73]. TPP can produce 3D nanostructures and trigger the polymerization process by applying laser pulses on photosensitive materials. This is a 3D prototype technology for precisely controllable sub-100 nm AM structures [73,74]. EBID uses a focused electron beam to decompose the precursor molecules into two parts: The volatile part is desorbed and discharged, and the nonvolatile part is left on the substrate to form a nanosized deposition layer [75–77]. These nanoscale AM technologies are expected to produce specific nanotopographies to kill bacteria by, for example, inducing excessive levels of strain through a mechanical process [45]. Ganjian et al. [54] used EBID to fabricate nanopillars with precisely controlled dimensions within the osteogenic range on silicon wafers (Figure 3). To the authors' knowledge, nanoscale AM technology has not been applied to Ti implants, and the printing of antimicrobial nanopatterns on Ti implants requires further exploration.

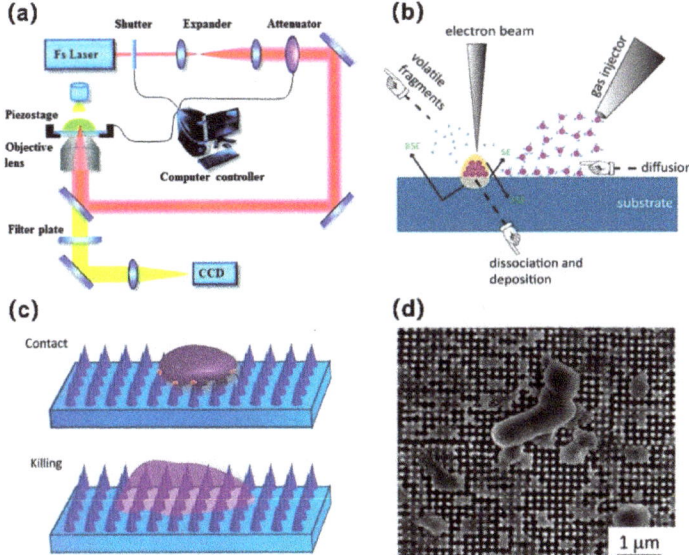

Figure 3. Schematic representation of (**a**) two-photon polymerization (TPP) (Reprinted with permission from ref. [78]. Copyright 2015 Royal Society of Chemistry) and (**b**) electron beam induced deposition (EBID); (**c**) schematic diagram showing the bactericidal behavior of a nanopattern structure, including deformation and being sunk on the nanopattern due to the penetration of the nanocolumn into the bacterial cell wall; (**d**) SEM image of damaged *Escherichia coli* bacteria on the surface of a nanopattern after 18 h of culture, observed from the above [54]. (CCD: Charge-coupled device).

3. AM Implants with Antimicrobial Loading

Oral or intravenous drug administration in the treatment of IAIs does not often work very well. This is due to the concentration of the drug in the blood, which is affected by the peak-and-valley effect. Additionally, drug release kinetics are often unpredictable, sometimes reaching toxic levels and sometimes falling below the therapeutic level. Moreover, inadequate blood perfusion restricts the blood supply in the bones, and only very small quantities of the drug reach the target site. Under these circumstances, utilizing the implant as a loader of the local drug delivery system is a promising method: It facilitates controlled and sustained release, a high local drug concentration, and minimizes the treatment period required [79,80]. Current implant drug loading methods generally include drug filling through implant design, a surface coating that releases drugs, and drug reservoirs through surface modification [81].

3.1. Hollow Implants

Ti implants with void volumes were designed to be filled with antimicrobial agents. These are known as hollow implants. The geometric freedom of AM technology facilitates the manufacturing of complex internal structures that would not be achievable with traditional methods. Park et al. [82] implanted hollow Ti implants perforated with microholes and loaded with a dexamethasone-based cartridge into the tibia of rabbits. The plasma pharmacokinetic behavior in these samples showed that the agent could be released continuously up to 7 weeks after implantation. Furthermore, a stainless-steel porous-wall hollow implant designed by Gimeno et al. [83] was filled with the synthetic antibiotic linezolid and was shown to exhibit good anti-infection properties in a model of tibia infection in sheep. It seems that drug-filled hollow implants may represent a novel approach to treat or prevent IAIs that do not require repeated injections or timely oral administration to maintain critical drug concentrations.

For the design of drug release routes, using microchannels for hollow implants is a typical method. The release profile can be predesigned by selecting the number of channels in order to achieve a rapid initial release and a slow, sustained, long-term release [84]. With the assistance of AM technology, hollow Ti implants with different channel orientations can be manufactured. It was found that channel orientation can affect the accuracy of the channel dimension [85,86], the back-pressure porosity of the injection material, the drug elution rate, and even the direction of drug release (Figure 4) [85]. Bezuidenhout et al. [87] designed a Ti alloy cube with channels. Polyethersulfone membrane discs were placed at the opening of each channel. It was able to control the release rate for the drug, and repeatable filling with antibiotics was possible through polyethersulfone membranes. Thus, antibiotic levels could be maintained above the minimum inhibitory concentration (MIC) for an extended period and drug resistance could be avoided. AM technology can also provide thin permeable walls with different porosities for hollow Ti implants. With the increase in porosity, the pattern of drug release profiles changes [88].

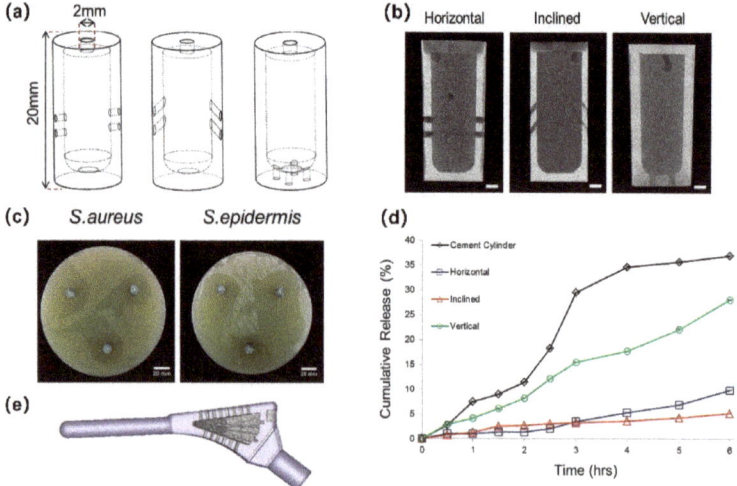

Figure 4. SLM Ti6Al4V implants containing a reservoir. Adapted with permission from ref. [85]. Copyright 2016 Elsevier.

(**a**) Schematics of implants with horizontal, inclined, and vertical pore channels; (**b**) coronal slice of antibiotic-loaded cement within implants; (**c**) zones of bacterial inhibition around implants with horizontally orientated pore channels showing directionality; (**d**) cumulative release of antibiotic from cement cylinders and cement-filled implants with different pore orientations; (**e**) re-engineered functional implant.

The filling material plays a pivotal role in the release spectrum, and its material type and solubility affect the reaction; for example, drugs with a higher water solubility exhibit a faster drug release rate [84]. The second factor is the state of the material that is filled. If the drug is directly filled, the elution involves one stage, and the drug is directly released from the channel. If the medium contains antibiotics, such as mesoporous silica particles or bone cement, the elution involves two stages: The drug is first desorbed from the medium and then released through the implant channels [85,89]. This two-stage eluting device has the ability to precisely control the drug release rate, because the drug medium and the hollow implant channel can be controlled independently.

The drug loading and drug eluting of these hollow Ti implants had promising antibacterial effects [85,88,90], which indicates that an optimized drug release profile could be achieved by rational structural design; however, a complete design scheme has not been proposed to date.

3.2. Porous Implants

3.2.1. Bioactive Coating

As a result of the increased surface area of porous implants, there is great potential for loading antimicrobial agents through coatings and surface treatments. Burton et al. [91] studied eight unit cell types of AM porous lattices and determined that the original Schwartz lattice geometry with 10% volume filling maintains the loading capacity, while allowing the maximum void volume in all lattice designs to load more antimicrobial agents. In addition to a homogeneous porous lattice structure, AM technology can also be used to manufacture implants with gradient porosity or surface blind pores. Sukhorukova et al. [92] used SLS to prepare a square blind holes network structure for loading antibiotics onto the surface of Ti plates. This had an obvious antibacterial effect, which was superior to the standard plate that contained a higher concentration of antibiotics.

Active coatings release preincorporated antimicrobial agents, such as antibiotics [93], inorganic antimicrobials [94,95] (e.g., silver (Ag), zinc (Zn), copper (Cu), gallium (Ga)), antiseptics [96], antimicrobial peptides [97], or certain types of metal oxides, to downregulate infection [45,98]. The biocompatibility of the implants should be ensured while considering the antimicrobial properties. Inspired by the extracellular matrix and proteins, organic and inorganic (hydroxyapatite) components, and composition combinations of living bone tissue, many bioactive materials can be used as candidates for antibacterial agents to coat implant surfaces [99]. Such coatings are not only effective in loading antimicrobial agents, but they also have the potential to promote tissue integration. These coatings include durable/biodegradable polymers, nanofibers, hydroxyapatite and gels [81], and titania nanotubes (TNT), which can mimic the nanoscale topology of bone and are among the most promising implant coatings. When antibacterial agents are mixed with these bioactive materials, the final drug release from the AM porous Ti implant surface is determined by the complex interaction between the coating characteristics, the drug properties, and the in vivo conditions [100].

There are many emerging surface treatment methods. During surface treatment and the coating of porous Ti implants, it is key to form a uniform surface over the entire specimen, reaching the entire inner surface. Some traditional coating techniques, such as plasma spraying, result in poor control over the thickness and surface topography and are not suitable for porous implants with complex geometries [99]. Therefore, chemical or electrochemical technology is a better choice, in order to reach the inner surface of the porous structure [45,101]. Technologies such as dipping, biomimetic deposition, chemical surface treatment using acidic and alkaline solutions, anodic oxidation, electrophoretic deposition, and plasma electrolytic oxidation are widely used in this field [45,99,102].

3.2.2. Nanometer Coating

In general, the nanomorphology of Ti surfaces with an antibacterial function is usually designed as nanotube and nanocoating forms [103]. TNT is easily adapted to AM porous structures and can be prepared on three-dimensional nonplanar surfaces [104]. A uniform layer of TNT can be formed on porous implants by liquid phase electrochemical treatment, e.g., anodizing [105,106] and micro-arc oxidation (MAO) [107–109]. In addition to the macroporous structure of AM technology, the microstructure of partially melted Ti microspheres on the surface of AM implants and TNT together constitutes a unique dual micro- to nanotopography (Figure 5) [110–112]. Biocompatibility and good corrosion resistance were demonstrated in TNT-coated Ti implants [113], and various promising results were reported in terms of improving the osteogenic activity around implants [111,114,115], especially regarding the dual micro- to nanotopography of AM porous scaffolds [111,116]. The hydrophilic surface of TNT has a positive effect on reducing bacterial adhesion, but the antibacterial performance of TNT is still poorer than that of mechanically polished samples, because the roughness of TNT affects its antibacterial properties [105].

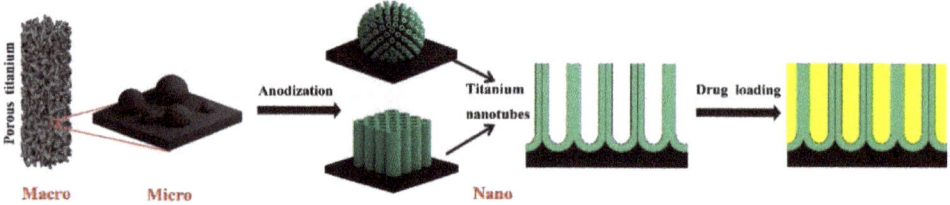

Figure 5. AM Ti implants with unique micro- and nanotopography, which can be used as a nano drug delivery system by combining AM technology and anodization. Adapted with permission from ref. [116]. Copyright 2020 Elsevier.

Nanotube structures with highly specific surface areas and pore volumes are both well organized and controllable and represent a good strategy for drug delivery and antibacterial activity [117]. Drug-eluting nanotubes significantly reduced bacterial adhesion to the surface when researchers filled the nanotubes with the antibiotic gentamicin [115]. Moreover, various parameters involved in anodizing (such as pore size and length) can precisely control the size of nanotubes [118], which, in turn, can control the drug release rate [119]. Drug-loaded TNT can be combined with other functional coatings, such as polymers [120] and hydrogels [116], which cover the openings of the TNTs and prolong the drug release time [121]. Maher et al. [122] proposed an innovative antimicrobial surface preparation method using SLM technology and electrochemical anodization to prepare the surfaces nanotopography of Ti alloys, thus promoting the nucleation reaction and growth of sharp nanospears through the hydrothermal process. The sharp triangular-shaped nanospears effectively destroyed bacteria by mechanically damaging their cell walls.

The majority of the research on nanocoatings focuses on metal nanoparticles [46,106,109] and nanocarriers, which aim to deliver antibacterial agents [123,124]. Metal nanoparticles have excellent antimicrobial effects [125]. Nanocoatings that do not contain antimicrobial agents are gradually being developed. Hu et al. [105] found that the composite effect of TNT and nanophase CaP on SLM Ti surfaces also exhibited good antibacterial activity. Furthermore, these antibacterial properties mainly come from the surface nanoroughness [105,123,126]. Rifai et al. [127] demonstrated that a nanodiamond (ND) coating can improve interface properties to inhibit bacterial colonization, and NDs coatings were applied on SLM Ti scaffolds using dip coating technology. Interestingly, it was found that controlling the embeddedness of nanoparticles is very helpful for maintaining the cell adhesion and osteogenic differentiation potential of TNT, especially when the nanoparticles do not completely cover the nanotubes [123].

4. Application of Antimicrobial Functionalization of AM Implants

4.1. Antibiotic

Antibiotics are the standard clinical tool for the local and systemic treatment of various infections caused by broad-spectrum pathogens. The concentration of local antibiotics is much higher than that achieved through parenteral administration. It can even be delivered to nonvascular areas, which can reduce the risk of antibiotic resistance and systemic toxicity [128,129]. Antibiotic implant materials, such as bone cement and temporary bone cement spacers, have exhibited effective antibacterial activity. AM porous Ti implants provide a variety of strategies for local antibiotic administration, among which surface coating technology is the most widely studied. In addition, from the perspective of implant structure design, antibiotic-added reservoirs are an important emerging method (see Section 3.1).

AM porous Ti implants have inherent advantages, such as their high porosity and large surface area, which provide adhesion sites for antibiotics. Moreover, they can be loaded with more antibiotics. Griseti et al. [129] found that the antibacterial effect of 3D porous Ti loaded with antibiotics lasted for 7 days, which was similar to that of early antibiotic bone cement. Smooth Ti alloy beads loaded with antibiotics demonstrated a limited inhibition effect before the second day. The change in the surface topology also improved the release profile of antibiotics and significantly increased the release amount [107,123].

There are several factors that need to be considered when choosing the type of antibiotic. The first is the antibacterial spectrum of the antibiotics, followed by its compatibility with the coating processing technology, and its stability and solubility, which determine the release profile. Gentamicin and vancomycin are the most widely used antibiotics for the prevention and treatment of IAIs. They are active against both Gram-negative and Gram-positive bacteria. Gentamicin and tobramycin, which have broad-spectrum antimicrobial properties and high temperature resistance, were approved by the United States Food and Drug Administration (FDA) for incorporation into bone cement for the treatment of prosthetic joint infections [130]. Vancomycin belongs to the glycopeptide antibiotics family and is effective against MRSA [131]. With the increase in bacterial resistance, and even the emergence of vancomycin-resistant strains [132], there is an urgent need for the application of novel antibiotics or combinations of antibiotics to treat refractory IAIs. Molina-Manso et al. tested the drug susceptibility of staphylococcal biofilms with a variety of antibiotics and found that rifampicin and tigecycline exhibited superior anti-biofilm activity to other antibiotics, which could be applied on the surface of implants to treat or prevent IAIs [133]. In addition, daptomycin [134,135] and minocycline [136] demonstrated efficacy against MRSA biofilms [137].

Current studies on antibiotic coatings focus on optimizing the release kinetics of eluting antibiotics. One important consideration is the duration of drug release, which must take into account both early and delayed IAIs. Another is the concentration of the released drug, which can help to avoid the development of bacterial resistance. In general, the release of antibiotics can be divided as follows: There is an initial outbreak period, during which the drug concentration reaches a high level; then, the release continues above MIC and stops before the development of antibiotic resistance (Figure 6). Stigter et al. [138] used the biomimetic coprecipitation method to incorporate antibiotics into the HA coating of titanium implants. Among the eight antibiotics studied, cephalosporins containing carboxylic groups were more strongly bound to the coating, with higher incorporation, a slower release rate, and more durable and effective antimicrobial activity.

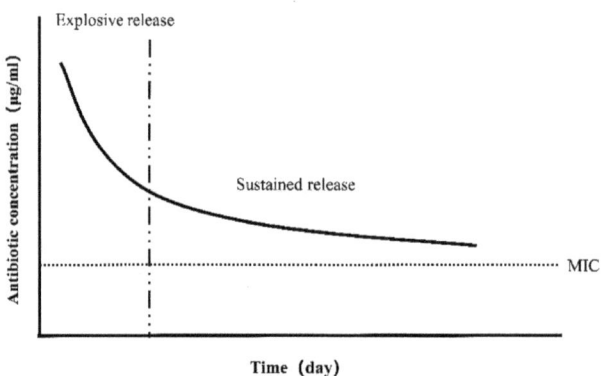

Figure 6. Ideal antibiotic release profile. The initial explosive release can kill bacteria effectively, and then the sustained release above the minimum inhibitory concentration (MIC) can maintain the antibacterial effect for a long period without antibiotic resistance.

Researchers have explored a variety of biocompatible coatings that stabilize adhesion to the Ti scaffold and load antibiotics. Common examples include dopamine [107], chitosan [139], gelatin [140], hyaluronic acid [141], and a variety of organic polymers [123,142–144]. Yavari et al. [140] applied multiple layers of gelatin- and chitosan-based coatings containing vancomycin and verified that the coating was almost completely degraded after 8 weeks. This timely biodegradation ensures the release of high doses of antibiotics during the perioperative period, while minimizing the risk of antibiotic resistance caused by long-term exposure to sub-MIC doses of antibiotics. Ghimire et al. [145] found that chitosan-fixed Ti implants significantly increased the sensitivity of adherent bacteria to antibiotics. Various innovative coatings, such as silk fibroin [146], bacterial cellulose [147], and phase-transited lysozyme [141], have also been proposed. Drug release from this conventional coating is continuous regardless of the occurrence of IAIs. In recent years, intelligent antimicrobial coatings have gradually emerged [148]. When the implant is invaded by bacteria and IAIs occurs, it is stimulated by temperature [149], pH [150], and electrical signals [151] to release antibiotics. Thus, the unnecessary release of antimicrobial agents in the uninfected state is avoided, and the risk of bacterial resistance is minimized.

Various processing and assembly methods have been explored that need to be suitable for customized scaffolds with complex shapes to ensure the continuous release of antibiotics. These include electrophoretic deposition (EPD) [128,143,146], electrospray deposition [123], covalent binding, layer-by-layer self-assembly [124,152], and electrospinning [142]. The advantage of EPD is that it can simply add a variety of antibacterial agents (such as antibiotics and nanoparticles) into different hydrogels and control the thickness and uniformity of the coating [143]. Given the hydrophilicity of antibiotics, EPD is a good approach. Bakhshandeh et al. [128] prepared chitosan and gelatin coatings in this way, and the antibiotics could be continuously released at a concentration higher than MIC for 21 days, thus achieving a long-term and highly effective bacteriostatic effect. Jahanmard et al. [142] applied antibiotic-loaded poly(ε-caprolactone) (PCL) and poly'1q'(lactic acid-co-glycolic acid) (PLGA) nanofiber coatings to lattice Ti implants by means of electrospinning. In this approach, the combination of specific drug–polymer interactions with bi-layer structures is crucial to prolong the inhibitory drug concentration, and, for the first time, it was shown that the antibacterial effect can last more than 6 weeks, preventing the early and delayed onset of IAI (Figure 7). Layer-by-layer self-assembly technology has great advantages in the stable fixation and continuous release of antibiotics. The self-assembled membrane prepared by Vaithilingam et al. [152] released less than 60% of the drug after 6 weeks.

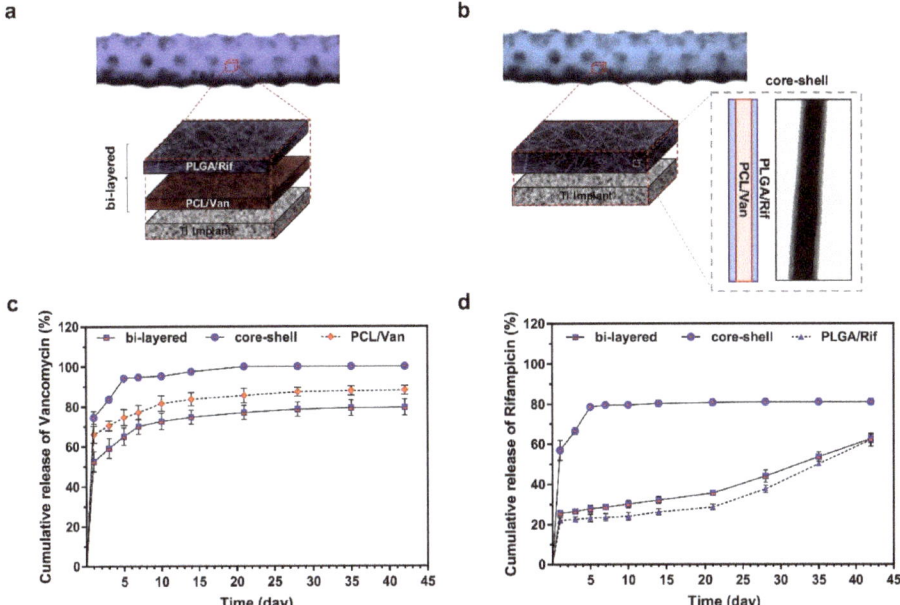

Figure 7. Various assembly and in vitro antibiotic release profiles of drug-loaded nanofiber-filled lattice implants. Reprinted with permission from ref. [142]. (**a**) The structure of the bi-layer nanofiber filled with PCL/Van as the inner layer and PLGA/Rif as the outer layer. (**b**) The structure of the core–shell nanofiber is composed of PCL/Van in the core and PLGA/Rif in the shell. The core–shell structure was verified by transmission electron microscope. Cumulative release of Van (**c**) and Rif (**d**) from three different nanofiber-filled lattice structures within 6 weeks. (Abbreviations: PCL, poly(ε-caprolactone); PLGA, poly'1q'(lactic-co glycolic acid); Van, vancomycin; Rif, rifampicin.).

The traditional view is that antibiotics are not cytotoxicity, and most antibiotic coatings demonstrate good biocompatibility [128]; however, Sukhorukova et al. [139] found that a standard dose of gentamicin (40 mg/mL) had short-term toxicity at an early stage and inhibited the proliferation of osteoblasts. Because the strong antimicrobial activity due to early rapid drug release compromises biocompatibility, caution is needed when using high concentrations of antibiotics.

4.2. Antibacterial Active Metal

4.2.1. Silver

In order to improve the antibacterial properties of AM implants, Ag has been widely studied for its broad-spectrum antibacterial activity and low toxicity to mammalian cells [153]. Table 1 shows the antibacterial effect of the Ag coating prepared on AM implants by different methods.

The toxic effects of Ag on microorganisms are attributed to the production of Ag ions [154]. Ag ions produce three main mechanisms of antibacterial action [155]: (1) Ag^+-induced direct membrane damage, through which Ag^+ can cause physical damage to the membrane and interact with sulfur-containing membrane proteins [156]; (2) through the reactive oxygen species (ROS) related to Ag^+, wherein the concentration of ROS is not related to the form of Ag, but is mainly related to the final concentration of Ag^+ [157]; (3) the cells uptake Ag^+ as a result of membrane perforation, resulting in the interruption of ATP production and the inhibition of DNA replication [155]. Compared with Gram-positive bacteria, Ag^+ has faster, longer, and more effective bactericidal effects on Gram-negative bacteria [106,108,158]. This may be due to the existence of a thick peptidoglycan layer in *S. aureus*, which can inhibit the transport of Ag ions through the cell membrane and has a low sensitivity [28]. Because of Ag's various antibacterial mechanisms, it is not easy to appear drug resistance, and it even has a strong antibacterial effect against MRSA [46].

To construct multifunctional porous implants, Ag ions are often combined with surface topological modification techniques to improve the osteogenic properties, as is the case for anodized TNT loaded with Ag [46,159]. Plasma electrolytic oxidation (PEO) can completely disperse and firmly adhere Ag nanoparticles (AgNPs) to the surface of the implant in a very short time by adding AgNPs into the electrolyte. AgNPs are encapsulated in an in-depth growth oxide layer to prevent the free circulation of AgNPs in the blood. The oxide layer causes AgNPs to become completely fixed, which can further prevent the potential nanotoxic effects [46]. Ag usually exists in the form of Ag_2O in PEO coatings. Gao et al. prepared TNT arrays embedded with Ag_2O nanoparticles on the surface of titanium using magnetron sputtering and anodic oxidation. Compared with the direct incorporation of Ag^+ into TNT, the Ag^+ release rate of Ag_2O was slower and the biocompatibility was better [106]. In addition to surface topological modification, composite biological coatings such as multifunctional hydrogels [160], polydopamine [108], chitosan [139], calcium phosphate [126], or silk protein [161] can be used to further enhance the biocompatibility of implants. Devlin-Mullin et al. [162] coated the surface of SLM 3D Ti scaffolds with an Ag nanolayer using atomic layer deposition, which produced good antibacterial properties and biocompatibility in vitro and in vivo. It also promoted angiogenesis and osteogenesis after implantation in rat tibia.

It is very important to improve antibacterial effects and prolong the anti-infection time, as it takes a long time (3 months) to achieve normal osseointegration. AM porous scaffolds and surface topological modification increase the depth of the reservoir, thereby increasing the ability to immobilize antimicrobial agents [108]. On this basis, compared with simple solution soaking [163], electrodeposition [159] and polydopamine-assisted coating [108] can form an Ag coating with stronger adhesion properties, providing Ag ion release for longer, and enhancing biocompatibility and the antibacterial properties. Shivaram et al. [159] prepared Ag coatings using electrodeposition and reported the most persistent in vitro release of Ag^+ to date (27 weeks), with the release of Ag^+ being within the potential toxicity limit of cells of 10 ppm (g/mL). In vivo experiments at 12 weeks showed good osteointegration properties and biocompatibility, but the long-term antimicrobial effects were not reported. Polydopamine (PDA) is a mussel-inspired multifunctional material, which can be deposited in situ onto TiO_2 via covalent cohesion. It can also chelate and reduce noble metal ions, inhibit the oxidative dissolution of AgNPs, and facilitate the long-term sustained dynamic release of Ag^+. In this way, the rigid TiO_2/PDA/Ag coating can be easily constructed on Ti [164]. Jia et al. [108] achieved an ultra-high loading capacity and a sustained release of Ag^+ through Mao, PDA, and Ag deposition. The antibacterial activity of scaffolds against planktonic/adherent bacteria (Gram-negative and Gram-positive) and even existing biofilms lasted for 12 weeks.

The synergistic effects of Ag and antibiotics on biofilm destruction have been observed. For instance, Ag can enhance the antibacterial activity of antibiotics. As Ag ions increase bacterial membrane permeability (even at sublethal concentrations), drug-resistant bacteria become sensitive to antibiotics [165]. AgNPs can promote free Ag^+ and antibiotics to kill biofilm bacteria through degrading the main components (polysaccharides, proteins, and nucleic acids) of the biofilm [108,166]. The special combination of Ag and antibiotics slows the release of the two drugs and avoids an initial sudden release. Furthermore, the release rate of each bactericide depends on the presence of other antibacterial components, such as Ag ions and antibiotics. The eradication of planktonic bacteria and adherent bacteria confirmed the synergistic effect of Ag and antibiotics. Moreover, reducing the drug concentration is beneficial in order to avoid the toxicity from Ag ions or antibiotics [128,158]. The slow release of Ag^+ provides very good antibacterial protection after the depletion of the antibiotic reservoirs [158].

Many in vitro antibacterial studies show that Ag ions have good antibacterial activity against bacteria and biofilms; however, Ag ions are somewhat toxic to mammalian cells, such as human mesenchymal stromal cells (hMSCs) [163], osteoblast-like cells [128], and macrophages [167]. The early explosion of Ag ions may lead to osteoblast toxicity; however, in the later stage, the toxicity is weakened when the Ag ion concentration is reduced, and the proliferation activity of osteoblasts is enhanced, which can offset the side effects in the early stage [108,161]. Two different outcomes were found when Ag-loaded chitosan-coated implants were implanted into the body. In the absence of infection or a low bacterial challenge, the host reaction can overcome the cytotoxicity of Ag, exhibiting a good antibacterial effect. When the host faces a high bacterial challenge, 0.1–1 µm Ag particles trigger inflammation and are released from the implant surface, which mediates local over-inflammation; kills neutrophils, so there is no antibacterial effect; and even aggravates infection-mediated bone remodeling [139]. Therefore, further in vivo studies are needed before Ag-coated implants will be ready for clinical use.

Table 1. The antibacterial effect of Ag coating prepared on AM implants.

Implant	AM Technology	Microorganism	Coating Technology	Result	Duration	Reference
Ti6Al4V	SLM	MRSA	PEO using electrolytes based on Ca/P species and AgNPs	Release of Ag ions; In vitro, antibacterial behavior against MRSA; Ex vivo, murine femoral infection model	4 weeks; 24 h; NR	[46,168,169]
Ti6Al4V	EBM	E. coli and S. aureus	MAO and PDA/Ag treatments	In vitro, antibacterial behavior against E. coli and S. aureus both planktonic and in biofilm	12 weeks	[108]
Ti6Al4V	EBM	S. aureus	Electrophoretic deposition of Ag and calcium phosphate nanoparticle layers	In vitro, antibacterial behavior against S. aureus	17 h	[126]
Ti	SLS	E. coli, S. aureus and N. crassa	TiCaPCON-Ag films by magnetron sputtering and loaded with gentamicin and amphotericin B	In vitro, antibacterial behavior against E. coli, S. aureus and N. crassa	3 days	[158]
Ti	LENS	NA	Anodized TNT followed by electrodeposition of Ag	Release of Ag ions	27 weeks	[159]
Ti	DMP	S. aureus	AgNO$_3$ were mixed with chitosan followed by EPD	In vitro, antibacterial behavior against S. aureus; In vivo, tibia intramedullary implant model inoculated with S. aureus	1 week; NR	[139]

Table 1. Cont.

Implant	AM Technology	Microorganism	Coating Technology	Result	Duration	Reference
Ti6Al4V	EBM	S. aureus	Hydrothermal growth of a titanate layer, on which nanosilver encapsulated silk fibrin multilayers were anchored through PDA-assisted, silk-on-silk self-assembly	In vitro, antibacterial activity against clinical pathogenic S. aureus both planktonic and in biofilm	6 weeks	[161]
Ti	SLM	S. epidermidis and MRSA	Atomic layer deposition of an Ag nanolayer	In vitro, antibacterial behavior against S. epidermidis and no antibacterial activity against MRSA; In vivo, vascularization and osseointegration tendency	4 days; NR	[162]
Ti	DMP	S. aureus	Anodized TNT followed by soaking in AgNO$_3$ solution	In vitro, antibacterial behavior against S. aureus	2 weeks	[163]
Ti	DMP	S. aureus	Vancomycin and AgNO$_3$ were mixed with the Chitosan/gelatin compound followed by EPD	In vitro, antibacterial behavior against S. aureus both planktonic and in biofilm	3 weeks	[128]

Abbreviations: SLM, selective laser melting; EBM, electron beam melting; SLS, selective laser sintering; LENS, laser engineered net shaping; DMP, direct metal printing; MRSA, methicillin-resistant S. aureus; PEO, plasma electrolytic oxidation; MAO, micro-arc oxidation; EPD, electrophoretic deposition; AgNPs, Ag nanoparticles; PDA, polydopamine; TNT, titania nanotubes; NR, never report.

4.2.2. Copper

Cu is a potential broad-spectrum inorganic antimicrobial agent. It is a necessary trace element in the human body and participates in the synthesis of enzymes. Cu is less cytotoxic than Ag [170] and can be metabolized by the human body. Therefore, Cu may be an effective substitute for Ag. It not only exhibits antibacterial activity against *E. coli* and *S. aureus* [171], but also demonstrates antibacterial and antibiofilm properties against the oral-specific bacteria *Streptococcus mutans* and *Porphyromonas gingivalis* [172], which is beneficial for using in dental materials. However, high doses of Cu can cause cytotoxicity. It is important to find the optimal concentration of Cu ions that can inhibit bacterial growth while avoiding cytotoxicity. Fowler et al. [173] studied the effect of Cu on the viability of

MC3T3 cells and *Staphylococcus epidermidis* in vitro, and the minimum inhibitory concentration of Cu ions in these two species ranged from 9×10^{-5} to 9×10^{-6} g/mL. As a result of species-specific and in vivo and in vitro differences, further studies are needed. At appropriate concentrations, Cu-modified implants also exhibited anti-inflammatory, proangiogenic, and osteogenic effects [174].

The antimicrobial mechanism of Cu can be divided into two elements: The direct mechanism and the indirect mechanism. The direct bactericidal mechanism involves Cu ions interacting with bacteria or biofilms. A large number of Cu ions flow into the bacteria, blocking the cell respiration chain and disturbing DNA synthesis. Then, a large amount of ROS are produced, leading to changes in the permeability of bacterial cell walls and leakage of the bacterial contents [175]. Cu inhibits the expression and transcription of positive biofilm regulators such as *sae* and *agr*, which inhibits biofilm formation [176]. The indirect bactericidal mechanism involves Cu activating the bactericidal ability mediated by macrophages. Cu can improve the ability of macrophages to uptake and kill bacteria, and it kills bacteria through the ROS pathway [177]. The concentration of copper ions required for the activation of macrophages is far lower than that required by the direct bactericidal mechanism, which can effectively help to avoid the cytotoxicity caused by high concentrations.

Similar to the method used for the preparation of the Ag coating, ion implantation [178], PEO [16], sol-gel [179], and electrodeposition [180] are applied in the preparation of copper coatings. For AM implants with complex geometries, PEO [16] and electrochemically assisted deposition [180] seem to be the superior methods. Interestingly, the application of copper as an antibacterial active substance in implants is not only through surface modification; copper-containing titanium alloy has also received much interest from researchers [181–183]. In the additive manufacturing process, Cu is added into the material by in situ alloying; thus, the process flow is simplified to a single-step process. In addition, it can help to avoid the wear and tear associated with surface coating and is expected to provide a lasting antibacterial effect. The antimicrobial properties of AM copper-containing titanium alloy are related to the Cu content, AM processing parameters, and heat treatment [181,183,184]. The majority of researchers consider 5 wt.% Cu to be appropriate for antibacterial function [184]. Furthermore, the in situ alloying of Cu increases the hardness and compressive strength of the material [185], which maintains good corrosion resistance in simulated body fluids [184].

In the preparation process, the phase transformation of antibacterial active metals inevitably occurs, such as during electrochemical treatments or powder processing at high temperatures, which may lead to the oxidation of the metal. Shimabukuro et al. [186] embedded Cu in the TiO_2 layer on the implant surface using MAO. They found that Cu existed in the form of Cu_2O and exerted an antibacterial effect. Therefore, it is necessary to further explore the antibacterial effects of metal oxides. The oxides of Cu are Cu_2O and CuO. Zhao et al. mixed Cu_2O nanoparticles of different concentrations into the ceramic oxide layer using MAO. They found that the addition of Cu_2O improved the antimicrobial performance of the MAO coating in a dose-dependent manner. It has also been demonstrated that Cu^+ is the key factor in terms of the antibacterial properties [187]. By comparing the minimum inhibitory concentrations and minimum bactericidal concentrations of Cu_2O and CuO against four kinds of periimplantitis-related bacteria, it can be inferred that Cu2O has a superior antibacterial activity to CuO [17].

4.2.3. Zinc

Compared with Ag, Zn has been less studied as an antimicrobial active substance, but its interesting biological effects have been gradually attracting attention. Zn is an essential trace element in the human body, which is involved in a variety of physiological processes, such as bone metabolism, cell signaling pathways, and immune regulation. Zn regulates the expression of bone morphogenetic protein gene, increases the activity of alkaline phosphatase, inhibits the bone resorption of osteoclasts, and induces the differentiation of osteoblasts, thus promoting bone formation. In addition, Zn ions and their nanoparticles have good antibacterial properties. Zn ions (positive charge) and the cell wall of bacteria (negative charge) are attracted to each other by electrostatic interaction, resulting in the destruction of the bacteria cell wall. Moreover, Zn ions enter the bacteria and interfere with DNA replication [188]. Not only does Zn itself have the ability to kill bacteria directly, but it can also inhibit bacterial infection by regulating host immune defenses [189]. Wang et al. prepared ZnO films on a titanium surface using magnetron sputtering. The results of the in vivo experiments demonstrated a stronger antibacterial effect than in the in vitro experiments, indicating that Zn further enhanced the antibacterial effect by regulating the immune system response in vivo.

Similar to Ag and Cu, electrochemical deposition [190], MAO [191], and layer-by-layer self-assembly [192] are used to introduce Zn onto the implant surface. Liu et al. [192] constructed a bio-multilayer structure containing Zn ions using the self-assembly technique, which promoted the biological activity and function of osteoblasts and produced antibacterial activity at the same time. Different from the inert surface of pure titanium, the Zn-containing surface has the ability to induce osteogenesis while avoiding IAIs, which is a promising modification method for biological implants.

In the Zn-doped TiO_2 layer prepared using EPO, Zn mainly exists in the form of ZnO. Surprisingly, when the MAO coating containing Zn was immersed in normal saline, the chemical state of the surface Zn changed from Zn^{2+} to ZnO, leading to an increase in antibacterial activity [191]. ZnO nanoparticles have attracted much attention because of their biocompatibility, low toxicity, chemical stability, antimicrobial activity, and their selective killing effect on normal cells and cancer cells. In particular, the small size effect of ZnO nanoparticles plays an important role in their antibacterial behavior, inhibiting the growth of a variety of bacteria [193,194].

4.2.4. Other Metals

Other metals such as gold (Au) [194], magnesium (Mg) [195], iron (Fe) [196], and their oxides also exhibit antibacterial effects. Au has stable chemical properties, good biocompatibility, and has the potential to promote the proliferation and differentiation of mesenchymal stem cells [197]. In addition, gold nanorods have special photophysical properties, which produce photothermal effects under the excitation of near-infrared light and exhibit antibacterial activity against a variety of bacterial strains [198]. The superparamagnetism of iron oxide nanoparticles makes targeted drug delivery possible. Moreover, it enables iron oxide nanoparticles to target the biofilm in the infected site and have a strong killing effect on drug-resistant strains [196].

4.2.5. Comparison

Ag, Cu, and Zn are the most commonly used inorganic antimicrobial agents. A comparison of their biological effects is necessary to improve their various applications. The MIC of Cu and Zn is much higher than that of Ag, with a difference of about two orders of magnitude. The antimicrobial activity of Ag is higher than that of Cu and Zn, i.e., Ag can produce a high level of antimicrobial activity at a lower concentration [16]. However, Ag is more likely to cause cytotoxicity, and its half maximal inhibitory concentration (IC_{50}s) to osteoblasts MC3T3-E1 is only 2.77 µM, while Cu and Zn show good biocompatibility in the appropriate concentration range, with IC_{50}s of 15.9 and 90.0 µM, respectively [199]. A combination of antimicrobial agents provides a way to solve these two problems. The combination of two or more antibacterial agents has a synergistic antibacterial effect and reduces the minimum inhibitory or bactericidal concentration [169,180,200]. For example, the combination of Ag and Zn can reduce the required concentration of Ag by two orders of magnitude, while maintaining the same antibacterial activity, greatly reducing cytotoxicity [169]. Strontium (Sr), a metal ion with osteogenic activity, was doped with Ag on the implant surface. It not only exhibited strong synergistic antibacterial behavior against drug-resistant strains, but also polarized macrophages (M2) through favorable immune regulation, thus further promoting the differentiation of preosteoblasts [168,201]. Shimabukuro et al. also investigated the biodegradation behavior of antibacterial active metal surfaces and found that their antibacterial properties had a time transient effect. When the antibacterial surface was soaked in normal saline for 28 days, the chemical states of the antibacterial active elements changed separately, leading to completely different changes in the antibacterial activity. The antimicrobial activity of the silver-coated specimens was gradually weakened, and Cu showed no significant change, while Zn showed enhanced antimicrobial activity [170].

Finally, it is important to note that although these antimicrobial active metals have shown excellent antimicrobial efficacy, they have not been approved as antimicrobial active ingredients in implants by the Food and Drug Administration (FDA) and Environmental Protection Agency (EPA) due to their potential cytotoxicity. Extreme caution should be exercised in the clinical application of these metal active ingredients until in vivo safety is further determined.

5. Conclusions

AM technology is increasingly being applied in orthopedics and stomatology, and the significant production freedom enables various innovative designs that cannot be achieved by traditional manufacturing methods. Through AM topological design, the mechanical properties and osteogenic properties of implants have been greatly improved. However, further optimization is needed to produce clinical applications. It is necessary to apply the surface modification technology of traditional solid titanium implants to the post-treatment of AM implants. Kirmanidou et al. [202] summarized a variety of surface modification methods to improve the mechanical properties, osteogenic properties, and antibacterial properties of titanium implants. Whether these surface modification methods are suitable for AM implants depends on whether they can be uniformly treated on surfaces with complex geometrical shapes, such as anodizing, acid/alkali treatments, and other liquid environments, in which surface modification methods are suitable. Sandblasting does not produce uniform modification of internal and external surfaces and is not suitable for AM implants with complex geometrical shapes.

In order to ensure the service life and suitability of AM Ti implants, antibacterial activity is not inconsequential. In this paper, various recent antimicrobial strategies applied to AM Ti implants are reviewed; however, the design of structural and antimicrobial coatings based on AM Ti implants remains a long way off. In the future, a balance between mechanical properties [203], microstructures and cell response [204], osteogenic properties, and antibacterial properties should be pursued in structural design. For antibacterial coatings, the release kinetics of antibacterial drugs might be further improved to suppress drug resistance and reduce the immune response caused by coating wear. The direct manipulation of AM technology to produce surfaces with inherent antimicrobial properties on AM Ti implants is also extremely promising. On the one hand, the antibacterial strategies of traditional Ti implants should be further applied to AM Ti implants, with emphasis on the development of surface treatment technologies suitable for complex geometries. On the other hand, novel and alternative strategies should be sought to combat IAIs, especially through the development of AM technology to improve antimicrobial performance and simplify the processing method. For the clinical application of these implants with bactericidal or bacteriostatic properties, and in order to improve the biocompatibility and prolong the functional life of implants, further research is needed. In particular, in vivo trials are necessary before clinical human trials can commence.

Funding: The work was funded by the National Natural Science Foundation of China (Grant No. 31971246).

Institutional Review Board Statement: Not applicable.

Informed Consent Statement: Not applicable.

Data Availability Statement: Not applicable.

Acknowledgments: We are very grateful to the National Natural Science Foundation of China (No. 31971246) Fund Committee for its support.

Conflicts of Interest: The authors declare no conflict of interest.

References

1. Schwarz, E.M.; Parvizi, J.; Gehrke, T.; Aiyer, A.; Battenberg, A.; Brown, S.A.; Callaghan, J.J.; Citak, M.; Egol, K.; Garrigues, G.E.; et al. 2018 International Consensus Meeting on Musculoskeletal Infection: Research Priorities from the General Assembly Questions. *J. Orthop. Res.* **2019**, *37*, 997–1006. [CrossRef]
2. Pulido, L.; Ghanem, E.; Joshi, A.; Purtill, J.J.; Parvizi, J. Periprosthetic joint infection: The incidence, timing, and predisposing factors. *Clin. Orthop. Relat. Res.* **2008**, *466*, 1710–1715. [CrossRef]
3. Kaplan, S.L. Recent lessons for the management of bone and joint infections. *J. Infect.* **2014**, *68* (Suppl. 1), S51–S56. [CrossRef] [PubMed]
4. Gristina, A.G.; Naylor, P.; Myrvik, Q. Infections from biomaterials and implants: A race for the surface. *Med. Prog. Technol.* **1988**, *14*, 205–224.
5. Ricciardi, B.F.; Muthukrishnan, G.; Masters, E.; Ninomiya, M.; Lee, C.C.; Schwarz, E.M. Staphylococcus aureus Evasion of Host Immunity in the Setting of Prosthetic Joint Infection: Biofilm and Beyond. *Curr. Rev. Musculoskelet. Med.* **2018**, *11*, 389–400. [CrossRef] [PubMed]
6. Costerton, J.W.; Montanaro, L.; Arciola, C.R. Biofilm in implant infections: Its production and regulation. *Int. J. Artif. Organs* **2005**, *28*, 1062–1068. [CrossRef] [PubMed]
7. Masters, E.A.; Trombetta, R.P.; de Mesy-Bentley, K.L.; Boyce, B.F.; Gill, A.L.; Gill, S.R.; Nishitani, K.; Ishikawa, M.; Morita, Y.; Ito, H.; et al. Evolving concepts in bone infection: Redefining "biofilm", "acute vs. chronic osteomyelitis", "the immune proteome" and "local antibiotic therapy". *Bone Res.* **2019**, *7*, 20. [CrossRef] [PubMed]
8. Bjarnsholt, T. The role of bacterial biofilms in chronic infections. *APMIS Suppl.* **2013**, 1–51. [CrossRef]
9. Qin, S.; Xu, K.; Nie, B.; Ji, F.; Zhang, H. Approaches based on passive and active antibacterial coating on titanium to achieve antibacterial activity. *J. Biomed. Mater. Res. A* **2018**, *106*, 2531–2539. [CrossRef] [PubMed]
10. Bekmurzayeva, A.; Duncanson, W.J.; Azevedo, H.S.; Kanayeva, D. Surface modification of stainless steel for biomedical applications: Revisiting a century-old material. *Mater. Sci. Eng. C* **2018**, *93*, 1073–1089. [CrossRef]
11. Chouirfa, H.; Bouloussa, H.; Migonney, V.; Falentin-Daudré, C. Review of titanium surface modification techniques and coatings for antibacterial applications. *Acta Biomater.* **2019**, *83*, 37–54. [CrossRef]
12. Sobolev, A.; Valkov, A.; Kossenko, A.; Wolicki, I.; Zinigrad, M.; Borodianskiy, K. Bioactive Coating on Ti Alloy with High Osseointegration and Antibacterial Ag Nanoparticles. *ACS Appl. Mater. Interfaces* **2019**, *11*, 39534–39544. [CrossRef]
13. Xie, C.-M.; Lu, X.; Wang, K.-F.; Meng, F.-Z.; Jiang, O.; Zhang, H.-P.; Zhi, W.; Fang, L.-M. Silver Nanoparticles and Growth Factors Incorporated Hydroxyapatite Coatings on Metallic Implant Surfaces for Enhancement of Osteoinductivity and Antibacterial Properties. *ACS Appl. Mater. Interfaces* **2014**, *6*, 8580–8589. [CrossRef]
14. Fazel, M.; Salimijazi, H.R.; Shamanian, M.; Minneboo, M.; Modaresifar, K.; van Hengel, I.A.J.; Fratila-Apachitei, L.E.; Apachitei, I.; Zadpoor, A.A. Osteogenic and antibacterial surfaces on additively manufactured porous Ti-6Al-4V implants: Combining silver nanoparticles with hydrothermally synthesized HA nanocrystals. *Mater. Sci. Eng. C Mater. Biol. Appl.* **2021**, *120*, 111745. [CrossRef]
15. Jacobs, A.; Renaudin, G.; Forestier, C.; Nedelec, J.M.; Descamps, S. Biological properties of copper-doped biomaterials for orthopedic applications: A review of antibacterial, angiogenic and osteogenic aspects. *Acta Biomater.* **2020**, *117*, 21–39. [CrossRef]
16. van Hengel, I.A.J.; Tierolf, M.; Fratila-Apachitei, L.E.; Apachitei, I.; Zadpoor, A.A. Antibacterial Titanium Implants Biofunctionalized by Plasma Electrolytic Oxidation with Silver, Zinc, and Copper: A Systematic Review. *Int. J. Mol. Sci.* **2021**, *22*, 3800. [CrossRef]
17. Vargas-Reus, M.A.; Memarzadeh, K.; Huang, J.; Ren, G.G.; Allaker, R.P. Antimicrobial activity of nanoparticulate metal oxides against peri-implantitis pathogens. *Int. J. Antimicrob. Agents* **2012**, *40*, 135–139. [CrossRef] [PubMed]
18. Tsikopoulos, K.; Sidiropoulos, K.; Kitridis, D.; Hassan, A.; Drago, L.; Mavrogenis, A.; McBride, D. Is coating of titanium implants effective at preventing *Staphylococcus aureus* infections? A meta-analysis of animal model studies. *Int. Orthop.* **2021**, *45*, 821–835. [CrossRef] [PubMed]
19. Attarilar, S.; Ebrahimi, M.; Djavanroodi, F.; Fu, Y.; Wang, L.; Yang, J. 3D Printing Technologies in Metallic Implants: A Thematic Review on the Techniques and Procedures. *Int. J. Bioprint.* **2021**, *7*, 306. [CrossRef] [PubMed]
20. Bose, S.; Traxel, K.D.; Vu, A.A.; Bandyopadhyay, A. Clinical significance of three-dimensional printed biomaterials and biomedical devices. *MRS Bull.* **2019**, *44*, 494–504. [CrossRef] [PubMed]
21. Zheng, C.; Attarilar, S.; Li, K.; Wang, C.; Liu, J.; Wang, L.; Yang, J.; Tang, Y. 3D-printed HA15-loaded beta-Tricalcium Phosphate/Poly (Lactic-co-glycolic acid) Bone Tissue Scaffold Promotes Bone Regeneration in Rabbit Radial Defects. *Int. J. Bioprint.* **2021**, *7*, 317. [CrossRef]
22. Maconachie, T.; Leary, M.; Lozanovski, B.; Zhang, X.Z.; Qian, M.; Faruque, O.; Brandt, M. SLM lattice structures: Properties, performance, applications and challenges. *Mater. Des.* **2019**, *183*, 18. [CrossRef]

23. Murr, L.E. Open-cellular metal implant design and fabrication for biomechanical compatibility with bone using electron beam melting. *J. Mech. Behav. Biomed. Mater.* **2017**, *76*, 164–177. [CrossRef] [PubMed]
24. Trevisan, F.; Calignano, F.; Aversa, A.; Marchese, G.; Lombardi, M.; Biamino, S.; Ugues, D.; Manfredi, D. Additive manufacturing of titanium alloys in the biomedical field: Processes, properties and applications. *J. Appl. Biomater. Funct. Mater.* **2018**, *16*, 57–67. [CrossRef] [PubMed]
25. Hafeez, N.; Liu, S.; Lu, E.; Wang, L.; Liu, R.; Lu, W.; Zhang, L.-C. Mechanical behavior and phase transformation of β-type Ti-35Nb-2Ta-3Zr alloy fabricated by 3D-Printing. *J. Alloys Compd.* **2019**, *790*, 117–126. [CrossRef]
26. Yang, Y.; Wang, G.; Liang, H.; Gao, C.; Peng, S.; Shen, L.; Shuai, C. Additive manufacturing of bone scaffolds. *Int. J. Bioprint* **2019**, *5*, 148. [CrossRef]
27. Revilla-León, M.; Meyer, M.J.; Özcan, M. Metal additive manufacturing technologies: Literature review of current status and prosthodontic applications. *Int. J. Comput. Dent.* **2019**, *22*, 55–67. [PubMed]
28. Maharubin, S.; Hu, Y.; Sooriyaarachchi, D.; Cong, W.; Tan, G.Z. Laser engineered net shaping of antimicrobial and biocompatible titanium-silver alloys. *Mater. Sci. Eng. C Mater. Biol. Appl.* **2019**, *105*, 110059. [CrossRef]
29. Mitun, D.; Balla, V.K.; Dwaipayan, S.; Devika, D.; Manivasagam, G. Surface properties and cytocompatibility of Ti-6Al-4V fabricated using Laser Engineered Net Shaping. *Mater. Sci. Eng. C Mater. Biol. Appl.* **2019**, *100*, 104–116. [CrossRef]
30. Mangano, F.; Chambrone, L.; van Noort, R.; Miller, C.; Hatton, P.; Mangano, C. Direct metal laser sintering titanium dental implants: A review of the current literature. *Int. J. Biomater.* **2014**, *2014*, 461534. [CrossRef]
31. Venkatesh, K.V.; Nandini, V.V. Direct metal laser sintering: A digitised metal casting technology. *J. Indian Prosthodont. Soc.* **2013**, *13*, 389–392. [CrossRef] [PubMed]
32. Sarker, A.; Leary, M.; Fox, K. Metallic additive manufacturing for bone-interfacing implants. *Biointerphases* **2020**, *15*. [CrossRef]
33. Razavykia, A.; Brusa, E.; Delprete, C.; Yavari, R. An Overview of Additive Manufacturing Technologies—A Review to Technical Synthesis in Numerical Study of Selective Laser Melting. *Materials* **2020**, *13*, 3895. [CrossRef] [PubMed]
34. Ataee, A.; Li, Y.; Song, G.; Wen, C. Metal Scaffolds Processed by Electron Beam Melting for Biomedical Applications. In *Metallic Foam Bone*; Woodhead Publishing: Cambridge, UK, 2017; pp. 83–110.
35. Dobbelstein, H.; Gurevich, E.L.; George, E.P.; Ostendorf, A.; Laplanche, G. Laser metal deposition of compositionally graded TiZrNbTa refractory high-entropy alloys using elemental powder blends. *Addit. Manuf.* **2019**, *25*, 252–262. [CrossRef]
36. Arabnejad, S.; Johnston, B.; Tanzer, M.; Pasini, D. Fully porous 3D printed titanium femoral stem to reduce stress-shielding following total hip arthroplasty. *J. Orthop. Res.* **2017**, *35*, 1774–1783. [CrossRef]
37. Glassman, A.H.; Bobyn, J.D.; Tanzer, M. New femoral designs: Do they influence stress shielding? *Clin. Orthop. Relat. Res.* **2006**, *453*, 64–74. [CrossRef]
38. Liang, H.; Yang, Y.; Xie, D.; Li, L.; Mao, N.; Wang, C.; Tian, Z.; Jiang, Q.; Shen, L. Trabecular-like Ti-6Al-4V scaffolds for orthopedic: Fabrication by selective laser melting and in vitro biocompatibility. *J. Mater. Sci. Technol.* **2019**, *35*, 1284–1297. [CrossRef]
39. Wang, X.; Xu, S.; Zhou, S.; Xu, W.; Leary, M.; Choong, P.; Qian, M.; Brandt, M.; Xie, Y.M. Topological design and additive manufacturing of porous metals for bone scaffolds and orthopaedic implants: A review. *Biomaterials* **2016**, *83*, 127–141. [CrossRef]
40. Cheng, A.; Humayun, A.; Boyan, B.D.; Schwartz, Z. Enhanced Osteoblast Response to Porosity and Resolution of Additively Manufactured Ti-6Al-4V Constructs with Trabeculae-Inspired Porosity. *3d Print Addit Manuf.* **2016**, *3*, 10–21. [CrossRef]
41. Karageorgiou, V.; Kaplan, D. Porosity of 3D biomaterial scaffolds and osteogenesis. *Biomaterials* **2005**, *26*, 5474–5491. [CrossRef]
42. Kuboki, Y.; Takita, H.; Kobayashi, D.; Tsuruga, E.; Inoue, M.; Murata, M.; Nagai, N.; Dohi, Y.; Ohgushi, H. BMP-induced osteogenesis on the surface of hydroxyapatite with geometrically feasible and nonfeasible structures: Topology of osteogenesis. *J. Biomed. Mater. Res.* **1998**, *39*, 190–199. [CrossRef]
43. Gotz, H.E.; Muller, M.; Emmel, A.; Holzwarth, U.; Erben, R.G.; Stangl, R. Effect of surface finish on the osseointegration of laser-treated titanium alloy implants. *Biomaterials* **2004**, *25*, 4057–4064. [CrossRef]
44. Bobbert, F.S.L.; Zadpoor, A.A. Effects of bone substitute architecture and surface properties on cell response, angiogenesis, and structure of new bone. *J. Mater. Chem B* **2017**, *5*, 6175–6192. [CrossRef]
45. Zadpoor, A.A. Additively manufactured porous metallic biomaterials. *J. Mater. Chem B* **2019**, *7*, 4088–4117. [CrossRef]
46. van Hengel, I.A.J.; Riool, M.; Fratila-Apachitei, L.E.; Witte-Bouma, J.; Farrell, E.; Zadpoor, A.A.; Zaat, S.A.J.; Apachitei, I. Selective laser melting porous metallic implants with immobilized silver nanoparticles kill and prevent biofilm formation by methicillin-resistant Staphylococcus aureus. *Biomaterials* **2017**, *140*, 1–15. [CrossRef] [PubMed]
47. Ranganathan, S.I.; Kohama, C.; Mercurio, T.; Salvatore, A.; Benmassaoud, M.M.; Kim, T.W.B. Effect of temperature and ultraviolet light on the bacterial kill effectiveness of antibiotic-infused 3D printed implants. *Biomed. Microdevices* **2020**, *22*, 59. [CrossRef] [PubMed]
48. Chen, X.; Gao, C.; Jiang, J.; Wu, Y.; Zhu, P.; Chen, G. 3D printed porous PLA/nHA composite scaffolds with enhanced osteogenesis and osteoconductivity in vivo for bone regeneration. *Biomed. Mater.* **2019**, *14*, 065003. [CrossRef]
49. Wang, Y.; Wang, S.; Zhang, Y.; Mi, J.; Ding, X. Synthesis of Dimethyl Octyl Aminoethyl Ammonium Bromide and Preparation of Antibacterial ABS Composites for Fused Deposition Modeling. *Polymers* **2020**, *12*, 2229. [CrossRef]
50. Zeng, W.; He, J.; Liu, F. Preparation and properties of antibacterial ABS plastics based on polymeric quaternary phosphonium salts antibacterial agents. *Polym. Adv. Technol.* **2019**, *30*, 2515–2522. [CrossRef]

51. Mania, S.; Ryl, J.; Jinn, J.R.; Wang, Y.J.; Michałowska, A.; Tylingo, R. The Production Possibility of the Antimicrobial Filaments by Co-Extrusion of the PLA Pellet with Chitosan Powder for FDM 3D Printing Technology. *Polymers* **2019**, *11*, 1893. [CrossRef] [PubMed]
52. Yang, F.; Zeng, J.; Long, H.; Xiao, J.; Luo, Y.; Gu, J.; Zhou, W.; Wei, Y.; Dong, X. Micrometer Copper-Zinc Alloy Particles-Reinforced Wood Plastic Composites with High Gloss and Antibacterial Properties for 3D Printing. *Polymers* **2020**, *12*, 621. [CrossRef]
53. Sarker, A.; Nhiem, T.; Rifai, A.; Brandt, M.; Tran, P.A.; Leary, M.; Fox, K.; Williams, R. Rational design of additively manufactured Ti6Al4V implants to control Staphylococcus aureus biofilm formation. *Materialia* **2019**, *5*. [CrossRef]
54. Ganjian, M.; Modaresifar, K.; Ligeon, M.R.O.; Kunkels, L.B.; Tumer, N.; Angeloni, L.; Hagen, C.W.; Otten, L.C.; Hagedoorn, P.-L.; Apachitei, I.; et al. Nature Helps: Toward Bioinspired Bactericidal Nanopatterns. *Adv. Mater. Interfaces* **2019**, *6*. [CrossRef]
55. Chen, M.; Zhang, E.; Zhang, L. Microstructure, mechanical properties, bio-corrosion properties and antibacterial properties of Ti-Ag sintered alloys. *Mater. Sci Eng. C Mater. Biol. Appl.* **2016**, *62*, 350–360. [CrossRef] [PubMed]
56. Oliver, J.D.; Eells, A.C.; Saba, E.S.; Boczar, D.; Restrepo, D.J.; Huayllani, M.T.; Sisti, A.; Hu, M.S.; Gould, D.J.; Forte, A.J. Alloplastic Facial Implants: A Systematic Review and Meta-Analysis on Outcomes and Uses in Aesthetic and Reconstructive Plastic Surgery. *Aesthetic Plast. Surg.* **2019**, *43*, 625–636. [CrossRef]
57. Hemmilä, M.; Karvonen, M.; Laaksonen, I.; Matilainen, M.; Eskelinen, A.; Haapakoski, J.; Puhto, A.P.; Kettunen, J.; Manninen, M.; Mäkelä, K.T. Survival of 11,390 Continuum cups in primary total hip arthroplasty based on data from the Finnish Arthroplasty Register. *Acta Orthop.* **2019**, *90*, 312–317. [CrossRef]
58. Laaksonen, I.; Lorimer, M.; Gromov, K.; Eskelinen, A.; Rolfson, O.; Graves, S.E.; Malchau, H.; Mohaddes, M. Trabecular metal acetabular components in primary total hip arthroplasty. *Acta Orthop.* **2018**, *89*, 259–264. [CrossRef] [PubMed]
59. Matharu, G.S.; Judge, A.; Murray, D.W.; Pandit, H.G. Trabecular Metal Versus Non-Trabecular Metal Acetabular Components and the Risk of Re-Revision Following Revision Total Hip Arthroplasty: A Propensity Score-Matched Study from the National Joint Registry for England and Wales. *J. Bone Jt. Surg. Am.* **2018**, *100*, 1132–1140. [CrossRef]
60. Pingueiro, J.; Piattelli, A.; Paiva, J.; Figueiredo, L.C.; Feres, M.; Shibli, J.; Bueno-Silva, B. Additive manufacturing of titanium alloy could modify the pathogenic microbial profile: An in vitro study. *Braz Oral Res.* **2019**, *33*, e065. [CrossRef]
61. Koutiri, I.; Pessard, E.; Peyre, P.; Amlou, O.; De Terris, T. Influence of SLM process parameters on the surface finish, porosity rate and fatigue behavior of as-built Inconel 625 parts. *J. Mater. Process. Technol.* **2018**, *255*, 536–546. [CrossRef]
62. Villapun, V.M.; Carter, L.N.; Gao, N.; Addison, O.; Webber, M.A.; Shepherd, D.E.T.; Andrews, J.W.; Lowther, M.; Avery, S.; Glanvill, S.J.; et al. A design approach to facilitate selective attachment of bacteria and mammalian cells to additively manufactured implants. *Addit. Manuf.* **2020**, *36*, 12. [CrossRef]
63. Sarker, A.; Tran, N.; Rifai, A.; Elambasseril, J.; Brandt, M.; Williams, R.; Leary, M.; Fox, K. Angle defines attachment: Switching the biological response to titanium interfaces by modifying the inclination angle during selective laser melting. *Mater. Des.* **2018**, *154*, 326–339. [CrossRef]
64. Ginestra, P.; Ferraro, R.M.; Zohar-Hauber, K.; Abeni, A.; Giliani, S.; Ceretti, E. Selective Laser Melting and Electron Beam Melting of Ti6Al4V for Orthopedic Applications: A Comparative Study on the Applied Building Direction. *Materials* **2020**, *13*, 5584. [CrossRef]
65. Xie, K.; Guo, Y.; Zhao, S.; Wang, L.; Wu, J.; Tan, J.; Yang, Y.; Wu, W.; Jiang, W.; Hao, Y. Partially Melted Ti6Al4V Particles Increase Bacterial Adhesion and Inhibit Osteogenic Activity on 3D-printed Implants: An In Vitro Study. *Clin. Orthop. Relat. Res.* **2019**, *477*, 2772–2782. [CrossRef]
66. Ginestra, P.; Ceretti, E.; Lobo, D.; Lowther, M.; Cruchley, S.; Kuehne, S.; Villapun, V.; Cox, S.; Grover, L.; Shepherd, D.; et al. Post Processing of 3D Printed Metal Scaffolds: A Preliminary Study of Antimicrobial Efficiency. *Procedia Manuf.* **2020**, *47*, 1106–1112. [CrossRef]
67. Szymczyk, P.; Junka, A.; Ziolkowski, G.; Smutnicka, D.; Bartoszewicz, M.; Chlebus, E. The ability of S. aureus to form biofilm on the Ti-6Al-7Nb scaffolds produced by Selective Laser Melting and subjected to the different types of surface modifications. *Acta Bioeng. Biomech.* **2013**, *15*, 69–76. [CrossRef]
68. Junka, A.F.; Szymczyk, P.; Secewicz, A.; Pawlak, A.; Smutnicka, D.; Ziolkowski, G.; Bartoszewicz, M.; Chlebus, E. The chemical digestion of Ti6Al7Nb scaffolds produced by Selective Laser Melting reduces significantly ability of Pseudomonas aeruginosa to form biofilm. *Acta Bioeng. Biomech.* **2016**, *18*, 115–120. [CrossRef]
69. Cox, S.C.; Jamshidi, P.; Eisenstein, N.M.; Webber, M.A.; Burton, H.; Moakes, R.J.A.; Addison, O.; Attallah, M.; Shepherd, D.E.T.; Grover, L.M. Surface Finish has a Critical Influence on Biofilm Formation and Mammalian Cell Attachment to Additively Manufactured Prosthetics. *ACS Biomater. Sci. Eng.* **2017**, *3*, 1616–1626. [CrossRef] [PubMed]
70. Szymczyk-Ziolkowska, P.; Hoppe, V.; Rusinska, M.; Gasiorek, J.; Ziolkowski, G.; Dydak, K.; Czajkowska, J.; Junka, A. The Impact of EBM-Manufactured Ti6Al4V ELI Alloy Surface Modifications on Cytotoxicity toward Eukaryotic Cells and Microbial Biofilm Formation. *Materials* **2020**, *13*, 2822. [CrossRef] [PubMed]
71. Wang, Q.; Zhou, P.; Liu, S.; Attarilar, S.; Ma, R.L.; Zhong, Y.; Wang, L. Multi-Scale Surface Treatments of Titanium Implants for Rapid Osseointegration: A Review. *Nanomaterials* **2020**, *10*, 1244. [CrossRef]
72. Skoric, L.; Sanz-Hernandez, D.; Meng, F.; Donnelly, C.; Merino-Aceituno, S.; Fernandez-Pacheco, A. Layer-by-Layer Growth of Complex-Shaped Three-Dimensional Nanostructures with Focused Electron Beams. *Nano Lett.* **2020**, *20*, 184–191. [CrossRef]
73. Liao, C.; Wuethrich, A.; Trau, M. A material odyssey for 3D nano/microstructures: Two photon polymerization based nanolithography in bioapplications. *Appl. Mater. Today* **2020**, *19*. [CrossRef]

74. Raimondi, M.T.; Eaton, S.M.; Nava, M.M.; Lagana, M.; Cerullo, G.; Osellame, R. Two-photon laser polymerization: From fundamentals to biomedical application in tissue engineering and regenerative medicine. *J. Appl Biomater. Funct. Mater.* **2012**, *10*, 55–65. [CrossRef] [PubMed]
75. van Dorp, W.F.; van Someren, B.; Hagen, C.W.; Kruit, P.; Crozier, P.A. Approaching the resolution limit of nanometer-scale electron beam-induced deposition. *Nano Lett.* **2005**, *5*, 1303–1307. [CrossRef] [PubMed]
76. Utke, I.; Hoffmann, P.; Melngailis, J. Gas-assisted focused electron beam and ion beam processing and fabrication. *J. Vac. Sci. Technol. B Microelectron. Nanometer Struct.* **2008**, *26*. [CrossRef]
77. Plank, H.; Winkler, R.; Schwalb, C.H.; Hutner, J.; Fowlkes, J.D.; Rack, P.D.; Utke, I.; Huth, M. Focused Electron Beam-Based 3D Nanoprinting for Scanning Probe Microscopy: A Review. *Micromachines* **2019**, *11*, 48. [CrossRef] [PubMed]
78. Xing, J.F.; Zheng, M.L.; Duan, X.M. Two-photon polymerization microfabrication of hydrogels: An advanced 3D printing technology for tissue engineering and drug delivery. *Chem Soc. Rev.* **2015**, *44*, 5031–5039. [CrossRef]
79. Kumeria, T.; Gulati, K.; Santos, A.; Losic, D. Real-time and in situ drug release monitoring from nanoporous implants under dynamic flow conditions by reflectometric interference spectroscopy. *ACS Appl Mater. Interfaces* **2013**, *5*, 5436–5442. [CrossRef]
80. Gulati, K.; Aw, M.S.; Findlay, D.; Losic, D. Local drug delivery to the bone by drug-releasing implants: Perspectives of nano-engineered titania nanotube arrays. *Ther. Deliv.* **2012**, *3*, 857–873. [CrossRef]
81. King, D.; McGinty, S. Assessing the potential of mathematical modelling in designing drug-releasing orthopaedic implants. *J. Control Release* **2016**, *239*, 49–61. [CrossRef]
82. Park, Y.S.; Cho, J.Y.; Lee, S.J.; Hwang, C.I. Modified titanium implant as a gateway to the human body: The implant mediated drug delivery system. *Biomed. Res. Int.* **2014**, *2014*, 801358. [CrossRef] [PubMed]
83. Gimeno, M.; Pinczowski, P.; Vazquez, F.J.; Perez, M.; Santamaria, J.; Arruebo, M.; Lujan, L. Porous orthopedic steel implant as an antibiotic eluting device: Prevention of post-surgical infection on an ovine model. *Int. J. Pharm.* **2013**, *452*, 166–172. [CrossRef] [PubMed]
84. Gimeno, M.; Pinczowski, P.; Perez, M.; Giorello, A.; Martinez, M.A.; Santamaria, J.; Arruebo, M.; Lujan, L. A controlled antibiotic release system to prevent orthopedic-implant associated infections: An in vitro study. *Eur. J. Pharm. Biopharm.* **2015**, *96*, 264–271. [CrossRef] [PubMed]
85. Cox, S.C.; Jamshidi, P.; Eisenstein, N.M.; Webber, M.A.; Hassanin, H.; Attallah, M.M.; Shepherd, D.E.T.; Addison, O.; Grover, L.M. Adding functionality with additive manufacturing: Fabrication of titanium-based antibiotic eluting implants. *Mater. Sci Eng. C Mater. Biol. Appl.* **2016**, *64*, 407–415. [CrossRef] [PubMed]
86. Hassanin, H.; Finet, L.; Cox, S.C.; Jamshidi, P.; Grover, L.M.; Shepherd, D.E.T.; Addison, O.; Attallah, M.M. Tailoring selective laser melting process for titanium drug-delivering implants with releasing micro-channels. *Addit. Manuf.* **2018**, *20*, 144–155. [CrossRef]
87. Bezuidenhout, M.B.; van Staden, A.D.; Oosthuizen, G.A.; Dimitrov, D.M.; Dicks, L.M. Delivery of antibiotics from cementless titanium-alloy cubes may be a novel way to control postoperative infections. *Biomed. Res. Int.* **2015**, *2015*, 856859. [CrossRef]
88. Bezuidenhout, M.B.; Booysen, E.; van Staden, A.D.; Uheida, E.H.; Hugo, P.A.; Oosthuizen, G.A.; Dimitrov, D.M.; Dicks, L.M.T. Selective Laser Melting of Integrated Ti6Al4V ELI Permeable Walls for Controlled Drug Delivery of Vancomycin. *ACS Biomater. Sci. Eng.* **2018**, *4*, 4412–4424. [CrossRef]
89. Perez, L.M.; Lalueza, P.; Monzon, M.; Puertolas, J.A.; Arruebo, M.; Santamaria, J. Hollow porous implants filled with mesoporous silica particles as a two-stage antibiotic-eluting device. *Int. J. Pharm.* **2011**, *409*, 1–8. [CrossRef]
90. Benmassaoud, M.M.; Kohama, C.; Kim, T.W.B.; Kadlowec, J.A.; Foltiny, B.; Mercurio, T.; Ranganathan, S.I. Efficacy of eluted antibiotics through 3D printed femoral implants. *Biomed. Microdevices* **2019**, *21*, 51. [CrossRef]
91. Burton, H.E.; Eisenstein, N.M.; Lawless, B.M.; Jamshidi, P.; Segarra, M.A.; Addison, O.; Shepherd, D.E.T.; Attallah, M.M.; Grover, L.M.; Cox, S.C. The design of additively manufactured lattices to increase the functionality of medical implants. *Mater. Sci. Eng. C Mater. Biol. Appl.* **2019**, *94*, 901–908. [CrossRef]
92. Sukhorukova, I.V.; Sheveyko, A.N.; Kiryukhantsev-Korneev, P.V.; Anisimova, N.Y.; Gloushankova, N.A.; Zhitnyak, I.Y.; Benesova, J.; Amler, E.; Shtansky, D.V. Two approaches to form antibacterial surface: Doping with bactericidal element and drug loading. *Appl. Surf. Sci.* **2015**, *330*, 339–350. [CrossRef]
93. Souza, J.G.S.; Bertolini, M.M.; Costa, R.C.; Nagay, B.E.; Dongari-Bagtzoglou, A.; Barão, V.A.R. Targeting implant-associated infections: Titanium surface loaded with antimicrobial. *iScience* **2021**, *24*, 102008. [CrossRef] [PubMed]
94. Chen, L.; Song, X.; Xing, F.; Wang, Y.; Wang, Y.; He, Z.; Sun, L. A Review on Antimicrobial Coatings for Biomaterial Implants and Medical Devices. *J. Biomed. Nanotechnol.* **2020**, *16*, 789–809. [CrossRef] [PubMed]
95. Rodriguez-Contreras, A.; Torres, D.; Guillem-Marti, J.; Sereno, P.; Pau-Ginebra, M.; Calero, J.A.; Maria Manero, J.; Ruperez, E. Development of novel dual-action coatings with osteoinductive and antibacterial properties for 3D-printed titanium implants. *Surf. Coat. Technol.* **2020**, *403*. [CrossRef]
96. Yadav, S.K.; Khan, G.; Mishra, B. Advances in patents related to intrapocket technology for the management of periodontitis. *Recent Pat. Drug Deliv.* **2015**, *9*, 129–145. [CrossRef] [PubMed]
97. Ma, M.; Kazemzadeh-Narbat, M.; Hui, Y.; Lu, S.; Ding, C.; Chen, D.D.; Hancock, R.E.; Wang, R. Local delivery of antimicrobial peptides using self-organized TiO_2 nanotube arrays for peri-implant infections. *J. Biomed. Mater. Res. A* **2012**, *100*, 278–285. [CrossRef] [PubMed]

98. Goodman, S.B.; Yao, Z.; Keeney, M.; Yang, F. The future of biologic coatings for orthopaedic implants. *Biomaterials* **2013**, *34*, 3174–3183. [CrossRef]
99. de Jonge, L.T.; Leeuwenburgh, S.C.; Wolke, J.G.; Jansen, J.A. Organic-inorganic surface modifications for titanium implant surfaces. *Pharm Res.* **2008**, *25*, 2357–2369. [CrossRef]
100. Barik, A.; Chakravorty, N. Targeted Drug Delivery from Titanium Implants: A Review of Challenges and Approaches. *Adv. Exp. Med. Biol.* **2020**, *1251*, 1–17. [CrossRef]
101. Zhang, L.-C.; Chen, L.-Y.; Wang, L. Surface Modification of Titanium and Titanium Alloys: Technologies, Developments, and Future Interests. *Adv. Eng. Mater.* **2020**, *22*. [CrossRef]
102. Han, C.; Yao, Y.; Cheng, X.; Luo, J.; Luo, P.; Wang, Q.; Yang, F.; Wei, Q.; Zhang, Z. Electrophoretic Deposition of Gentamicin-Loaded Silk Fibroin Coatings on 3D-Printed Porous Cobalt-Chromium-Molybdenum Bone Substitutes to Prevent Orthopedic Implant Infections. *Biomacromolecules* **2017**, *18*, 3776–3787. [CrossRef]
103. Liu, J.; Liu, J.; Attarilar, S.; Wang, C.; Tamaddon, M.; Yang, C.; Xie, K.; Yao, J.; Wang, L.; Liu, C.; et al. Nano-Modified Titanium Implant Materials: A Way Toward Improved Antibacterial Properties. *Front. Bioeng. Biotechnol.* **2020**, *8*, 576969. [CrossRef]
104. Popat, K.C.; Eltgroth, M.; LaTempa, T.J.; Grimes, C.A.; Desai, T.A. Titania nanotubes: A novel platform for drug-eluting coatings for medical implants? *Small* **2007**, *3*, 1878–1881. [CrossRef] [PubMed]
105. Hu, X.; Xu, R.; Yu, X.; Chen, J.; Wan, S.; Ouyang, J.; Deng, F. Enhanced antibacterial efficacy of selective laser melting titanium surface with nanophase calcium phosphate embedded to TiO_2 nanotubes. *Biomed. Mater.* **2018**, *13*, 045015. [CrossRef] [PubMed]
106. Gao, A.; Hang, R.; Huang, X.; Zhao, L.; Zhang, X.; Wang, L.; Tang, B.; Ma, S.; Chu, P.K. The effects of titania nanotubes with embedded silver oxide nanoparticles on bacteria and osteoblasts. *Biomaterials* **2014**, *35*, 4223–4235. [CrossRef] [PubMed]
107. Zhang, T.; Zhou, W.; Jia, Z.; Wei, Q.; Fan, D.; Yan, J.; Yin, C.; Cheng, Y.; Cai, H.; Liu, X.; et al. Polydopamine-assisted functionalization of heparin and vancomycin onto microarc-oxidized 3D printed porous Ti_6Al_4V for improved hemocompatibility, osteogenic and anti-infection potencies. *Sci. China-Mater.* **2018**, *61*, 579–592. [CrossRef]
108. Jia, Z.; Xiu, P.; Xiong, P.; Zhou, W.; Cheng, Y.; Wei, S.; Zheng, Y.; Xi, T.; Cai, H.; Liu, Z.; et al. Additively Manufactured Macroporous Titanium with Silver-Releasing Micro-/Nanoporous Surface for Multipurpose Infection Control and Bone Repair—A Proof of Concept. *ACS Appl Mater. Interfaces* **2016**, *8*, 28495–28510. [CrossRef]
109. Jia, Z.; Li, M.; Xiu, P.; Xu, X.; Cheng, Y.; Zheng, Y.; Xi, T.; Wei, S.; Liu, Z. A novel cytocompatible, hierarchical porous Ti6Al4V scaffold with immobilized silver nanoparticles. *Mater. Lett.* **2015**, *157*, 143–146. [CrossRef]
110. Maher, S.; Kaur, G.; Lima-Marques, L.; Evdokiou, A.; Losic, D. Engineering of Micro- to Nanostructured 3D-Printed Drug-Releasing Titanium Implants for Enhanced Osseointegration and Localized Delivery of Anticancer Drugs. *ACS Appl. Mater. Interfaces* **2017**, *9*, 29562–29570. [CrossRef] [PubMed]
111. Gulati, K.; Prideaux, M.; Kogawa, M.; Lima-Marques, L.; Atkins, G.J.; Findlay, D.M.; Losic, D. Anodized 3D-printed titanium implants with dual micro- and nano-scale topography promote interaction with human osteoblasts and osteocyte-like cells. *J. Tissue Eng. Regen. Med.* **2017**, *11*, 3313–3325. [CrossRef]
112. Xue, T.; Attarilar, S.; Liu, S.; Liu, J.; Song, X.; Li, L.; Zhao, B.; Tang, Y. Surface Modification Techniques of Titanium and its Alloys to Functionally Optimize Their Biomedical Properties: Thematic Review. *Front. Bioeng. Biotechnol.* **2020**, *8*, 603072. [CrossRef]
113. Nair, M.; Elizabeth, E. Applications of Titania Nanotubes in Bone Biology. *J. Nanosci. Nanotechnol.* **2015**, *15*, 939–955. [CrossRef]
114. Ahn, T.K.; Lee, D.H.; Kim, T.S.; Jang, G.C.; Choi, S.; Oh, J.B.; Ye, G.; Lee, S. Modification of Titanium Implant and Titanium Dioxide for Bone Tissue Engineering. *Adv. Exp. Med. Biol.* **2018**, *1077*, 355–368. [CrossRef]
115. Popat, K.C.; Eltgroth, M.; Latempa, T.J.; Grimes, C.A.; Desai, T.A. Decreased Staphylococcus epidermis adhesion and increased osteoblast functionality on antibiotic-loaded titania nanotubes. *Biomaterials* **2007**, *28*, 4880–4888. [CrossRef]
116. He, P.; Zhang, H.; Li, Y.; Ren, M.; Xiang, J.; Zhang, Z.; Ji, P.; Yang, S. 1α,25-Dihydroxyvitamin D3-loaded hierarchical titanium scaffold enhanced early osseointegration. *Mater. Sci. Eng. C Mater. Biol. Appl.* **2020**, *109*, 110551. [CrossRef] [PubMed]
117. Losic, D.; Simovic, S. Self-ordered nanopore and nanotube platforms for drug delivery applications. *Expert Opin. Drug. Deliv.* **2009**, *6*, 1363–1381. [CrossRef] [PubMed]
118. Maher, S.; Mazinani, A.; Barati, M.R.; Losic, D. Engineered titanium implants for localized drug delivery: Recent advances and perspectives of Titania nanotubes arrays. *Expert Opin. Drug. Deliv.* **2018**, *15*, 1021–1037. [CrossRef] [PubMed]
119. Kang, H.-J.; Kim, D.J.; Park, S.-J.; Yoo, J.-B.; Ryu, Y.S. Controlled drug release using nanoporous anodic aluminum oxide on stent. *Thin Solid Film.* **2007**, *515*, 5184–5187. [CrossRef]
120. Kumeria, T.; Mon, H.; Aw, M.S.; Gulati, K.; Santos, A.; Griesser, H.J.; Losic, D. Advanced biopolymer-coated drug-releasing titania nanotubes (TNTs) implants with simultaneously enhanced osteoblast adhesion and antibacterial properties. *Colloids Surf. B Biointerfaces* **2015**, *130*, 255–263. [CrossRef] [PubMed]
121. Li, Y.; Yang, Y.; Li, R.; Tang, X.; Guo, D.; Qing, Y.; Qin, Y. Enhanced antibacterial properties of orthopedic implants by titanium nanotube surface modification: A review of current techniques. *Int. J. Nanomed.* **2019**, *14*, 7217–7236. [CrossRef]
122. Maher, S.; Wijenayaka, A.R.; Lima-Marques, L.; Yang, D.; Atkins, G.J.; Losic, D. Advancing of Additive-Manufactured Titanium Implants with Bioinspired Micro- to Nanotopographies. *ACS Biomater. Sci. Eng.* **2021**, *7*, 441–450. [CrossRef]
123. Im, S.Y.; Kim, K.M.; Kwon, J.S. Antibacterial and Osteogenic Activity of Titania Nanotubes Modified with Electrospray-Deposited Tetracycline Nanoparticles. *Nanomaterials* **2020**, *10*, 1093. [CrossRef]
124. Han, L.; Wang, M.; Sun, H.; Li, P.; Wang, K.; Ren, F.; Lu, X. Porous titanium scaffolds with self-assembled micro/nano-hierarchical structure for dual functions of bone regeneration and anti-infection. *J. Biomed. Mater. Res. A* **2017**, *105*, 3482–3492. [CrossRef]

Review

β-Ti Alloys for Orthopedic and Dental Applications: A Review of Progress on Improvement of Properties through Surface Modification

Longfei Shao [1,†], Yiheng Du [1,†], Kun Dai [1,†], Hong Wu [2], Qingge Wang [2,*], Jia Liu [3,*], Yujin Tang [3,*] and Liqiang Wang [1,*]

1. State Key Laboratory of Metal Matrix Composites, School of Material Science and Engineering, Shanghai Jiao Tong University, No. 800 Dongchuan Road, Shanghai 200240, China; shaolongfei@sjtu.edu.cn (L.S.); 03170813@sjtu.edu.cn (Y.D.); daikun1989@sjtu.edu.cn (K.D.)
2. State Key Lab of Powder Metallurgy, Central South University, Changsha 410083, China; hwucsu@csu.edu.cn
3. Affiliated Hospital of Youjiang Medical University for Nationalities, Baise 533000, China
* Correspondence: wendymewqg@163.com (Q.W.); liujia0111@live.cn (J.L.); tangyujin196709@163.com (Y.T.); wang_liqiang@sjtu.edu.cn (L.W.)
† The authors contributed equally to this work as first authors.

Abstract: Ti and Ti alloys have charming comprehensive properties (high specific strength, strong corrosion resistance, and excellent biocompatibility) that make them the ideal choice in orthopedic and dental applications, especially in the particular fabrication of orthopedic and dental implants. However, these alloys present some shortcomings, specifically elastic modulus, wear, corrosion, and biological performance. Beta-titanium (β-Ti) alloys have been studied as low elastic modulus and low toxic or non-toxic elements. The present work summarizes the improvements of the properties systematically (elastic modulus, hardness, wear resistance, corrosion resistance, antibacterial property, and bone regeneration) for β-Ti alloys via surface modification to address these shortcomings. Additionally, the shortcomings and prospects of the present research are put forward. β-Ti alloys have potential regarding implants in biomedical fields.

Keywords: beta titanium alloy; elastic modulus; wear resistance; corrosion property; surface modification; osseointegration

1. Introduction

Titanium and its alloys have been widely used as bone implants in clinical dentistry and orthopedics, especially CP Ti (α) and Ti-6Al-4V (TC4, α + β) [1]. However, the development of CP Ti and TC4 was restricted because of high elastic modulus and toxic element vanadium (V). Low elastic modulus and non-toxic β-Ti alloys were designed to solve these problems. β-Ti alloys majorly represented a β-phase-dominated microstructure after annealing and air cooling to room temperature with the BCC form of titanium (called beta). The alloying elements in the titanium matrix can be one or more of these metals, including molybdenum (Mo), vanadium (V), niobium (Nb), tantalum (Ta), zirconium (Zr), manganese (Mn), iron (Fe), chromium (Cr), cobalt (Co), nickel (Ni), and copper (Cu) [2–4]. The combination of different types and contents of elements leads to a variety of β-Ti alloys with distinct properties; the resultant β-Ti alloys usually have excellent formability and facile welding characteristics [5,6].

As shown in Table 1, these Ti alloys have been applied as implants in clinical surgery. Nevertheless, the low wear resistance of Ti alloys became the new issue. Researchers developed varieties with modified processing and technology to enhance the wear resistance and endowed β-Ti alloys with antibacterial properties and bone regeneration.

Table 1. The uses, advantages, and disadvantages of Ti and its alloys.

Materials	Type	Advantages	Disadvantages	Applications	Clinical Surgery	Ref.
CP Ti	α	Good biocompatibility	Low strength and poor wear resistance	Dental implants	√	[7]
Ti–3Al–2.5V	α + β	Good strength and corrosion resistance	Toxicity elements (Al, V)	Dental implants	√	[8]
Ti–6Al–4V (TC4)	α + β	Excellent strength and corrosion resistance	High elastic modulus, toxicity elements (Al, V), and poor wear resistance	Bone fixation plates and stem of artificial hip joints	√	[1]
Ti–6Al–7Nb	α + β	Good wear resistance	Toxicity element (Al)	Dental prostheses knee, wrist, and femoral stems, fasteners, fixation plates, and screws	√	[9]
Ti–5Al–2.5Fe	α + β	Good wear resistance	Toxicity element (Al)	Hip prostheses	√	[1]
Ti–2.5Al–2.5Mo–2.5Zr (TAMZ)	α + β	High compatibility, toughness, fatigue resistance	Toxicity element (Al)	Hip stems, endosseous, subperiosteal, or transosteal implants in dentistry	√	[10]
Ti–12Mo–6Zr–2Fe (TMZF)	β	Low elastic modulus, high fracture toughness, good wear resistance, and corrosion resistance	Head-neck taper fretting and corrosion, flexural rigidity	Femoral neck shaft, acetabular implant, and femoral stems	√	[11]
Ti–13Nb–13Zr	near β	Low elastic modulus, low density, paramagnetic properties, low thermal conductivity	Low hardness and resistance	Head and acetabulum of hip endoprostheses	√	[12]
Ti–24Nb–4Zr–8Sn (Ti2448)	β	High biocompatibility and mechanical properties	Low wear resistance	Artificial hip joints and dental roots	√	[13,14]
Ti–15Mo	β	More biocompatible, lower modulus, better processability	Lower strength	Femoral hip implant components	√	[15]
Ti–28Nb–24.5Zr	β	Low elastic modulus, high strength and toughness, excellent mechanical properties and biocompatibility	Poor wear property	Surgical and orthopedic implants	√	[16]

The elastic modulus is one of the most specific mechanical characteristics and among the vitally prominent physical indicators that substantially affect the long-term stability of implants in medical applications. In this regard, the maximum elastic modulus of human bone is 15–30 GPa, while the elastic modulus of CP Ti (100 GPa) and TC4 (112 GPa) is far higher than that of human bone [1]. The elastic modulus of β-Ti is more than 50 GPa, which is higher than that of human bone [17–20]. Because of the mismatch in stiffness, most of the stress is concentrated on the implants. Human bone grows and rebuilds during the whole process. Thus, the bone density may decrease due to a lack of load in the long-term surrounding the implants, which causes bone resorption or nonunions and a stress shielding phenomenon while affecting usual recovery [21]. On the other hand, low elastic modulus material will severely deform after being stressed and lose its supporting effect [22]. Moreover, achieving an intimate attachment of implanted Ti alloys to the biological tissue is challenging and may lead to implant fracture and fall off the tissue. Thus, it is necessary to explore the relationship between the elastic modulus and chemical composition, hardness, and porosity. To overcome the issue regarding stress shielding, different chemical compositions and microstructures of β-Ti alloys were developed; some of them have been used in the biomedical field (Table 1), and some are still at the research stage. In Figure 1, some researchers reported an approach to control the elastic modulus by designing the material microstructures. Hardness improvement without affecting other properties is of particular importance in β-Ti alloys for implants [23,24].

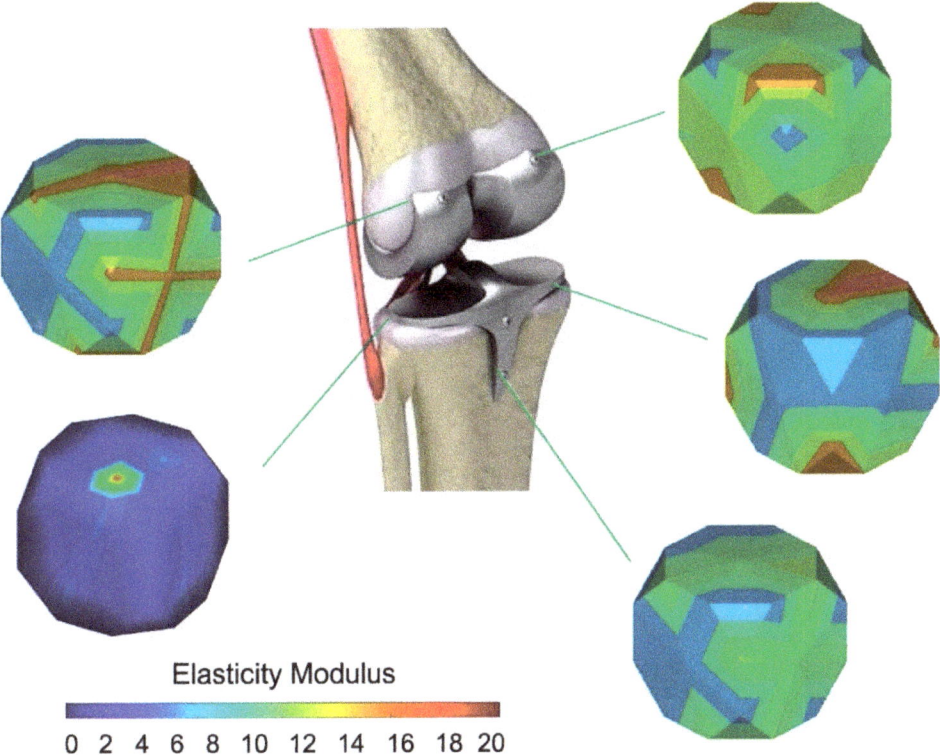

Figure 1. Design of the elastic modulus by changing the material microstructures. Reproduced with permission from [25], copyright Elsevier, 2020.

The Ti alloys always present low wear resistance. The high coefficients of friction and large wear loss are considered as the negative factors of the tribological behaviors. Although there is high corrosion resistance, the protective oxide layer on the surface of Ti alloys can be destroyed during infection in body fluid with low oxygen content and biomolecules. The wear and tear of the prosthetic components (femoral head and cup of a hip implant) creates metallic wear debris with the size of 0.05 μm that leads to adverse cellular responses, toxicity, and inflammation, ultimately causing osteolysis, implant loosening, or the formation of a pseudo-tumor [26–29]. The wear debris percentage is about 4%–5% of all the implant failure cases.

The electrochemical performance (corrosion resistance) of Ti and its alloys' implants has a decisive role in human health. In the body fluid environment, the toxic ions (for example, Co, Cr, Ni, Al, and V ions) release into the surrounding tissues and even organs by body fluids via corrosion, increasing the risk of cytotoxic and even genotoxic and allergic responses in medical implantation. The corrosion of objects may produce adverse effects, reduce the implant life, and endanger the safety of human life. In addition, the implants may cause failure due to corrosion-fatigue under the effect of cyclically load. The reason for corrosion failure is related to implant design. The number and size of defects (including porosity, grain, and inclusions) decides the quality of implants. Due to the micro-current effects between the different phases and almost non-toxic elements, β-Ti alloys have broad prospects in biomedicine compared with other Ti alloys [29].

Another crucial requirement for long-term implant stability is related to their favorable biological response. In this regard, bacterial infection is one of the main reasons for the failure of Ti alloys in human implants [30]. It is reported that the implants' failure rate reaches 0.5%–5% due to bacterial infection [29]. Regarding bacterial infection, particular pathogens result in biofilms and microbial reproduction on the surface of implants after adhering, colonizing, and proliferating, leading to bone destruction [31]. The biofilm is hard to remove, which results in revision surgeries. In order to promote the biomedical application of β-Ti alloys, it is of crucial importance to increase the antibacterial abilities of β-Ti alloys.

The excellent bone regeneration decides the stability and service life of implants. Generally, Ti and Ti alloys are bio-inert [32]. The factors affecting the bone regeneration process are the patient's age and bone quality, and anatomical location [29]. The elastic modulus, hardness, wear resistance, and corrosion resistance may cause low bone-implant contact then implant failure.

Hence, these properties are mutually influencing and closely connected. The stress shielding, toxicity, and poor wear resistance are difficult to overcome [33]. Developing novel β-Ti alloys without toxic elements is the only choice to solve these problems. The elastic modulus of β-Ti is lower than other Ti alloys implants. Moreover, the wear resistance can be improved by surface modification.

There are many literature studies that have reported a variety of methods for addressing these shortcomings, such as substrate (surface) modification methods and deposition of surface coatings. The common surface modification techniques involve in friction stir processing (FSP), ultrasonic nanocrystal surface modification (UNSM), laser surface treatment (LST), surface mechanical attrition treatment (SMAT), and equal channel angular pressing (ECAP). The surface coating deposition methods include chemical vapor deposition (CVD), acid and alkali treatment [34], evaporation, sputtering, and laser cladding [35,36]. The exclusive focus of the present review is the literature on the surface modification methods.

In the recent literature, there are many review articles on Ti alloys and on Ti alloys for biomedical applications via the surface modification method to improve their properties [1,37,38]. The previous reviews did not focus exclusively on orthopedic and dental applications and did not focus exclusively on substrate (surface) modification methods but also included coating deposition methods.

The purpose of the review is to perform a critical review of the literature on the surface modification methods, with an emphasis on the improvement in the hardness, wear resistance, friction coefficient, corrosion resistance, antibacterial activity, and bone regeneration performance of β-Ti alloys in orthopedic and dental applications. Moreover, the review also includes a brief discussion of the shortcomings of this body of literature.

2. Elastic Modulus

In the field of orthopedic implants, elastic modulus is one of the most crucial physical indicators. The maximum elastic modulus of human bone is 30 GPa, while the elastic modulus of Ti alloy is generally above 80 GPa. A high elastic modulus of a metallic implant will cause stress shielding and affect normal recovery [39]. The "stress shielding" effect is associated with the disproportional load distribution between a bone and an adjacent implant due to the elastic modulus mismatch. The metallic implant materials are usually considerably stiffer as compared to bones. Eventually, the elastic modulus mismatch may lead to bone resorption and loosening or failure of the implant. The elastic modulus of β-Ti alloys matches with cortical bone. In addition, porous Ti alloys are found to potential choices to develop low elastic modulus Ti alloys [40–43]. The elastic modulus of common β-Ti alloys is summarized in Figure 2.

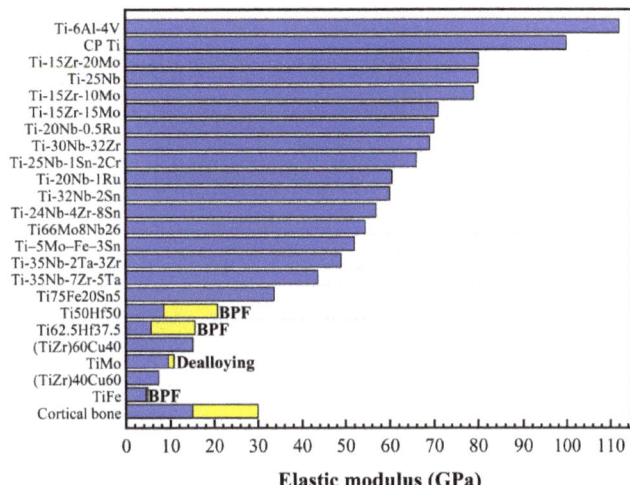

Figure 2. The elastic modulus of different composition and modified β-Ti alloys (BPF -impregnated by bisphenol F epoxy resin). Reprinted with permission from ref. [40,44–50]. Copyright Elsevier, 2018, 2019, 2020, and 2021.

Modern surface modification treatments, such as dealloying, have been proven to reduce the elastic modulus and nonporous structure [51,52]. Okulov et al. [53] synthesized the Ti–Mg interpenetrating phase composite material with a low elastic modulus by liquid metal dealloying. The elastic modulus (17.6 GPa) of the composite material was several times lower than those of the individual component phase, namely Mg (45 GPa) and Ti (110 GPa). In another study, they [43] developed a design strategy of light-weight nano-/microporous alloys by selective corrosion in liquid metal. The elastic modulus could be adjusted between 4.4 and 24 GPa, which affords matching to bone.

In fact, low elastic modulus material deforms easily after being stressed [46]. Thus, Ti alloys are difficult to closely adhere to the biological tissue. Eventually, the implants break and fall off from the tissue. It is vital to keep the elastic modulus in an appropriate range.

3. Hardness

There are many ways to measure the hardness of materials with their specific mechanical meanings, such as Brinell, Vickers, and nano-indentation methods. As shown in Figure 3, the hardness of unstrengthened β-Ti alloy is much higher than human materials, such as bones and femurs [24]. The combination of high hardness, low elastic modulus, and excellent biocompatibility is desirable but difficult to achieve simultaneously, so it is of great importance to improve the hardness without reducing other properties.

Heat treatment plays a vital part in the improvement of hardness. It is reported that Ti–xNb–3Zr–2Ta alloys (x = 33, 31, 29, 27, 25) (wt.%) were carried out by water quenching and air cooling, respectively. The hardness value of both the water quenching and air-cooling group enhanced significantly with the reduction in niobium content. However, there are two different mechanisms to increase the hardness between water quenching and air cooling. For the water quenching group, with the volume fractions of the martensite phase, the hardness increased. For the air cooling group, the improvement of hardness was attributed to the lattice distortion of the β matrix [54].

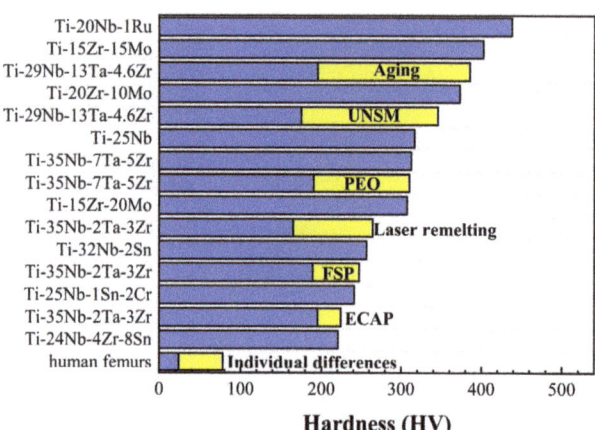

Figure 3. The hardness of different composition and modified β-Ti alloys. Reprinted with permission from ref. [45,55–60]. Copyright Elsevier, 2003, 2013, 2015, 2019, and 2021.

FSP, as a novel solid-state surface modification technique, is an attractive method for inducing localized thermomechanical effects [61]. In order to modify the surface of the newly developed Ti-35Nb-2Ta-3Zr (wt.%) alloy, Wang et al. [56] explored the microhardness value of deformation pass from single pass to three passes at the same rotation speed. The microhardness increased from 189 HV to 208 HV via FSP. After three passes, the microhardness was up to 247 HV. In addition, the ECAP method can produce considerable uniform plastic strains to refine grains and enhance mechanical properties. It is reported that the hardness values of β Ti–35Nb–3Zr–2Ta biomedical alloy with refined grains and uniform microstructures after four passes reached 216 HV. Moreover, a low elastic modulus of about 59 GPa was obtained [59].

Niinomi et al. [62] reported that TNTZ could effectively inhibit bone atrophy and enhance bone remodeling in vivo. Additionally, TNTZ exhibits a low elastic modulus that is approximately half of TC4 [63,64]. However, due to its relatively low wear resistance, TNTZ cannot satisfy the requirements for biomedical implants [65]. Hence, there is a substantial motivation to improve the wear resistance of TNTZ by increasing the hardness while maintaining its relatively low elastic modulus. The TNTZ samples after aging exhibited varying hardness values at different temperatures. Within a definite temperature range, the hardness increased as the temperature rose.

The UNSM technique as a branch of the physical methods is controlled by a computer numerical system. Therefore, innovative processed surfaces with complex geometric shapes and microstructures can be created via accurate parameter selection to improve the properties of surfaces [60]. Besides, researchers investigating the performance of coatings on the surface of Ti alloys have proposed various schemes to compensate for some of the shortcomings of Ti alloys [66,67]. Kheradmandfard et al. [68] processed the TNTZ alloy with the UNSM technique to enhance its hardness. The result showed that the hardness at the surface improved from 195 Hv to 385 Hv; by moving away from the surface, the rate of increase in surface hardness decreases. However, the reason for the hardness increment by UNSM is attributed to grain refinement and a working hardening phenomenon. It was seen that the samples have a thin diffusion layer with a weak bonding to the substrate after diffusion [67].

Chauhan et al. [69] used laser surface heat treatment (LST) to modify the surface microstructure of VT3-1 α-β-Ti alloy. LST was carried out at various laser powers (100–250 W) and scanning speeds of 150–500 mm/min. EBSD)analysis of the laser-affected zone exhibited a distinct microstructure across the depth of the specimen, and it changed with laser power and scanning speed. At a lower scanning speed (150 mm/min) and high laser power

(200 W), an almost complete β-containing microstructure was developed at the surface with a high hardness level of 750 Hv. Obtaining dendritic microstructure and homogenous elemental distribution is helpful in hardness improving.

A laser was also used in the study of remelting on Ti-35Nb-2Ta-3Zr by Zhang et al. [58]. It was found that the average microhardness value increased from 165 Hv up to 264 Hv, which could be ascribed to the surface of the material changing from an equiaxed crystal to a needle-like structure after laser remelting.

Most of research is concentrated on improving the hardness of β-Ti alloys by thermo-mechanical effects. The use of more advanced digital control systems to accurately select parameters for processing, which includes heat treatment and complex microstructure building, may be a further development direction.

4. Wear Resistance

Generally, medical Ti alloys show poor wear properties, and it is easy to become worn in the human body [70–72]. Prosthetic components' wear will produce pieces, causing adverse cellular behavior and inflammatory reactions, eventually leading to implant loosening [17]. The presence of particle corrosion and wear products in the surrounding tissues of implants may cause osteoporosis and eventually lead to the failure of implants [73]. Therefore, it is important to clarify the wear mechanism and find the solution to increase wear resistance.

The wear mechanisms of Ti alloys mainly include oxidation wear, adhesive wear, abrasive wear, and layered wear. However, different loading sliding velocities, matrix material, and ambient temperature conditions will change the wear mechanism [74–76]. According to Archard's laws, the sliding wear resistance is proportional to the alloy hardness [77]. Additionally, the friction coefficient can also affect the wear resistance of the material. Generally, a low friction coefficient is beneficial to the wear resistance of the material. Some researchers designed β-Ti via modification processes to decrease the friction coefficient on the material surface, as shown in Table 2.

Table 2. Summary of preparation process and tribological performance.

Material	Preparation Process	Coefficient of Friction Before	Coefficient of Friction After	Wear Loss Before	Wear Loss After	Mechanism	Ref.
Ti-24Nb-38Zr-2Mo	cold crucible levitation melting + cold rolling + solution treatment	1.25	1.10	1.3 mg	0.9 mg	plowing and some indication of abrasive wear	[78]
Ti-24Nb-38Zr-2Mo-0.15c	cold crucible levitation melting + cold rolling + solution treatment	1.20	0.90	1.0 mg	0.5 mg	plowing and some indication of abrasive wear	[78, 79]
Ti-35Nb-2Ta-3Zr	hybrid surface modification	0.6	0.15	/	/	abrasive wear + adhesive wear	[31]
Ti-13Zr-13Nb-0.5B	melting + heat treatment + hot rolling + solid solution + water quenching	0.42	0.4	/	/	microcutting; abrasive wear	[73]
Ti-29Nb-13Ta-4.6Zr	Picosecond + laser processing	/	/	0.00102 mm^3	0.00014 mm^3	wear debris containment effect and loading pressure	[79, 80]
Ti-25Nb-3Zr-2Sn-3Mo	vacuum induction nitriding	0.65	0.25	0.2 mm^3	0.0007 mm^3	abrasive wear (rod-shaped TiN0.3 phase and soft β matrix)	[80]
Ti-5Al-5Mo-5V-3Cr-0.5Fe	vacuum arc melting + forging + stress relaxation	0.57	0.45	0.073 mm^3	0.034 mm^3	small amount of adhesive wear and slight abrasion wear	[81]
Ti-29Nb-13Ta-4.6Zr	solution-treated + water quenching + UNSM	/	/	0.00102 mm^3	0.00014 mm^3	increased surface hardness + nanoscale lamellar grains + α precipitates	[82]
Ti-30Nb-4Sn	laser nitriding	0.70	0.18	/	/	three-body abrasive wear	[83]
Ti-10V-2Fe-3Al	hot-rolled + heat treatment	0.6	0.6	/	/	oxidation + superelasticity	[74]

In addition to the above factors, wear property can be evaluated by wear resistance indices (H/E and H^3/E^2_{eff}). The Ti-Nb-Sn-Cr alloy was designed during adjusting Sn and Cr. After measurement and calculation, Ti-25Nb-1Sn-2Cr displayed the highest H/E (0.03327) and H^3/E^2_{eff} (0.00261 GPa) [84]. Tong et al. [78] reported the difference in the wear resistance between β-type Ti–24Nb–38Zr–2Mo and CP Ti. The samples exhibited an excellent wear resistance compared with CP Ti because of the Nb2O5 oxide-containing passivation film. Based on that, Majumdar et al. [73] designed Ti-13Zr-13Nb-0.5B and evaluated the wear rate and wear mechanism. The wear rate after furnace cooling and aging treatment was the lowest. The main wear mechanism was microcutting in a dry condition, while the wear mechanism was abrasive wear in the case of bovine serum. Liu et al. [31] developed the hybrid coating on Ti-35Nb-2Ta-3Zr substrate and improved the wear resistance. It involved abrasive wear and adhesive wear by analyzing the friction coefficient and the wear morphology. It is reported [82] that the UNSM treatment on a new β-type TNTZ alloy resulted in a nanostructure surface layer fabrication. The wear volume of the UNSM-treated (1.02×10^{-3} mm^3) sample was more than seven times higher than that of the untreated (0.14×10^{-3} mm^3) one.

The issue of poor wear resistance of Ti alloys needs to be solved as soon as possible. At present, the design of alloys based on the chemical composition is a good approach to improve the wear resistance. In addition, several technologies have been developed to prepare a modified layer with high wear resistance. The wear resistance has improved, and the friction coefficient has decreased in the meantime. However, the layers via some modified processes are easy to peel off because of a weak bonding force.

5. Corrosion Resistance

Some specific Ti alloys are not good for the human body because of the release of toxic ions [85]. This phenomenon has a negative effect on bone repair and body health. Moreover, the study has proven that TC4 showed toxicity or lead to intoxication after implanting for 3 years [86]. The patient showed sensory–motor axonal neuropathy and hearing loss, which was related to high concentrations of V in the blood (6.1 µg/L) and urine (56.0 µg/L). Moreover in another study, neurotoxic of aluminum (Al) was demonstrated [87]. Besides, some other elements (such as Ni and Cr) can cause bone resorption, mobility, and breaking. Although TC4 presents excellent corrosion resistance, it is not the best choice for implants.

β-Ti alloys are widely studied due to their non-cytotoxic nature (without toxic alloying elements) and excellent corrosion resistance [88,89]. However, the applicability of these alloys should be further verified as it is necessary to understand and improve the corrosion resistance of β-Ti alloys. Surface modification technologies are widely utilized to additionally improve the corrosion resistance of β-Ti alloys. Usually, the electrolytes are Hank's, Ringer's, and SBF solution in experiments of corrosion resistance. In addition to the modulation of alloys' chemical composition, the common surface treatment involves plasma electrolytic oxidation (PEO), plasma injection, vapor deposition, sputtering, alkali treatment, and other technologies, which can fabricate oxide layers on the alloy surface. The electrochemical behavior of Ti alloys mainly depends on E_b and I_{pass}. Thus, higher E_b, higher E_p, and lower I_{pass} in the polarization curves exhibit great corrosion resistance.

It is reported that the corrosion current densities (3–4 nA/cm^2) and the film resistances (10^5 Ω·cm^2) of the low-cost Ti-4.7Mo-4.5Fe are close to those of TC4 in Ringer's solution [90]. Great corrosion resistance can be attributed to the stability of the oxide film on the Ti-4.7Mo-4.5Fe surface. The Ti-13Nb-13Zr alloy exhibited similar corrosion resistance compared with TC4 because the source of the Ti-13Nb-13Zr and TC4 corrosion resistances is the same; both Ti alloys produced TiO$_2$ film in Hank's solution [91].

Jin et al. [92] fabricated TiNbZrFe alloy via surface mechanical attrition treatment and reached the nanocrystalline size of 10–30 nm. Due to the fabricated stable and dense passive layer on the nanocrystallized surface of the TiNbZrFe alloy, the corrosion resistance improved significantly.

Prakash et al. [93] studied a new method of surface modification on β-Ti alloy Ti-35Nb-7Ta-5Zr using HA hybrid electrical discharge machining. The method can deposit the biomimetic nanoporous HA layer in situ. The deposited layer was composed of titanium, niobium, tantalum, zirconium, oxygen, calcium, and phosphorus. As shown in Figure 4, the samples treated with the electrical discharge machining (EDM) technique showed superior and higher corrosion resistance than the untreated samples in Ringer's simulated body fluid. Moreover, the HA-deposited bio-ceramic layer indicated excellent corrosion resistance.

Chen et al. [94] successfully developed the PEO coating on the Ti-39Nb-6Zr alloy and studied the electrochemical corrosion and wear behavior of the Ti-39Nb-6Zr alloy before and after modification in PBS solution. The results showed that, after PEO surface treatment, the corrosion and wear resistance of the Ti-39Nb-6Zr alloy in the PBS solution was significantly improved. Wang et al. [95] conducted a systematic study on the MAO treatment of Ti-35Nb-2Ta-3Zr alloy in sodium silicate electrolyte. The results showed that the Ti-35Nb-2Ta-3Zr alloy had good film-forming properties, and a layer of porous-structure-nested film is formed on the surface of Ti-35Nb-2Ta-3Zr alloy. The relative content of the biologically active anatase phase was much higher than that of the rutile phase, which can effectively improve the deposition ability of HA. In addition, since the surface of Ti-35Nb-2Ta-3Zr treated by micro-arc oxidation has a Nb_2O_5 phase, an oxide film with the denser and thicker condition is formed, leading to excellent corrosion resistance. Gu et al. [96] also developed Ti-35Nb-2Ta-3Zr alloy by means of surface modification. They used fabricated TiO_2/Ti-35Nb-2Ta-3Zr anticorrosion micro/nano-composites with different amounts of TiO_2 particles via FSP. The refined surface microstructure could increase the compactness of the surface oxide films that eventually affect the corrosion performance. Material with the most TiO_2 content has Icorr magnitude five times less than the substrate; hence, the corrosion resistance can be further improved by the TiO_2 micro/nano-composite layer. Figure 5 displays the potentiodynamic polarization curves and EIS patterns of the substrate and TiO_2 micro/nano-composite layers of Ti-35Nb-2Ta-3Zr substrate and FSPed in Hank's solution. Icorr reduced with the increasing amount of TiO_2 incorporation at the same rotation speed. The capacitances of the outer and inner layer are less than the substrate, which was attributed to the surface topography and thickness of the passive film.

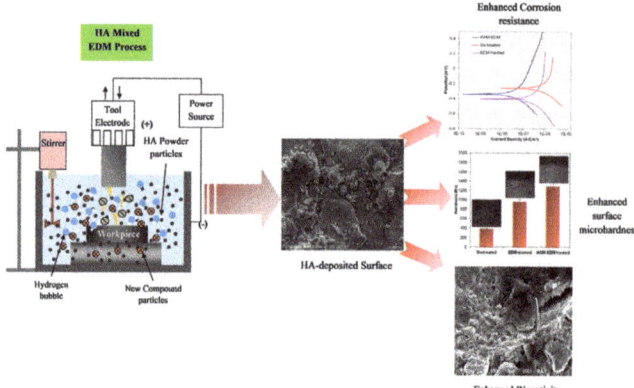

Figure 4. The enhancement of corrosion resistance, hardness, and bioactivity on HA-deposited surface (surface modification of β-phase Ti implant by hydroaxyapatite mixed electric discharge machining to enhance the corrosion resistance and in-vitro bioactivity). Reproduced with permission from [93], copyright Elsevier, 2017.

Liu et al. [97] proposed a unique method that combined alkali treatment with natural cross-linking agent proanthocyanidin to form a sub-micron porous structure (pore size 100–300 nm). The dense inner oxide layer mainly provided enhanced corrosion resistance for the Ti2448 alloy stent. Pina et al. [98] found that Ti30Nb4Sn alloys containing Sn have excellent chemical resistance. Bahl et al. [99] studied the tribo-electrochemical behavior of different β-Ti alloys sintered by powder metallurgy. According to the results, the active and passive dissolution rates increased by attaining the Sn content of 2% and 4%, leading to enhanced mechanically activated corrosion under the friction corrosion condition.

Diomidis et al. [100] tested the open circuit potential and anodic current of Ti-13Nb-13Zr and Ti-29Nb-13Ta-4.6Zr alloys. Ti-13Nb-13Zr and Ti-29Nb-13Ta-4.6Zr showed the ability to restore their passive state during the fretting process. The excellent potential of these β-Ti alloys to restore their passive state during the fretting process was related to the interaction of the mechanical properties of the passive surface layer and its contact pressure. In addition, they also studied the effect of the bovine serum albumin addition, hyaluronic acid, and triglyceride dipolyphosphate on the fretting corrosion of the Ti-12.5Mo alloy. The results showed that the addition of the sliding film leads to a decrease in the friction coefficient, the elastic adjustment of displacement, and the wear rate of the alloy.

However, the human body environment is very complicated. In addition to electrochemical corrosion, biomedical β-Ti alloy implants also need to satisfy the friction corrosion conditions. In addition, the material quality, the effect of mechanical force, the chemical composition of the medium, and the alloy itself will affect the β-Ti alloy. Under such circumstances, the biomedical β-Ti alloy needs further research on corrosion resistance [85].

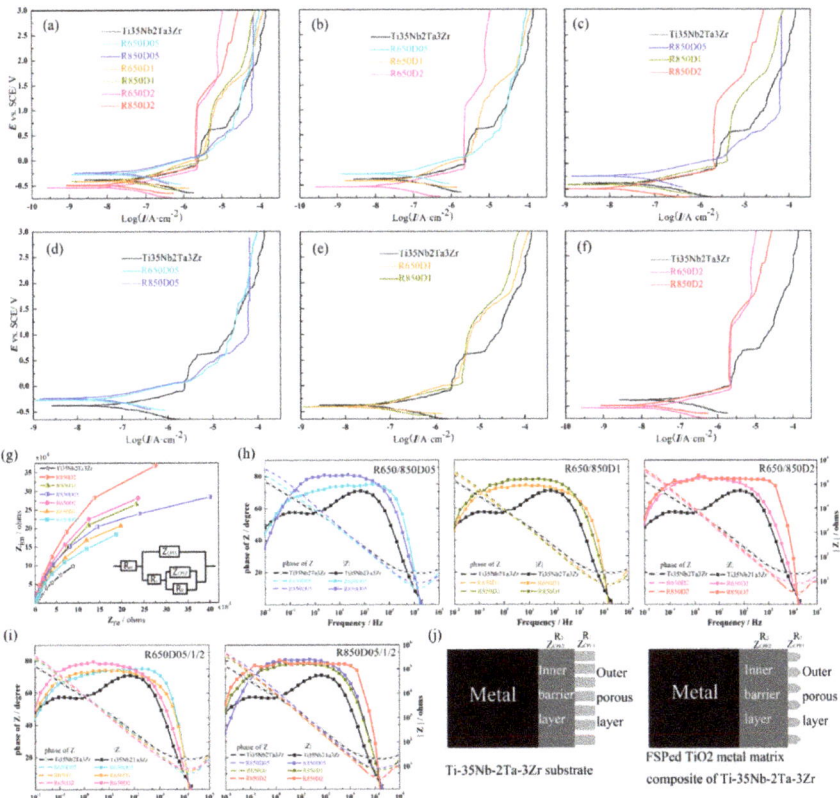

Figure 5. The corrosion behavior of the substrate and TiO$_2$ micro/nano-composite layers in Hank's solution: (**a**–**h**) potentiodynamic polarization curves, (**g**–**j**) EIS pattern: (**g**) Nyquist diagram with equivalent circuit, (**h**) Bode diagrams of TiO$_2$ micro/nano-composite layers at the different rotation speeds, (**i**) Bode diagrams of TiO$_2$ micro/nano-composite layers at the different amount of TiO$_2$ added, (**j**) schematic diagrams of the substrate and TiO$_2$ micro/nano-composite layers. Reproduced with permission from [96], copyright Elsevier, 2019.

6. Biological Response

6.1. Antibacterial Property

Antibacterial property refers to the ability of a material to remain effective under the action of bacteria or microorganisms. High antibacterial performance is one of the vital requirements in human implant materials [101]. Bacterial infection has been indicated as one of the main factors in the failure of metallic implants [102]. β-Ti alloys have been widely used in medical and surgical implants, but they are currently facing the challenge of implant-related infections [83]. Hence, it is crucial to improve the antibacterial property of the β-Ti alloys.

Chang et al. [103] utilized the fully automated fiber laser system (Micro Laser Systems) for laser nitriding of Ti-30Nb-4Sn in the open air. A set of samples was prepared by varying the duty cycle from 5% to 100%. In Figure 6, the studied laser-nitrided samples are displayed, including DC5-DC100. The results showed a significant difference in the total bacterial adherence/biofilm formation on the surface coverage of the untreated samples and the laser-nitrided samples. The untreated sample had the highest biofilm coverage of 29.8%, and the total biofilm coverage was significantly decreased and reached 2.9% (DC60).

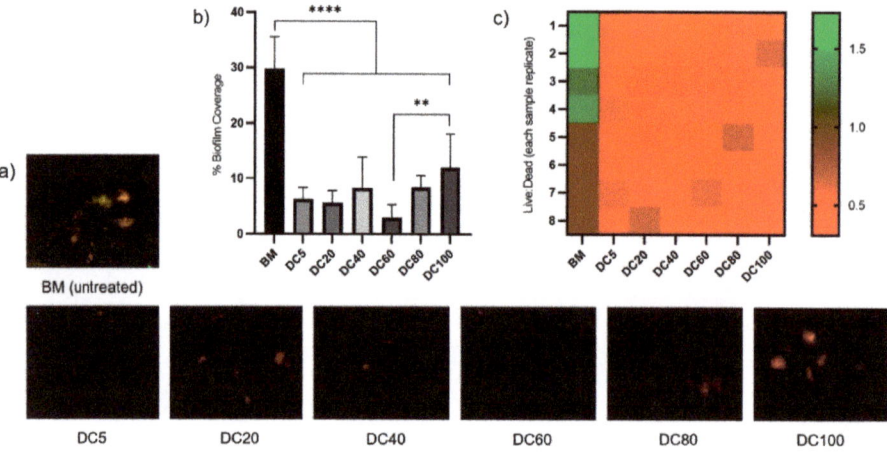

Figure 6. Bacterial coverage results for the untreated and laser-nitrided surfaces after 24h culture with S. aureus. (**a**) Representative images showing bacterial adherence/biofilm coverage on each sample. Images were obtained by live/dead staining and fluorescence microscopy. Live (viable) bacteria are stained green, and dead (non-viable bacteria) stained red. (**b**) Percentage biofilm coverage on each treated sample and untreated sample as determined by image analysis (* denotes significant difference, **: $p < 0.01$. ****: $p < 0.0001$). (**c**) Ratio of live: dead bacteria on each image analyzed for each sample and untreated sample control and data displayed as a heat-map. The scale shows samples with a greater proportion of live bacteria represented in green and greater proportion of dead bacteria represented in red. Reproduced with permission from [103], copyright Elsevier, 2020.

A laser nitriding treatment was performed on the Ti-35Nb-7Zr-6Ta surface [104]. Biological studies of the laser-nitrided surfaces included in vitro culture for 24 h using mesenchymal stem cell (MSC) fluorescence staining and *Staphylococcus aureus* (*S. aureus*) live/dead staining. The bacteria coverage percentage was decreased from more than 5% to less than 1%, so it can be claimed that the laser-nitrided surfaces (laser power: 45 W) led to a significant antibacterial effect.

Shi et al. [105] fabricated a low elastic modulus Ti-13Nb-13Zr-5Cu alloy with a good antibacterial property (against *S. aureus* > 90%) as the precipitation of the Ti_2Cu phase. Liu et al. [106] prepared SLA–TiCu by sandblasting and large-grits etching (SLA) to avoid implant-related infection and promote early bone integration. Studies had shown that the antibacterial rate of SLA–TiCu surface increased from 36.6% to 99.9% against *S. aureus*.

Most β-Ti alloys do not present an antibacterial property. After being implanted in the human body, the implants are prone to infection, which may cause implant failure, especially in the early stage after implantation, causing pain and an economic burden to the patients. Therefore, it is of great significance to endow the antibacterial property of β-Ti alloys (incorporation of antibacterial particles) via surface modification.

6.2. Bone Regeneration

In recent years, the surface morphology modification effect on the biological activity of titanium implants has been extensively studied. Topography investigations from the bionics perspective revealed that the more similar micro-topography of the implant surface to the human bone structure led to more effectiveness in the biological activity improvement of the implant. More and more shreds of evidence show that the surface microenvironment structure can affect cell growth, which, in turn, affects osseointegration [107]. Micromorphology has been identified as a potential stimulating factor for stem cell osteogenic differentiation and new bone formation. It was reported that Ti, Zr, Nb, Ta, Au, Mo, and Sn are highly biocompatible elements [108]. Hence, subsequent surface processing of β-Ti alloys that includes high biocompatible elements can effectively improve bone regeneration performance. Researchers have proposed various surface modification strategies to produce micron or sub-micron surface topography on titanium implants, such as sandblasting, acid etching, anodizing, spin-coating, sputtering deposition, etc.

Scholars applied some exceptional processes in combination with the characteristics of the alloy itself to directly construct the micro-topography on the surface without introducing other active materials to improve the bone regeneration performance [109–111]. Micro-scale surface topographies can enhance bio-mechanical interlocking and increase bone-anchoring. Kheradmandfard et al. [68] fabricated a gradient nanostructure layer on the TNTZ surface by UNSM technology. The microstructure of the top surface layer was composed of nanoflakes with a width of 60–200 nm. The results showed that the nanostructured titanium surface induced enhanced cell adhesion, osteoblast differentiation, and improved osseointegration toward mesenchymal stem cells. The micro-pattern will assist the TNTZ implants in attaining high biological activity and bone regeneration performance.

Li et al. [112] prepared hierarchical nanostructures on the surface of Ti-24Nb-4Zr-7.9Sn alloy through anodic oxidation, forming a nanoscale bone-like structure and nanotubes. They observed the behavior of bone marrow stromal cells on the surface of the sample through in vitro culture. They also conducted histological analysis after implanting the modified materials in vivo to evaluate the biocompatibility and osseointegration of the implant surface. The results showed a generation of a hierarchical structure with nano-scale bone-like layers on the surface of Ti-24Nb-4Zr-7.9Sn. The BIC value was 65.48 ± 8.0% compared with 33.21 ± 6.05% of untreated specimen, and the bone area reached 55.28 ± 14.92% compared with 20.64 ± 10.28% of untreated. The Ti-24Nb-4Zr-7.9Sn surface showed high biocompatibility with bone marrow mesenchymal stem cells in vitro and osseointegration in vivo.

Fu et al. [113] found that the behavior of murine mesenchymal stem cells (MSCs) and murine preosteoblastic cells (MC3T3-E1) is regulated by the cell-material interface. They used selective laser melting and alkaline heat treatment techniques on titanium implants and constructed ordered-micro and disordered-nano patterned structures. Compared with untreated specimens, the expression level of osteogenic genes in the treated implants increased by 3.43 times. The optimal bone formation structure had a steady wave structure (horizontal direction: ridge, 2.7 μm; grooves, 5.3 μm; and vertical direction: distance, 700 μm) with the appropriate density of nano-branches (6.0 per μm^2). Ordered grooves provided direction guidance for cells, and disordered branches influenced the shape of cells by maintaining nanostructures with different spacing and densities. Micro-nano patterned structures can provide biophysical clues to guide the development of cell phenotypes, including cell size, shape, and direction, thereby affecting cell survival, growth, and differentiation processes. Therefore, the superimposed isotropic and anisotropic cues, ordered-micro and disordered-nano patterned structures, can promote integrin α5, integrin β1, cadherin 2, Runx2, and Opn by activating Wnt/β-catenin signaling and Ocn, thereby further transferring and changing the cell shape and inducing nuclear orientation (as shown in Figure 7). The osteogenic differentiation induced by the ordered-micro and disordered-nano pattern structure is related to Wnt/β-catenin signaling and was further proved by the common Wnt signaling inhibitor Dickkopf1. Ordered micro-topography and disordered nano-topography pattern structures can lead to osteogenic differentiation in vitro and bone regeneration in vivo.

The osseointegration ability of β-Ti alloys increases in comparison to the material before modification, which is essential for implants. The surface microenvironment structure plays an important role in cell growth, which, in turn, affects osseointegration. Many modified methods, including sandblasting, anodizing, and so on, can provide microstructure and even nanostructure. Additionally, the osseointegration behaviors of β-Ti alloys increase in comparison to the material before modification, which is essential for implants. There is a trend of constructing the micro-topography directly on the alloy surface without introducing other active materials. Furthermore, the principle of the complex microstructure improving bone regeneration requires further study.

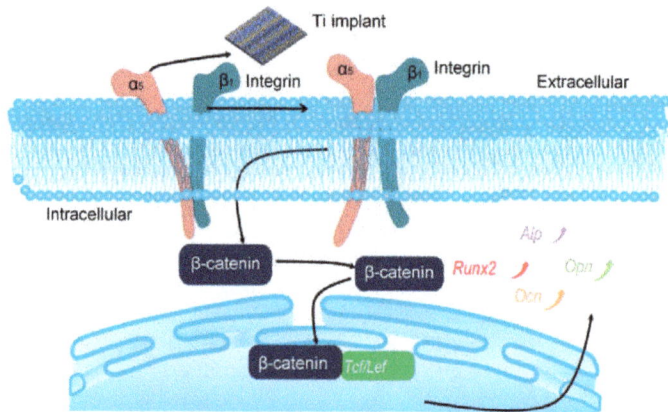

Figure 7. Schematic illustration of patterned Ti for osteogenic differentiation. Function 1: Activation of integrin α5β1 heterodimers and regulation of cell morphology. Function 2: Stimulation of MSCs and MC3T3-E1-related gene and β-catenin expressions to promote osteogenic differentiation. Reproduced with permission from [113], copyright Elsevier, 2020.

7. Shortcomings and Prospects

β-Ti alloys show notable potential in orthopedic and dental applications, mainly as low modulus and low or non-toxic elements. However, low wear resistance is a challenge for β-Ti alloys, which causes abrasive particles running in the body fluid environment. Additionally, corrosion behavior may lead to infection and implant failure. Besides, despite excellent biocompatibility, antibacterial ability and bone regeneration are also necessary with β-Ti alloys for bone recovery. Researchers have studied and developed various novel β-Ti alloys, but there is a lack of systematic optimization. The traditional methods of selecting chemical compositions have high time costs. This results in most of the β-Ti alloys still being in the fundamental research stage rather than clinical surgery. Thus, a general standard is necessary to be formulated. Moreover, the relationship of hardness, friction behavior, and corrosion resistance is not clarified yet. It seems that the new strategy may be to explain the relationship between these characteristics by computer simulation. The single and comprehensive model of the simulated environment needs to be designed to reduce unnecessary animal experiments in vivo. In short, β-Ti alloys present considerable prospects in the future for orthopedic and dental applications.

8. Conclusions

The performance of the β-Ti alloy consists of more adaptation to the required demands of biomedical materials compared to other alloys, but it is still far from the requirements of biomedical materials. It is necessary to develop novel targeted β-Ti alloys rapidly. The computer simulation to search for and optimize the composition of Ti alloys may be a suitable approach to solve the issue. The utilization of surface modification techniques can improve the mechanical properties, wear resistance, corrosion resistance, and biological responses of β-Ti alloys. However, the specific relationship between these aspects is complicated and needs to be clarified by considering the deeper mechanisms rather than just adjusting the parameters during the processes. The improvements of β-Ti alloys deserve continuous attention. The successful development of β-Ti alloys is beneficial for clinical surgery in the orthopedic and dental fields.

Author Contributions: Conceptualization, J.L. and Y.T.; investigation, Q.W.; resources, J.L. and Y.T.; data curation, Q.W.; writing—original draft preparation, L.S., Y.D. and K.D.; writing—review and editing, Q.W.; visualization, L.W.; supervision, H.W. and L.W.; project administration, L.W. All authors have read and agreed to the published version of the manuscript.

Funding: The authors would like to acknowledge the financial supports of National Natural Science Foundation under (Grant No.51674167,51831011 and 5191101889), Shanghai Science and Technology Project: 20S31900100 Medical Engineering Cross Research Foundation of Shanghai Jiao Tong University (Grant No. YG2021QN61). The authors would like to acknowledge the institutes of Guangxi Key Laboratory of basic and translational research of Bone and Joint Degenerative Diseases, and Guangxi Biomedical Materials and Engineering Research Center for Bone and Joint Degenerative Diseases.

Institutional Review Board Statement: Not applicable.

Informed Consent Statement: Not applicable.

Data Availability Statement: Not applicable.

Acknowledgments: The authors sincerely thank Shokouh Attarilar for language polishing of this paper.

Conflicts of Interest: The authors declare no conflict of interest.

References

1. Zhang, L.C.; Chen, L.Y. A Review on Biomedical Titanium Alloys: Recent Progress and Prospect. *Adv. Eng. Mater.* **2019**, *21*, 1801215. [CrossRef]
2. Raji, S.A.; Popoola AP, I.; Pityana, S.L.; Popoola, O.M. Characteristic effects of alloying elements on β solidifying titanium aluminides: A review. *Heliyon* **2020**, *6*, e04463. [CrossRef]
3. Rossi, M.C.; Amado, J.M.; Tobar, M.J.; Vicente, A.; Yañez, A.; Amigó, V. Effect of alloying elements on laser surface modification of powder metallurgy to improve surface mechanical properties of beta titanium alloys for biomedical application. *J. Mater. Res. Technol.* **2021**, *14*, 1222–1234. [CrossRef]
4. Pitchi, C.S.; Priyadarshini, A.; Sana, G.; Narala, S.K.R. A review on alloy composition and synthesis of β-Titanium alloys for biomedical applications. *Mater. Today Proc.* **2020**, *26*, 3297–3304. [CrossRef]
5. Goldberg, J. An Evaluation of Beta Titanium Alloys for Use in Orthodontic Appliances. *J. Dent. Res.* **1979**, *58*, 593–599. [CrossRef]
6. Liu, S.; Liu, W.; Liu, J.; Liu, J.; Zhang, L.; Tang, Y.; Zhang, L.-C.; Wang, L. Compressive properties and microstructure evolution in NiTiNb alloy with mesh eutectic phase. *Mater. Sci. Eng. A* **2021**, *801*, 140434. [CrossRef]
7. Geetha, M.; Singh, A.K.; Asokamani, R.; Gogia, A.K. Ti based biomaterials, the ultimate choice for orthopaedic implants—A review. *Prog. Mater. Sci.* **2009**, *54*, 397–425. [CrossRef]
8. Bolzoni, L.; Ruiz-Navas, E.M.; Gordo, E. Investigation of the factors influencing the tensile behaviour of PM Ti–3Al–2.5V alloy. *Mater. Sci. Eng. A* **2014**, *609*, 266–272. [CrossRef]
9. Semlitsch, M.F.; Weber, H.; Streicher, R.M.; Schön, R. Joint replacement components made of hot-forged and surface-treated Ti-6Al-7Nb alloy. *Biomaterials* **1992**, *13*, 781–788. [CrossRef]
10. He, G.; Liu, H.; Tan, Q.; Ni, J. Diffusion bonding of Ti–2.5Al–2.5Mo–2.5Zr and Co–Cr–Mo alloys. *J. Alloys Compd.* **2011**, *509*, 7324–7329. [CrossRef]
11. Kwon, Y.M.; An, S.; Yeo, I.; Tirumala, V.; Chen, W.; Klemt, C. Radiographic Risk Factors Associated With Adverse Local Tissue Reaction in Head-Neck Taper Corrosion of Primary Metal-on-Polyethylene Total Hip Arthroplasty. *J. Am. Acad. Orthop. Surg.* **2021**, *29*, 353–360. [CrossRef]

12. Sak, A.; Moskalewicz, T.; Zimowski, S.; Cieniek, L.; Dubiel, B.; Radziszewska, A.; Kot, M.; Lukaszczyk, A. Influence of polyetheretherketone coatings on the Ti-13Nb-13Zr titanium alloy's bio-tribological properties and corrosion resistance. *Mater. Sci. Eng. C Mater. Biol. Appl.* **2016**, *63*, 52–61. [CrossRef]
13. Liu, Y.J.; Li, S.J.; Wang, H.L.; Hou, W.T.; Hao, Y.L.; Yang, R.; Sercombe, T.B.; Zhang, L.C. Microstructure, defects and mechanical behavior of beta-type titanium porous structures manufactured by electron beam melting and selective laser melting. *Acta Mater.* **2016**, *113*, 56–67. [CrossRef]
14. Liu, Y.; Li, S.; Hou, W.; Wang, S.; Hao, Y.; Yang, R.; Sercombe, T.B.; Zhang, L.-C. Electron Beam Melted Beta-type Ti–24Nb–4Zr–8Sn Porous Structures With High Strength-to-Modulus Ratio. *J. Mater. Sci. Technol.* **2016**, *32*, 505–508. [CrossRef]
15. Ho, W.F.; Ju, C.P.; Chern Lin, J.H. Structure and properties of cast binary Ti–Mo alloys. *Biomaterials* **1999**, *20*, 2115–2122. [CrossRef]
16. Zha, S. Study on the Microstructure and Property of New β-Ti28Nb24.5Zr Alloy for Biomedic Applications. Master's Thesis, Tianjin University, Tianjin, China, 2006.
17. Zhu, C.; Lv, Y.; Qian, C.; Ding, Z.; Jiao, T.; Gu, X.; Lu, E.; Wang, L.; Zhang, F. Microstructures, mechanical, and biological properties of a novel Ti-6V-4V/zinc surface nanocomposite prepared by friction stir processing. *Int. J. Nanomed.* **2018**, *13*, 1881. [CrossRef]
18. Wang, L.; Xie, L.; Lv, Y.; Zhang, L.-C.; Chen, L.; Meng, Q.; Qu, J.; Zhang, D.; Lu, W. Microstructure evolution and superelastic behavior in Ti-35Nb-2Ta-3Zr alloy processed by friction stir processing. *Acta Mater.* **2017**, *131*, 499–510. [CrossRef]
19. Hafeez, N.; Liu, J.; Wang, L.; Wei, D.; Tang, Y.; Lu, W.; Zhang, L.-C. Superelastic response of low-modulus porous beta-type Ti-35Nb-2Ta-3Zr alloy fabricated by laser powder bed fusion. *Addit. Manuf.* **2020**, *34*, 101264. [CrossRef]
20. Raza, D.; Kumar, G.; Uzair, M.; Singh, M.K.; Sultan, D.; Kumar, R. Development and heat treatment of β-phase titanium alloy for orthopedic application. *Mater. Today Proc.* **2021**. [CrossRef]
21. Karre, R.; Dey, S.R. Progress in Development of Beta Titanium Alloys for Biomedical Applications. In *Encyclopedia of Smart Materials*; Elsevier: Amsterdam, The Netherlands, 2019; Volume 5, pp. 512–527. [CrossRef]
22. Liu, S.; Liu, J.; Wang, L.; Ma, R.L.W.; Zhong, Y.; Lu, W.; Zhang, L.C. Superelastic behavior of in-situ eutectic-reaction manufactured high strength 3D porous NiTi-Nb scaffold. *Scr. Mater.* **2020**, *181*, 121–126.
23. Xie, K.Y.; Wang, Y.; Zhao, Y.; Chang, L.; Wang, G.; Chen, Z.; Cao, Y.; Liao, X.; Lavernia, E.J.; Valiev, R.Z.; et al. Nanocrystalline beta-Ti alloy with high hardness, low Young's modulus and excellent in vitro biocompatibility for biomedical applications. *Mater Sci. Eng. C Mater. Biol. Appl.* **2013**, *33*, 3530–3536. [CrossRef]
24. Singleton, R.C.; Pharr, G.M.; Nyman, J.S. Increased tissue-level storage modulus and hardness with age in male cortical bone and its association with decreased fracture toughness. *Bone* **2021**, *148*, 115949. [CrossRef]
25. Callioglu, S.; Acar, P. Design of beta-Titanium microstructures for implant materials. *Mater. Sci. Eng. C Mater. Biol. Appl.* **2020**, *110*, 110715. [CrossRef] [PubMed]
26. Dearnley, P.A.; Dahm, K.L.; Çimenoğlu, H. The corrosion–wear behaviour of thermally oxidised CP-Ti and Ti–6Al–4V. *Wear* **2004**, *256*, 469–479. [CrossRef]
27. Niinomi, M.; Kuroda, D.; Fukunaga, K.-I.; Morinaga, M.; Kato, Y.; Yashiro, T.; Suzuki, A. Corrosion wear fracture of new β type biomedical titanium alloys. *Mater. Sci. Eng. A* **1999**, *263*, 193–199. [CrossRef]
28. Kaur, S.; Ghadirinejad, K.; Oskouei, R.H. An Overview on the Tribological Performance of Titanium Alloys with Surface Modifications for Biomedical Applications. *Lubricants* **2019**, *7*, 65. [CrossRef]
29. Bahl, S.; Suwas, S.; Chatterjee, K. Comprehensive review on alloy design, processing, and performance of β Titanium alloys as biomedical materials. *Int. Mater. Rev.* **2020**, *66*, 114–139. [CrossRef]
30. Torrento, J.E.; Grandini, C.R.; Sousa, T.S.P.; Rocha, L.A.; Gonçalves, T.M.; Sottovia, L.; Rangel, E.C.; Cruz, N.C.; Correa, D.R.N. Bulk and surface design of MAO-treated Ti-15Zr-15Mo-Ag alloys for potential use as biofunctional implants. *Mater. Lett.* **2020**, *269*, 127661. [CrossRef]
31. Liu, S.; Wang, Q.; Liu, W.; Tang, Y.; Liu, J.; Zhang, H.; Liu, X.; Liu, J.; Yang, J.; Zhang, L.C.; et al. Multi-scale hybrid modified coatings on titanium implants for non-cytotoxicity and antibacterial properties. *Nanoscale* **2021**, *13*, 10587–10599. [CrossRef] [PubMed]
32. Kandavalli, S.R.; Wang, Q.; Ebrahimi, M.; Gode, C.; Djavanroodi, F.; Attarilar, S.; Liu, S. A Brief Review on the Evolution of Metallic Dental Implants: History, Design, and Application. *Front. Mater.* **2021**, *8*, 140. [CrossRef]
33. Attarilar, S.; Yang, J.; Ebrahimi, M.; Wang, Q.; Liu, J.; Tang, Y.; Yang, J. The Toxicity Phenomenon and the Related Occurrence in Metal and Metal Oxide Nanoparticles: A Brief Review From the Biomedical Perspective. *Front. Bioeng. Biotechnol.* **2020**, *8*, 822. [CrossRef]
34. Çaha, I.; Alves, A.C.; Rocha, L.A.; Toptan, F. A Review on Bio-functionalization of β-Ti Alloys. *J. Bio-Tribo-Corros.* **2020**, *6*, 1–31. [CrossRef]
35. Wang, Q.; Zhou, P.; Liu, S.; Attarilar, S.; Ma, R.L.; Zhong, Y.; Wang, L. Multi-Scale Surface Treatments of Titanium Implants for Rapid Osseointegration: A Review. *Nanomater* **2020**, *10*, 1244. [CrossRef]
36. Wang, Q.; Wu, L.; Liu, S.; Cao, P.; Yang, J.; Wang, L. Nanostructured Titanium Alloys Surface Modification Technology for Antibacterial and Osteogenic Properties. *Curr. Nanosci.* **2021**, *17*, 175–193. [CrossRef]
37. Zhang, L.-C.; Chen, L.-Y.; Wang, L. Surface Modification of Titanium and Titanium Alloys: Technologies, Developments, and Future Interests. *Adv. Eng. Mater.* **2020**, *22*, 1901258. [CrossRef]
38. Kurup, A.; Dhatrak, P.; Khasnis, N. Surface modification techniques of titanium and titanium alloys for biomedical dental applications: A review. *Mater. Today Proc.* **2021**, *39*, 84–90. [CrossRef]

39. Liang, S. Review of the Design of Titanium Alloys with Low Elastic Modulus as Implant Materials. *Adv. Eng. Mater.* **2020**, *22*, 2000555. [CrossRef]
40. Okulov, I.V.; Okulov, A.V.; Soldatov, I.V.; Luthringer, B.; Willumeit-Romer, R.; Wada, T.; Kato, H.; Weissmuller, J.; Markmann, J. Open porous dealloying-based biomaterials as a novel biomaterial platform. *Mater. Sci. Eng. C Mater. Biol. Appl.* **2018**, *88*, 95–103. [CrossRef] [PubMed]
41. Luthringer, B.J.; Ali, F.; Akaichi, H.; Feyerabend, F.; Ebel, T.; Willumeit, R. Production, characterisation, and cytocompatibility of porous titanium-based particulate scaffolds. *J. Mater. Sci. Mater. Med.* **2013**, *24*, 2337–2358. [CrossRef] [PubMed]
42. Prashanth, K.; Zhuravleva, K.; Okulov, I.; Calin, M.; Eckert, J.; Gebert, A. Mechanical and Corrosion Behavior of New Generation Ti-45Nb Porous Alloys Implant Devices. *Technologies* **2016**, *4*, 33. [CrossRef]
43. Okulov, I.V.; Weissmuller, J.; Markmann, J. Dealloying-based interpenetrating-phase nanocomposites matching the elastic behavior of human bone. *Sci. Rep.* **2017**, *7*, 20. [CrossRef]
44. Okulov, A.V.; Volegov, A.S.; Weissmüller, J.; Markmann, J.; Okulov, I.V. Dealloying-based metal-polymer composites for biomedical applications. *Scr. Mater.* **2018**, *146*, 290–294. [CrossRef]
45. Liu, J.; Tang, Y.; Liu, J.; Wang, L. Research progress in titanium alloy in the field of orthopaedic implants. *J. Mater. Eng.* **2021**, *49*, 11–15.
46. Okulov, I.V.; Okulov, A.V.; Volegov, A.S.; Markmann, J. Tuning microstructure and mechanical properties of open porous TiNb and TiFe alloys by optimization of dealloying parameters. *Scr. Mater.* **2018**, *154*, 68–72. [CrossRef]
47. Berger, S.A.; Okulov, I.V. Open Porous α + β Titanium Alloy by Liquid Metal Dealloying for Biomedical Applications. *Metals* **2020**, *10*, 1450. [CrossRef]
48. Li, P.; Zhang, H.; Tong, T.; He, Z. The rapidly solidified β-type Ti–Fe–Sn alloys with high specific strength and low elastic modulus. *J. Alloys Compd.* **2019**, *786*, 986–994. [CrossRef]
49. Xu, Y.; Gao, J.; Huang, Y.; Rainforth, W.M. A low-cost metastable beta Ti alloy with high elastic admissible strain and enhanced ductility for orthopaedic application. *J. Alloys Compd.* **2020**, *835*, 155391. [CrossRef]
50. Li, P.; Ma, X.; Tong, T.; Wang, Y. Microstructural and mechanical properties of β-type Ti–Mo–Nb biomedical alloys with low elastic modulus. *J. Alloys Compd.* **2020**, *815*, 152412. [CrossRef]
51. Guo, X.; Zhang, C.; Tian, Q.; Yu, D. Liquid metals dealloying as a general approach for the selective extraction of metals and the fabrication of nanoporous metals: A review. *Mater. Today Commun.* **2021**, *26*, 102007. [CrossRef]
52. Okulov, I.V.; Joo, S.H.; Okulov, A.V.; Volegov, A.S.; Luthringer, B.; Willumeit-Romer, R.; Zhang, L.; Madler, L.; Eckert, J.; Kato, H. Surface Functionalization of Biomedical Ti-6Al-7Nb Alloy by Liquid Metal Dealloying. *Nanomater* **2020**, *10*, 1479. [CrossRef]
53. Okulov, I.V.; Wilmers, J.; Joo, S.-H.; Bargmann, S.; Kim, H.S.; Kato, H. Anomalous compliance of interpenetrating-phase composite of Ti and Mg synthesized by liquid metal dealloying. *Scr. Mater.* **2021**, *194*, 113660. [CrossRef]
54. Zhu, Y.; Wang, X.; Wang, L.; Fu, Y.; Qin, J.; Lu, W.; Zhang, D. Influence of forging deformation and heat treatment on microstructure of Ti–xNb–3Zr–2Ta alloys. *Mater. Sci. Eng. C* **2012**, *32*, 126–132. [CrossRef]
55. Zhang, T.; Liu, C.-T. Design of titanium alloys by additive manufacturing: A critical review. *Adv. Powder Mater.* **2021**. [CrossRef]
56. Wang, L.; Qu, J.; Chen, L.; Meng, Q.; Zhang, L.-C.; Qin, J.; Zhang, D.; Lu, W. Investigation of Deformation Mechanisms in β-Type Ti-35Nb-2Ta-3Zr Alloy via FSP Leading to Surface Strengthening. *Metall. Mater. Trans. A* **2015**, *46*, 4813–4818. [CrossRef]
57. Niinomi, M. Fatigue performance and cyto-toxicity of low rigidity titanium alloy, Ti–29Nb–13Ta–4.6Zr. *Biomaterials* **2003**, *24*, 2673–2683. [CrossRef]
58. Zhang, T.; Fan, Q.; Ma, X.; Wang, W.; Wang, K.; Shen, P.; Yang, J.; Wang, L. Effect of laser remelting on microstructural evolution and mechanical properties of Ti-35Nb-2Ta-3Zr alloy. *Mater. Lett.* **2019**, *253*, 310–313. [CrossRef]
59. Lin, Z.; Wang, L.; Xue, X.; Lu, W.; Qin, J.; Zhang, D. Microstructure evolution and mechanical properties of a Ti-35Nb-3Zr-2Ta biomedical alloy processed by equal channel angular pressing (ECAP). *Mater. Sci. Eng. C Mater. Biol. Appl.* **2013**, *33*, 4551–4561. [CrossRef]
60. Liu, R.; Yuan, S.; Lin, N.; Zeng, Q.; Wang, Z.; Wu, Y. Application of ultrasonic nanocrystal surface modification (UNSM) technique for surface strengthening of titanium and titanium alloys: A mini review. *J. Mater. Res. Technol.* **2021**, *11*, 351–377. [CrossRef]
61. Wang, L.; Wang, Y.; Huang, W.; Liu, J.; Tang, Y.; Zhang, L.; Fu, Y.; Zhang, L.-C.; Lu, W. Tensile and superelastic behaviors of Ti-35Nb-2Ta-3Zr with gradient structure. *Mater. Des.* **2020**, *194*, 108961. [CrossRef]
62. Niinomi, M.; Nakai, M. Titanium-Based Biomaterials for Preventing Stress Shielding between Implant Devices and Bone. *Int. J. Biomater.* **2011**, *2011*, 836587. [CrossRef]
63. Kuroda, D.; Niinomi, M.; Morinaga, M.; Kato, Y.; Yashiro, T. Design and mechanical properties of new β type titanium alloys for implant materials. *Mater. Sci. Eng. A* **1998**, *243*, 244–249. [CrossRef]
64. Niinomi, M.; Nakai, M.; Hieda, J. Development of new metallic alloys for biomedical applications. *Acta Biomater.* **2012**, *8*, 3888–3903. [CrossRef] [PubMed]
65. Lee, Y.-S.; Niinomi, M.; Nakai, M.; Narita, K.; Cho, K.; Liu, H. Wear transition of solid-solution-strengthened Ti–29Nb–13Ta–4.6Zr alloys by interstitial oxygen for biomedical applications. *J. Mech. Behav. Biomed. Mater.* **2015**, *51*, 398–408. [CrossRef]
66. Makuch, N.; Kulka, M.; Dziarski, P.; Przestacki, D. Laser surface alloying of commercially pure titanium with boron and carbon. *Opt. Lasers Eng.* **2014**, *57*, 64–81. [CrossRef]
67. Gao, Q.; Yan, H.; Qin, Y.; Zhang, P.; Guo, J.; Chen, Z.; Yu, Z. Laser cladding Ti-Ni/TiN/TiW+TiS/WS2 self-lubricating wear resistant composite coating on Ti-6Al-4V alloy. *Opt. Laser Technol.* **2019**, *113*, 182–191. [CrossRef]

68. Kheradmandfard, M.; Kashani-Bozorg, S.F.; Kim, C.L.; Hanzaki, A.Z.; Pyoun, Y.S.; Kim, J.H.; Amanov, A.; Kim, D.E. Nanostructured beta-type titanium alloy fabricated by ultrasonic nanocrystal surface modification. *Ultrason. Sonochem.* **2017**, *39*, 698–706. [CrossRef] [PubMed]
69. Chauhan, A.S.; Jha, J.S.; Telrandhe, S.; Srinivas, V.; Gokhale, A.A.; Mishra, S.K. Laser surface treatment of α-β titanium alloy to develop a β -rich phase with very high hardness. *J. Mater. Process. Technol.* **2021**, *288*, 116873. [CrossRef]
70. Molinari, A.; Straffelini, G.; Tesi, B.; Bacci, T. Dry sliding wear mechanisms of the Ti6Al4V alloy. *Wear* **1997**, *208*, 105–112. [CrossRef]
71. Cui, W.-F.; Niu, F.-J.; Tan, Y.-L.; Qin, G.-W. Microstructure and tribocorrosion performance of nanocrystalline TiN graded coating on biomedical titanium alloy. *Trans. Nonferrous Met. Soc. China* **2019**, *29*, 1026–1035. [CrossRef]
72. Graves, A.; Norgren, S.; Wan, W.; Singh, S.; Kritikos, M.; Xiao, C.; Crawforth, P.; Jackson, M. On the mechanism of crater wear in a high strength metastable β titanium alloy. *Wear* **2021**, *484*, 203998. [CrossRef]
73. Majumdar, P.; Singh, S.B.; Chakraborty, M. Wear properties of Ti–13Zr–13Nb (wt.%) near β titanium alloy containing 0.5wt.% boron in dry condition, Hank's solution and bovine serum. *Mater. Sci. Eng. C* **2010**, *30*, 1065–1075. [CrossRef]
74. Mehdi, M.; Farokhzadeh, K.; Edrisy, A. Dry sliding wear behavior of superelastic Ti–10V–2Fe–3Al β-titanium alloy. *Wear* **2016**, *350*, 10–20. [CrossRef]
75. Weng, Z.; Gu, K.; Cui, C.; Cai, H.; Liu, X.; Wang, J. Microstructure evolution and wear behavior of titanium alloy under cryogenic dry sliding wear condition. *Mater. Charact.* **2020**, *165*, 110385. [CrossRef]
76. Li, X.X.; Zhou, Y.; Ji, X.L.; Li, Y.X.; Wang, S.Q. Effects of sliding velocity on tribo-oxides and wear behavior of Ti–6Al–4V alloy. *Tribol. Int.* **2015**, *91*, 228–234. [CrossRef]
77. Huang, C.; Zhang, Y.; Vilar, R.; Shen, J. Dry sliding wear behavior of laser clad TiVCrAlSi high entropy alloy coatings on Ti–6Al–4V substrate. *Mater. Des.* **2012**, *41*, 338–343. [CrossRef]
78. Tong, X.; Sun, Q.; Zhang, D.; Wang, K.; Dai, Y.; Shi, Z.; Li, Y.; Dargusch, M.; Huang, S.; Ma, J.; et al. Impact of scandium on mechanical properties, corrosion behavior, friction and wear performance, and cytotoxicity of a beta-type Ti–24Nb-38Zr-2Mo alloy for orthopedic applications. *Acta Biomater.* **2021**, *134*, 791–803. [CrossRef] [PubMed]
79. Nakai, M.; Iwasaki, T.; Ueki, K. Differences in the effect of surface texturing on the wear loss of β-type Ti–Nb–Ta–Zr and (α+β)-type Ti–6Al–4V ELI alloys in contact with zirconia in physiological saline solution. *J. Mech. Behav. Biomed. Mater.* **2021**, *124*, 104808. [CrossRef]
80. Jiang, X.; Dai, Y.; Xiang, Q.; Liu, J.; Yang, F.; Zhang, D. Microstructure and wear behavior of inductive nitriding layer in Ti–25Nb–3Zr–2Sn–3Mo alloys. *Surf. Coat. Technol.* **2021**, *427*, 127835. [CrossRef]
81. Hua, K.; Zhang, Y.; Zhang, F.; Kou, H.; Li, X.; Wu, H.; Wang, H. Microstructure refinement and enhanced wear-resistance modulated by stress relaxation processing in a metastable β titanium alloy. *Mater. Charact.* **2021**, *181*, 111505. [CrossRef]
82. Kheradmandfard, M.; Kashani-Bozorg, S.F.; Lee, J.S.; Kim, C.-L.; Hanzaki, A.Z.; Pyun, Y.-S.; Cho, S.-W.; Amanov, A.; Kim, D.-E. Significant improvement in cell adhesion and wear resistance of biomedical β-type titanium alloy through ultrasonic nanocrystal surface modification. *J. Alloys Compd.* **2018**, *762*, 941–949. [CrossRef]
83. Zhang, Y.; Chu, K.; He, S.; Wang, B.; Zhu, W.; Ren, F. Fabrication of high strength, antibacterial and biocompatible Ti–5Mo–5Ag alloy for medical and surgical implant applications. *Mater. Sci. Eng. C Mater. Biol. Appl.* **2020**, *106*, 110165. [CrossRef] [PubMed]
84. Jawed, S.F.; Rabadia, C.D.; Liu, Y.J.; Wang, L.Q.; Li, Y.H.; Zhang, X.H.; Zhang, L.C. Mechanical characterization and deformation behavior of β-stabilized Ti–Nb–Sn–Cr alloys. *J. Alloys Compd.* **2019**, *792*, 684–693. [CrossRef]
85. Dias Corpa Tardelli, J.; Bolfarini, C.; Candido Dos Reis, A. Comparative analysis of corrosion resistance between beta titanium and Ti–6Al–4V alloys: A systematic review. *J. Trace Elem. Med. Biol.* **2020**, *62*, 126618. [CrossRef] [PubMed]
86. Moretti, B.; Pesce, V.; Maccagnano, G.; Vicenti, G.; Lovreglio, P.; Soleo, L.; Apostoli, P. Peripheral neuropathy after hip replacement failure: Is vanadium the culprit? *Lancet* **2012**, *379*, 1676. [CrossRef]
87. Mirza, A.; King, A.; Troakes, C.; Exley, C. Aluminium in brain tissue in familial Alzheimer's disease. *J. Trace Elem. Med. Biol.* **2017**, *40*, 30–36. [CrossRef]
88. Bansal, P.; Singh, G.; Sidhu, H.S. Improvement of surface properties and corrosion resistance of Ti13Nb13Zr titanium alloy by plasma-sprayed HA/ZnO coatings for biomedical applications. *Mater. Chem. Phys.* **2021**, *257*, 123738. [CrossRef]
89. Vlcak, P.; Fojt, J.; Koller, J.; Drahokoupil, J.; Smola, V. Surface pre-treatments of Ti-Nb-Zr-Ta beta titanium alloy: The effect of chemical, electrochemical and ion sputter etching on morphology, residual stress, corrosion stability and the MG-63 cell response. *Results Phys.* **2021**, *28*, 104613. [CrossRef]
90. Abd-elrhman, Y.; Gepreel, M.A.H.; Abdel-Moniem, A.; Kobayashi, S. Compatibility assessment of new V-free low-cost Ti–4.7Mo–4.5Fe alloy for some biomedical applications. *Mater. Des.* **2016**, *97*, 445–453. [CrossRef]
91. Assis, S.L.D.; Wolynec, S.; Costa, I. Corrosion characterization of titanium alloys by electrochemical techniques. *Electrochim. Acta* **2006**, *51*, 1815–1819. [CrossRef]
92. Jin, L.; Cui, W.-F.; Song, X.; Liu, G.; Zhou, L. Effects of surface nanocrystallization on corrosion resistance of β-type titanium alloy. *Trans. Nonferrous Met. Soc. China* **2014**, *24*, 2529–2535. [CrossRef]
93. Prakash, C.; Uddin, M.S. Surface modification of β-phase Ti implant by hydroaxyapatite mixed electric discharge machining to enhance the corrosion resistance and in-vitro bioactivity. *Surf. Coat. Technol.* **2017**, *326*, 134–145. [CrossRef]
94. Chen, L.; Wei, K.; Qu, Y.; Li, T.; Chang, B.; Liao, B.; Xue, W. Characterization of plasma electrolytic oxidation film on biomedical high niobium-containing β-titanium alloy. *Surf. Coat. Technol.* **2018**, *352*, 295–301. [CrossRef]

95. Wang, C.; Ma, F.; Liu, P.; Chen, J.; Liu, X.; Zhang, K.; Li, W.; Han, Q. The influence of alloy elements in Ti 6Al 4V and Ti 35Nb 2Ta 3Zr on the structure, morphology and properties of MAO coatings. *Vacuum* **2018**, *157*, 229–236. [CrossRef]
96. Gu, H.; Ding, Z.; Yang, Z.; Yu, W.; Zhang, W.; Lu, W.; Zhang, L.-C.; Wang, K.; Wang, L.; Fu, Y.-f. Microstructure evolution and electrochemical properties of TiO_2/Ti-35Nb-2Ta-3Zr micro/nano-composites fabricated by friction stir processing. *Mater. Des.* **2019**, *169*, 107680. [CrossRef]
97. Liu, C.-F.; Li, S.-J.; Hou, W.-T.; Hao, Y.-L.; Huang, H.-H. Enhancing corrosion resistance and biocompatibility of interconnected porous β-type Ti-24Nb-4Zr-8Sn alloy scaffold through alkaline treatment and type I collagen immobilization. *Appl. Surf. Sci.* **2019**, *476*, 325–334. [CrossRef]
98. Pina, V.G.; Dalmau, A.; Devesa, F.; Amigo, V.; Munoz, A.I. Tribocorrosion behavior of beta titanium biomedical alloys in phosphate buffer saline solution. *J. Mech. Behav. Biomed. Mater.* **2015**, *46*, 59–68. [CrossRef]
99. Bahl, S.; Das, S.; Suwas, S.; Chatterjee, K. Engineering the next-generation tin containing beta titanium alloys with high strength and low modulus for orthopedic applications. *J. Mech. Behav. Biomed. Mater.* **2018**, *78*, 124–133. [CrossRef]
100. Diomidis, N.; Mischler, S.; More, N.S.; Roy, M.; Paul, S.N. Fretting-corrosion behavior of β titanium alloys in simulated synovial fluid. *Wear* **2011**, *271*, 1093–1102. [CrossRef]
101. Yuan, Z.; He, Y.; Lin, C.; Liu, P.; Cai, K. Antibacterial surface design of biomedical titanium materials for orthopedic applications. *J. Mater. Sci. Technol.* **2021**, *78*, 51–67. [CrossRef]
102. Cai, D.; Zhao, X.; Yang, L.; Wang, R.; Qin, G.; Chen, D.-F.; Zhang, E. A novel biomedical titanium alloy with high antibacterial property and low elastic modulus. *J. Mater. Sci. Technol.* **2021**, *81*, 13–25. [CrossRef]
103. Chang, X.; Smith, G.C.; Quinn, J.; Carson, L.; Chan, C.W.; Lee, S. Optimization of anti-wear and anti-bacterial properties of beta TiNb alloy via controlling duty cycle in open-air laser nitriding. *J. Mech. Behav. Biomed. Mater.* **2020**, *110*, 103913. [CrossRef]
104. Lubov Donaghy, C.; McFadden, R.; Kelaini, S.; Carson, L.; Margariti, A.; Chan, C.-W. Creating an antibacterial surface on beta TNZT alloys for hip implant applications by laser nitriding. *Opt. Laser Technol.* **2020**, *121*, 105793. [CrossRef]
105. Shi, A.; Cai, D.; Hu, J.; Zhao, X.; Qin, G.; Han, Y.; Zhang, E. Development of a low elastic modulus and antibacterial Ti-13Nb-13Zr-5Cu titanium alloy by microstructure controlling. *Mater. Sci. Eng. C Mater. Biol. Appl.* **2021**, *126*, 112116. [CrossRef]
106. Liu, H.; Liu, R.; Ullah, I.; Zhang, S.; Sun, Z.; Ren, L.; Yang, K. Rough surface of copper-bearing titanium alloy with multifunctions of osteogenic ability and antibacterial activity. *J. Mater. Sci. Technol.* **2020**, *48*, 130–139. [CrossRef]
107. Kaur, M.; Singh, K. Review on titanium and titanium based alloys as biomaterials for orthopaedic applications. *Mater. Sci. Eng. C Mater. Biol. Appl.* **2019**, *102*, 844–862. [CrossRef]
108. Sidhu, S.S.; Singh, H.; Gepreel, M.A.H. A review on alloy design, biological response, and strengthening of β-titanium alloys as biomaterials. *Mater. Sci. Eng. C Biomim. Mater. Sens. Syst.* **2021**, *121*, 111661. [CrossRef]
109. Zhao, D.; Liang, H.; Han, C.; Li, J.; Liu, J.; Zhou, K.; Yang, C.; Wei, Q. 3D printing of a titanium-tantalum Gyroid scaffold with superb elastic admissible strain, bioactivity and in-situ bone regeneration capability. *Addit. Manuf.* **2021**, *47*, 102223. [CrossRef]
110. Liang, H.; Zhao, D.; Feng, X.; Ma, L.; Deng, X.; Han, C.; Wei, Q.; Yang, C. 3D-printed porous titanium scaffolds incorporating niobium for high bone regeneration capacity. *Mater. Des.* **2020**, *194*, 108890. [CrossRef]
111. Jirka, I.; Vandrovcova, M.; Frank, O.; Tolde, Z.; Plsek, J.; Luxbacher, T.; Bacakova, L.; Stary, V. On the role of Nb-related sites of an oxidized beta-TiNb alloy surface in its interaction with osteoblast-like MG-63 cells. *Mater. Sci. Eng. C Mater. Biol. Appl.* **2013**, *33*, 1636–1645. [CrossRef] [PubMed]
112. Li, X.; Chen, T.; Hu, J.; Li, S.; Zou, Q.; Li, Y.; Jiang, N.; Li, H.; Li, J. Modified surface morphology of a novel Ti-24Nb-4Zr-7.9Sn titanium alloy via anodic oxidation for enhanced interfacial biocompatibility and osseointegration. *Colloids Surf. B Biointerfaces* **2016**, *144*, 265–275. [CrossRef]
113. Fu, J.; Liu, X.; Tan, L.; Cui, Z.; Liang, Y.; Li, Z.; Zhu, S.; Zheng, Y.; Kwok Yeung, K.W.; Chu, P.K.; et al. Modulation of the mechanosensing of mesenchymal stem cells by laser-induced patterning for the acceleration of tissue reconstruction through the Wnt/beta-catenin signaling pathway activation. *Acta Biomater.* **2020**, *101*, 152–167. [CrossRef] [PubMed]

Review

A Review of Effects of Femtosecond Laser Parameters on Metal Surface Properties

Hongfei Sun, Jiuxiao Li *, Mingliang Liu, Dongye Yang * and Fangjie Li

School of Materials Engineering, Shanghai University of Engineering Science, Shanghai 201620, China
* Correspondence: lijiuxiao@126.com (J.L.); dongye.yang@sues.edu.cn (D.Y.)

Abstract: As a laser technology, the femtosecond laser is used in biomedical fields due to its excellent performance—its ultrashort pulses, high instantaneous power, and high precision. As a surface treatment process, the femtosecond laser can prepare different shapes on metal surfaces to enhance the material's properties, such as its wear resistance, wetting, biocompatibility, etc. Laser-induced periodic surface structures (LIPSSs) are a common phenomenon that can be observed on almost any material after irradiation by a linearly polarized laser. In this paper, the current research state of LIPSSs in the field of biomedicine is reviewed. The influence of laser parameters (such as laser energy, pulse number, polarization state, and pulse duration) on the generation of LIPSSs is discussed. In this paper, the applications of LIPSSs by femtosecond laser modification for various purposes, such as in functional surfaces, the control of surface wettability, the surface colonization of cells, and the improvement of tribological properties of surfaces, are reviewed.

Keywords: laser-induced periodic surface structures; femtosecond laser processing; functional surfaces; application

1. Introduction

In 1954, Charles Towns et al. [1] made the first maser, the precursor of the laser. It opened the door to a series of astonishing inventions and discoveries. D. E. Spence et al. [2] obtained the first laser with titanium-doped sapphire as the gain medium in 1991, which is considered to be the first femtosecond laser of real significance. Femtosecond lasers are used in various fields, such as information, environment, medicine, defense, and industry, because of their short pulses, high energy, and high peak power [3–13].

The femtosecond laser (fs-laser) has good application prospects in the biomedical field. There are various ways to improve the biocompatibility of medical implants, such as changing the alloy composition [14], designing porous structures [15], and various processes [16,17]. As a surface treatment process, the femtosecond laser can prepare different shapes on metal surfaces to enhance the material's properties, such as wear resistance, wetting, and biocompatibility [18–20]. Surface modification technologies include mechanical methods (e.g., friction stirring [21,22], burnishing [23]), chemical methods (e.g., anodic oxidation [24], chemical vapor deposition [25]) and physical methods (e.g., thermal spraying [26], physical vapor deposition [27]). Compared with traditional surface modification technology, laser surface modification has outstanding advantages, such as high precision, flexibility, versatility, etc.

In this paper, the application of femtosecond laser surface modification is reviewed in various fields and for various purposes, including in patterns, the coloration of functional surfaces, the control of surface wettability, the surface colonization of cells, and the improvement of tribological properties of nanostructured metal surfaces, and we explore the connection between femtosecond laser parameters and patterns to provide a reference for future applications.

Citation: Sun, H.; Li, J.; Liu, M.; Yang, D.; Li, F. A Review of Effects of Femtosecond Laser Parameters on Metal Surface Properties. *Coatings* **2022**, *12*, 1596. https://doi.org/10.3390/coatings12101596

Academic Editors: Angela De Bonis and Anton Ficai

Received: 31 July 2022
Accepted: 22 September 2022
Published: 21 October 2022

Publisher's Note: MDPI stays neutral with regard to jurisdictional claims in published maps and institutional affiliations.

Copyright: © 2022 by the authors. Licensee MDPI, Basel, Switzerland. This article is an open access article distributed under the terms and conditions of the Creative Commons Attribution (CC BY) license (https:// creativecommons.org/licenses/by/ 4.0/).

2. Laser-Induced Periodic Stripe Structure (LIPSS) with Femtosecond Laser

A laser-induced periodic surface structure (LIPSS) is a surface relief composed of periodic lines that can be observed on almost any material after irradiation of a linearly polarized laser beam, especially when using ultrashort laser pulses of durations in the range of picoseconds to femtoseconds [28–36].

A lot of work has been performed to study the formation mechanism of the femtosecond laser on LIPSSs. A LIPSS can be classified according to the characteristic ratio of its spatial periods (Λ) to the irradiation wavelength (λ) and the polarization direction of the linear laser beam used to produce them [37–39]. Figure 1a provides a general classification of LIPSSs observed on irradiation with femtosecond laser pulses. The period of low spatial frequency LIPSSs (LSFL) is slightly equal to or less than the laser wavelength. They are perpendicular (LSFL-I) or parallel (LSFL-II) to the polarization direction of the laser. In contrast, the period of high spatial frequency LIPSSs (HSFL) is smaller than half of the irradiation wavelength and may be formed as deep surface gratings (HSFL-I, depth-to-period aspect ratio A > 1) or as shallow surface gratings (HSFL-II, depth-to-period aspect ratio A < 1). Figure 1b [37] provides an LSFL-I type structure on a Ti6Al4V surface after femtosecond laser irradiation. The double arrows indicate the direction of the laser beam polarization. The LSFL structure has a period of $\Lambda LSFL \sim 620 \pm 80$ nm and is perpendicular to the polarization direction of the laser beam. Figure 1c shows the HSFL-II structure formed on the laser irradiation surface [40]. The HSFL structures have periods of less than 100 nm ($\Lambda HSFL \sim 80 \pm 20$ nm) and are parallel to the direction of the laser beam polarization.

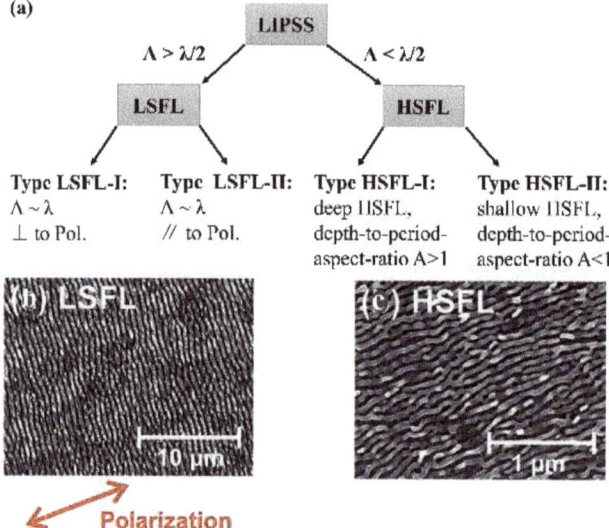

Figure 1. (a) General classification of fs-laser-induced periodic surface structures. SEM micrographs [37] of near wavelength LSFL-I (b) and sub-wavelength HSFL-II (c) on Ti6Al4V surfaces after irradiation with fs-laser in the air (pulse duration τ = 30 fs, center wavelength λ = 790 nm, pulse repetition frequency 1 kHz). The double arrows in (b) mark the direction of laser beam polarization.

The current theories on the formation of LIPSSs can be divided into two classes, i.e.: (i) Electromagnetic theories describing the deposition of optical energy into a solid. By introducing the η of efficacy factors, the researchers analyzed the interaction of electromagnetic radiation with microscopic rough surfaces through theoretical and experimental combinations [41]. (ii) Matter reorganization theory, which is based on the redistribution of the surface matter (Figure 2 [38,42]). The researchers believe that HSFL is formed through the self-organization of irradiated materials and is related to the surface instability caused by atomic diffusion and surface erosion effects [43–46]. The difference between both classes can be summarized as follows: electromagnetic scattering and absorption effects sow the spatial signature of the structure during laser irradiation, and the reorganization of matter takes longer. Figure 2a shows that static thermal melting or ablation from a sample occurs on shorter time scales. The laser beam is marked in green. Figure 2b shows a dynamic response via self-organization from a laser-produced instability.

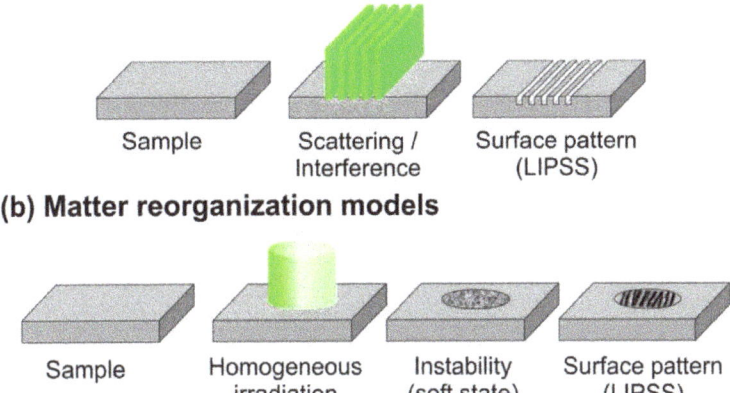

Figure 2. Ref. [42] Fundamental processes occur during LIPSS formation according to (**a**) electromagnetic models and (**b**) matter reorganization models. The laser irradiation is marked in green.

3. Laser Parameters That Control LIPSS

Studies have shown that laser peak fluence, the number of laser pulses, laser polarization state, pulse duration, and the processing environment are all key parameters affecting LIPSS [47–52].

3.1. Laser Fluence

The laser fluence has a large impact on the morphology of LIPSSs; different types of LIPSS are obtained by varying the laser fluence on the same material [49–51,53,54]. J. Bonse [55] obtained both LSFL and HSFL on the surface of titanium by varying the laser fluence, and the laser fluence affects the periodicity of LIPSSs. Georg Schnell et al. [50] report the formation of nano- and micro-structures on Ti6Al4V evoked by different scanning strategies and fluences with an fs-laser. Figure 3 shows the SEM microstructure images of the femtosecond laser pulses of different energy fluences. As shown in Figure 3a [50], the surface morphology is LIPSS when the laser fluence is 0.14 J/cm^2, the surface topography is micron spacing grooves when the laser fluence is 0.86 J/cm^2, and the surface morphology is cones and micro craters at a laser fluence of 4.76 J/cm^2. Shi-zhen Xu et al. [56] explore the influence of laser scan fluence on the formation of micro/nanostructures on the surfaces of fused silica. At a fixed laser scan speed (1.7 mm/s), the HFSL was observed at a low fluence region (1.8–2.5 J/cm^2). A transition from HSFL to LSFL occurred when a critical energy fluence threshold (2.5 J/cm^2) was exceeded (Figure 3b). The phenomenon can help to form process design guidelines to tailor large-scale surfaces with self-organized features, and can be used in future studies. For most materials, the periodicity of LIPSSs increases as the laser fluence increases. Better surface topography can be obtained at a low laser fluence approximate to the material ablation threshold. To achieve high efficiency for industrial applications, the ablation rate is increased by increasing the laser fluence. However, the processing quality is significantly reduced due to the thermal damage caused by the highly effective penetration depth. To avoid adverse effects on the sample, a suitable fluence is an advantageous condition to realize cold processing.

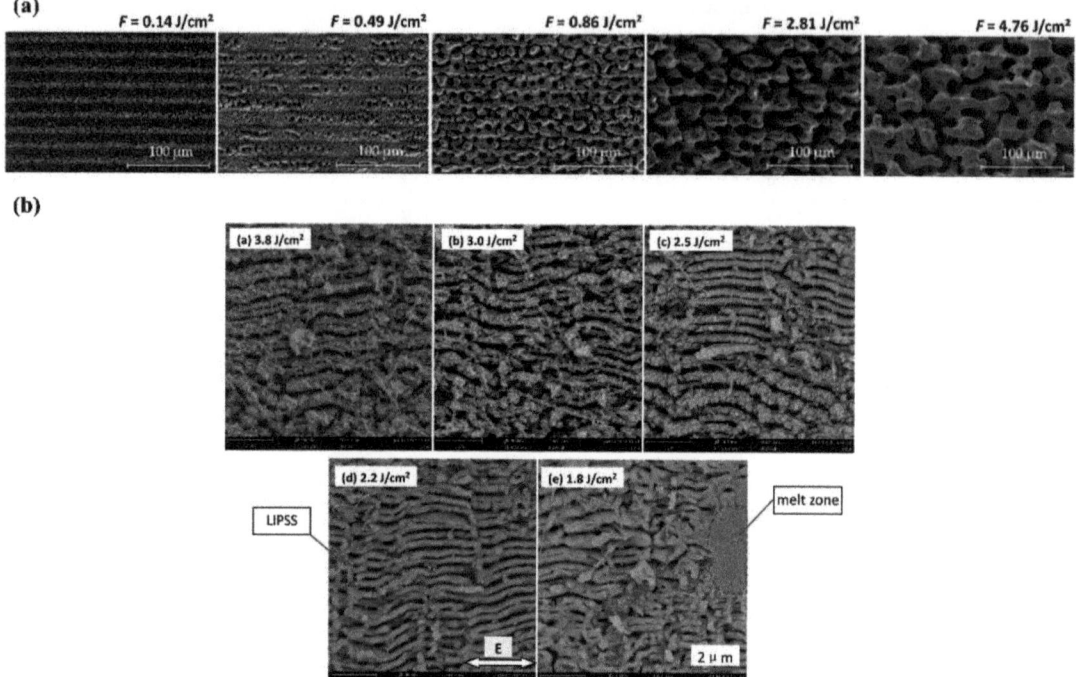

Figure 3. (a) Ref. [50] SEM images of structured surfaces with increasing laser fluence. (b) Ref. [56] Microstructures of femtosecond laser irradiation regions with different fluence.

3.2. Number of Pulses

The higher the number of laser pulses, the easier it is to obtain a more regular LIPSS. Evangelos Skoulas et al. [57] studied the effect of pulse number on the formation of LIPSSs. As the number of pulses increases, the surface roughness increases, and the period of LIPSS decreases. At the same time, as the number of pulses or the laser fluence increases, the depth of the pit and the height of the microstructure increase. Xu Ji et al. [58] prepared nanoholes on the silicon surface by a femtosecond laser. Figure 4 provides the SEM images of the depth of the surface pit and the height of the microstructure with the different numbers of pulses. As shown in Figure 4a, a shallow modified zone is formed with pulses. As shown in Figure 4b, when N is 4, the rectangular nanoholes were created on the silicon surface. As the number of pulses increases, the energy is absorbed more efficiently along the direction of laser polarization. This results in two rows of nanohole chains, forming LSFL, as shown in Figure 4c. For the pulse number, N = 8, the nanoholes become larger and deeper, as shown in Figure 4d. The deeper and larger nanoholes can be created on the surface by increasing the pulse number and fluences. When the pulse number increases to 10, most of the HSFL is broken, as shown in Figure 4e. Rao Li et al. [59] obtained the femtosecond laser-induced damage threshold (LIDT) by measuring the damage morphology under different energies and pulse numbers of the femtosecond laser. For the multi-pulse radiation, the LIDT of the thin film decreases as the number of pulses increases due to the accumulation effect. To obtain a high-quality periodic structure, it is necessary to accurately measure the laser damage threshold of the material.

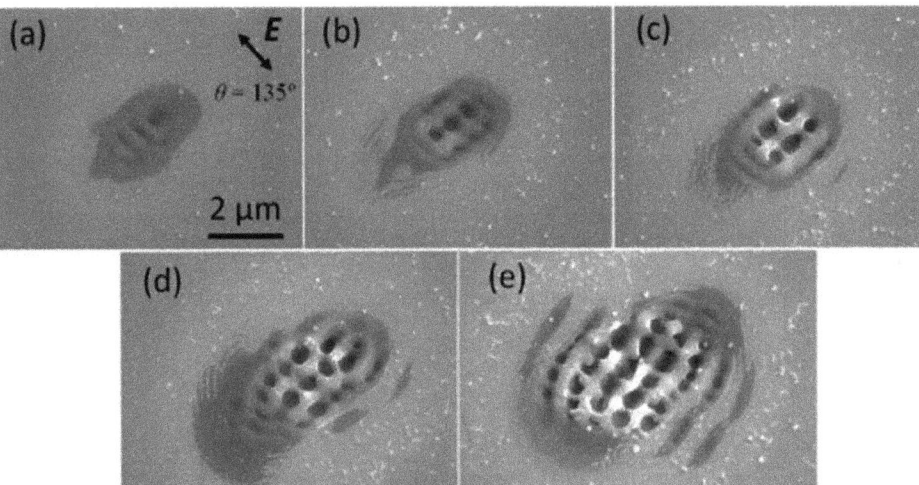

Figure 4. Ref. [58] SEM images of periodic structures induced on the surface of a silicon wafer by femtosecond laser at a fixed laser fluence (F = 0.22 J/cm^2) with different pulse number (N): (**a**) N = 2, (**b**) N = 4, (**c**) N = 6, (**d**) N = 8 and (**e**) N = 10, respectively. The arrow shows the direction of laser polarization.

3.3. Polarization States

The structure characteristics of the material surface after femtosecond laser modification are related to the polarization state of the laser beam [60,61]. The orientation and shape of the laser-induced periodic structure are determined by the polarization of the incident light. For example, circularly polarized beams can acquire triangular periodic structures [62–64], and elliptically polarized beams can acquire spherical nanoparticles [65–67]. When a linearly polarized laser beam is applied, the ripple direction is perpendicular or parallel to the polarization direction of the incident laser beam. Zhang Hao et al. [48] used the finite-difference time-domain method (FDTD) to study the surface morphology of LIPSSs under various polarization states (linear, circular, radial and azimuthal). The surface morphology simulated using circular polarization lasers is consistent with the triangular LIPSSs and spherical nanoparticles reported in the literature [62,68]. Evangelos Skoulas et al. [57] obtained a nanoscale controllable periodic structure on the nickel surface by laser direct writing with radial and azimuthal polarization beams, which mimicked the placoid structures found in the skin of sharks. Figure 5 shows the characteristic surface morphologies attained in SEM micrographs obtained at a scanning speed v = 0.5 mm/s and a laser fluence F = 0.24 J/cm^2, for linear Gaussian (a,b), radial (c,d) and azimuthal (e,f) cylindrical vector beams, respectively. The images (b,d,f) are higher magnifications of areas of the red dashed squares. As shown in Figure 5a, linear laser direct writing obtains LIPSSs on the surface. Figure 5b shows how the radial and azimuthal beams were irradiated to obtain a rhombus-like structure.

3.4. Pulse Duration

Pulse duration is a relevant parameter in laser processing, and different laser systems (e.g., nanosecond, picosecond, femtosecond) obtain different surface morphologies. Sun Yuanyuan et al. [69] used a continuous laser, nanosecond laser, and femtosecond laser to modify the surface of a ferromanganese alloy. The results show that the effect of the continuous laser and nanosecond laser on the material is mainly melted generation. Surface grooves in the micron range can be obtained using nanosecond lasers. Femtosecond laser ablation generates LIPSSs on the surface of the material without altering the crystal structure. LSFL can be obtained under laser irradiation with a nanosecond pulse duration or longer, while HSFL with periods much smaller than λ is only suitable for the irradiation of ultrashort pulsed lasers in the range of picoseconds to femtoseconds. Sungkwon Shin et al. [70] ablated the Invar sheets with a laser with different pulse durations, and the results show that the femtosecond laser treatment obtained high precision micro-holes with no thermal damage (i.e., Figure 6). Figure 6a–c shows SEM images of laser pulses irradiating Invar, corresponding to the pulse durations of 10 ns, 15 ps, and 300 fs, respectively. In the ns laser processing with a laser fluence of 5 J/cm^2, a pulse repetition rate of 50 Hz, and a wavelength of 248 nm, the surface is observed to produce burrs. High-precision micro-holes with no thermal damage at the edges were obtained by fs laser processing with F = 0.29 J/cm^2, f = 200 kHz, and λ = 1035 nm.

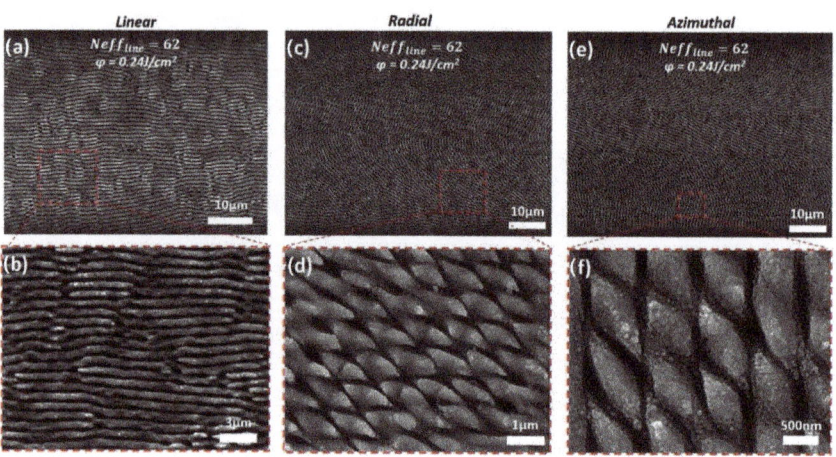

Figure 5. Ref. [57] The SEM images depicting line scans produced by linearly (**a**,**b**), radially (**c**,**d**), and azimuthally polarized (**e**,**f**) beams, respectively, at v = 0.5 mm/s, and F = 0.24 J/cm^2. The images (**b**,**d**,**f**) are higher magnifications of an area inside the red dashed squares and reveal the biomimetic shark skin-like morphology of the processed areas.

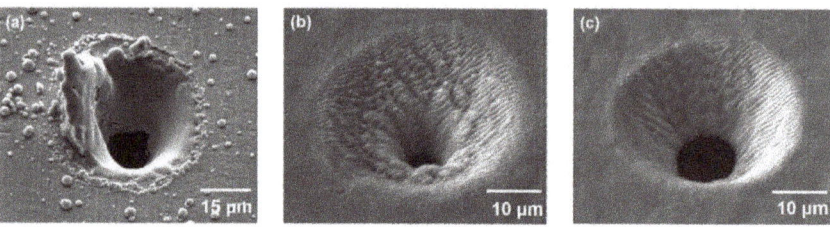

Figure 6. Ref. [70] SEM images of micro-holes on Invar were processed at three pulse durations of (**a**) 10 ns, (**b**) 15 ps, and (**c**) 300 fs, respectively. The ns laser parameters are F = 5 J/cm^2, f = 50 Hz, and λ = 248 nm. For the ps and fs laser, F = 0.29 J/cm^2, f = 200 kHz, and λ = 1035 nm.

3.5. Ambient Medium

In addition to the above laser-related parameters, the ambient medium around the sample also has a significant impact on the surface morphology of laser processing. Zhiduo Xin et al. [71] reported the results of femtosecond laser texturing and femtosecond laser nitriding experiments on Ti6Al4V. After femtosecond laser texturing, as shown in Figure 7a,c, cuboid structures of 125 × 125 × 130 mm^3 were formed on the surface. After nitriding, as shown in Figure 7b,d, a uniform crack-free TiN coating was prepared on the top of the textured structures with a thickness of 40–60 mm. As shown in Figure 7e, the morphology analysis shows that only slight height variations are introduced into the textured structures by femtosecond laser nitriding. Vadim Yalishev et al. [72] reported the surface morphology changes and wettability of titanium processed by femtosecond lasers in both the air and a vacuum. The results show that the laser texture obtained under vacuum conditions can form a permanent superhydrophilic surface. Yang Yang et al. [73] studied the microstructure of the titanium action of femtosecond lasers in three different liquid environments. Cavities and islands were observed on the sample surface. After femtosecond laser modification in the supersaturated Hydroxyapatite (HA, $Ca_{10}(PO_4)_6(OH)_2$) suspension, the biocompatible element Ca-P is firmly deposited on the surface. Thus, the corresponding functional surface can be obtained by changing the ambient medium, which also provides a new way to understand the ablation mechanism of the femtosecond laser.

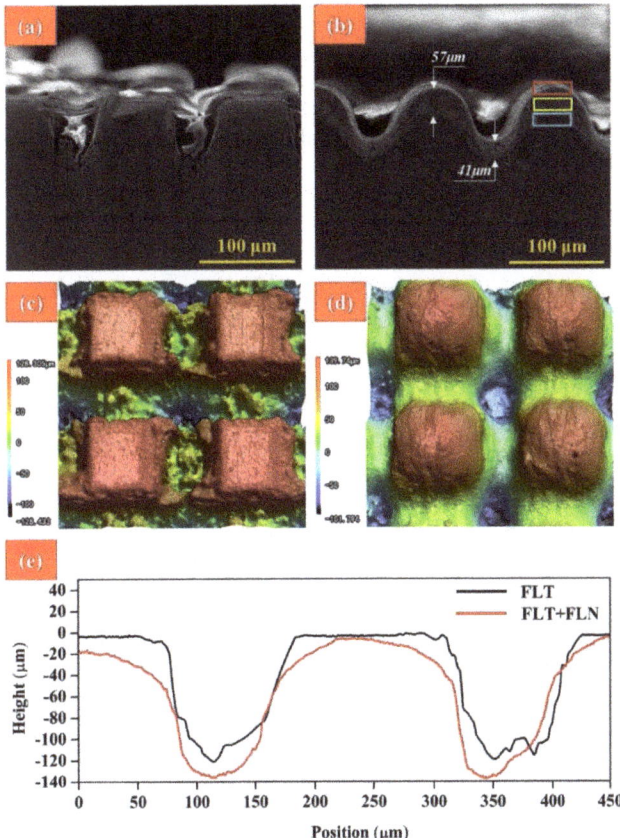

Figure 7. Ref. [71] The surface morphology and cross-section microstructure of the samples pro−cessed by Femtosecond laser texturing (FLT) and FLT+Femtosecond laser nitriding (FLN); (**a**) SEM image of FLT, (**b**) SEM image of FLT+FLN, (**c**) 3D morphology of the cross-section of FLT, (**d**) 3D morphology of the cross-section of FLT+FLN, (**e**) Profile curves corresponding to (**c,d**).

In summary, LIPSSs (ripples) can be obtained when the fluence of the laser is slightly greater than the ablation threshold of the material. Increasing the laser energy will obtain the surface morphology of grooves, pits, etc. In addition, the orientation and shape of LIPSSs are affected by the polarization state of the laser. An increase in the number of pulses will make the surface pits deeper. The laser processing environment is also one of the important parameters that affects surface morphology. Although the relevant parameters that affect the formation of LIPSSs have been reported, there is still a lack of a general algorithm to control the regularity of LIPSSs. In the future, artificial intelligence (AI) and algorithms will discover and control the regularity of LIPSSs.

4. Application of LIPSS

Surface texturing by laser irradiation can change various materials' properties and create multifunctional surfaces [74–78]. Materials can be better applied by customizing functional surfaces.

4.1. Structural Color

One of the most obvious applications of LIPSSs is optics. Since their period is in the same range as the radiation wavelength of visible light, they can effectively act as a diffraction grating, producing a "structural color". B. Dusser et al. [79] studied how to change the direction of the ripples to transmit information onto metal surfaces, creating a portrait of Vincent van Gogh on stainless steel surfaces (Figure 8A). Wang Chao et al. [80] prepared LIPSSs on a Ti6Al4V surface by laser irradiation, and observed differences in the laser texture color under natural light, and the surface color changes with the changes in the laser parameters (Figure 8B). Figure 8B(a) shows an optical image of the sample after laser irradiation, which includes "nine-squares" and "JLU". Figure 8B(b) lists the laser parameters corresponding to each square. As shown in Figure 8B(c), when captured in a dark environment, the difference in colors in the "nine squares" is evident. Moreover, as shown in Figure 8B(d), when changing the shooting angle, the "JLU" could present various colors. The results in Figure 8 show that the LIPSS has potential applications for Ti6Al4V surface coloring. Different colors can be observed by changing the laser parameters to regulate the period and direction of the LIPSS, as well as the incident light and the viewing angle [81–85]. High-quality and regular LIPSSs are prepared in large areas on metal surfaces, making it possible to apply them to optical sensors, anti-counterfeiting, decoration, and laser marking, etc.

4.2. Wetting Behavior

The wetting behavior of LIPSSs has attracted the attention of many researchers. In general, the wettability of liquids to solid surfaces depends on three major factors: (1) the surface energy of the solids and liquids, (2) the viscosity of the liquids, and (3) the surface morphology of the solids. Surface topography can significantly affect the contact angle of droplets placed on the surface. Figure 9 shows that the surface morphology has a great effect on surface roughness and contact angle. The variation in the contact angles (θ_M) measured for 15 samples irradiated at different laser fluences is presented in Figure 9a. As the laser fluence increases, the contact angle increases. Figure 9b shows different surface morphologies and the increase in contact angle of the water droplets on different surface structures. The water contact angle measurement shows that the femtosecond laser treatment of Au turns its originally hydrophilic surface ($\theta_M \sim 74°$) into a hydrophobic surface ($\theta_M \sim 108°$). The θ_M measurements indicate that as the surface nano/microstructures increase, the θ_M significantly increases as well. Numerous studies [32,35,86–89] have shown that bioinspired surfaces with superwettability can be prepared using ultrashort pulse lasers. Alexandre Cunha et al. [90] generated hydrophilic surface textures on the surface of Ti–6Al–4V alloys by femtosecond laser processing. They show that the surface treatment of metal surfaces with femtosecond lasers is an effective technique for improving surface wettability. A. Y. Vorobyev et al. [91] prepared superhydrophobic and self-cleaning multifunctional surfaces using femtosecond laser pulses. Research by Erin Liu et al. [92] demonstrates that femtosecond fiber lasers can form layered structures on metal surfaces, demonstrating superhydrophobic, self-cleaning, and light-trapping properties. Sohail A. Jalil et al. [93] investigated the surface structure of femtosecond laser-induced gold (Au) and its effect on hydrophobicity. The result shows that the femtosecond laser processing turns originally hydrophilic Au into a superhydrophobic surface. It can be seen that surfaces with superwettability have a significant impact on other fields, such as for sensors, thermal management, biomedicine, etc. The long-term stability of LIPSSs' surface wetting properties (e.g., hydrophobicity or hydrophilicity) in applications will be a popular topic in the future.

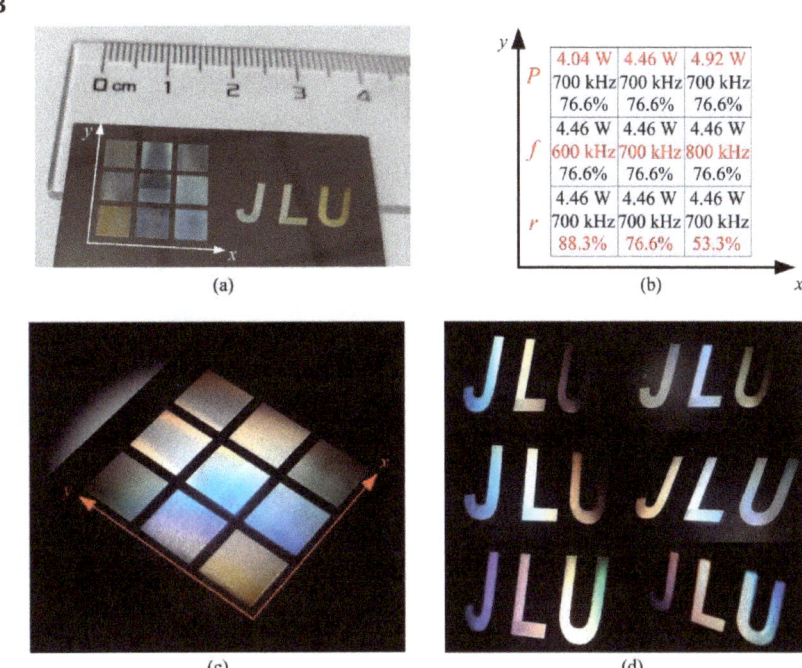

Figure 8. (**A**) Ref. [79] A portrait of Vincent van Gogh on stainless steel surfaces. (**B**) Ref. [80] (a) Optical images of the sample after laser irradiation, and (b) the corresponding laser parameters (P, Laser power; f, Laser repetition frequency; r, Pulse overlap rate between two adjacent scanning lines). (c) shows the optical images "nine-squares" captured in the dark environment, and (d) shows the color change in "JLU" when changing the shooting angle.

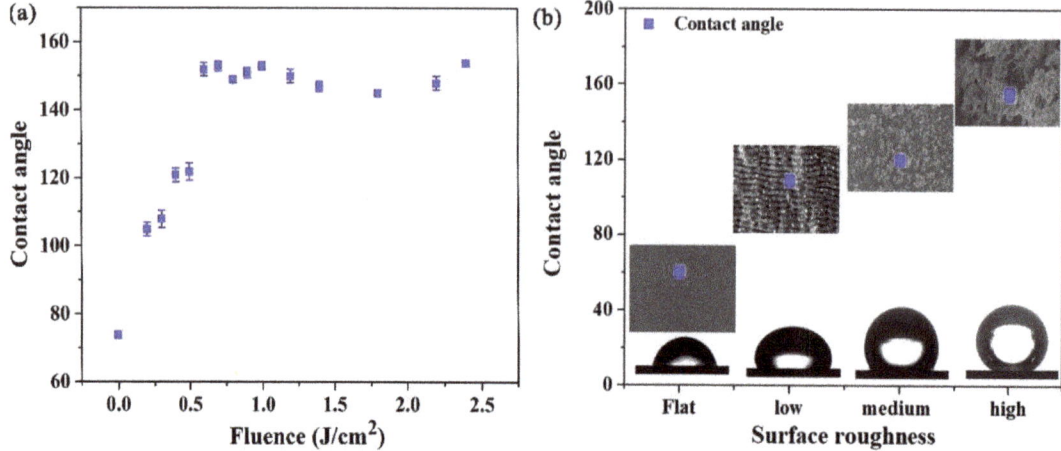

Figure 9. Ref. [93] (**a**) The measured contact angle values as a function of laser fluence. (**b**) The contact angle values measured on the initial surface roughness at low fluence. Corresponding surface morphologies are depicted in the insets.

4.3. Biomedical Applications

Another promising application area for LIPSS is biomedicine, which can inhibit the formation of bacterial biofilm and affect cell growth. Laser texturing has been used in the biomedical field as a method of altering surface morphology to potentially improve osseointegration [94–97]. Research [98–101] has shown that different surface topographies have a great influence on cell growth. Kai Borcherding et al. [102] described the adhesion and shape of osteoblast-like cells (MG-63) after laser treatment of titanium alloys. Compared to pure titanium, the cell viability was improved on the structured surface, indicating good cytocompatibility. Alexandre Cunha et al. [94] prepared three types of surface textures by femtosecond laser: LIPSSs, nanopillars, and microcolumns covered with LIPSSs. Compared with the polished reference group, the cell area and adhesion area of human mesenchymal stem cells on the surface of the laser-treated titanium alloy are reduced. Xiao Luo et al. [103] applied femtosecond laser irradiation to produce three types of nano-ripples on the surface of pure titanium, and to investigate their anti-bacterial behavior and their biocompatibility. The three types of nano-ripples include LIPSSs (type 1 textures), nano-ripples interrupted by grooves (type 2 textures), and columns with overlapping LIPSS (type 3 textures). The control group is the mechanical polishing group. The results shows that three types of nano-ripples can prevent bacterial colonization and biofilm formation. As demonstrated in Figure 10a, the staining of F-actin and the nucleus shows the adhesion states of rat mesenchymal stem cells on the substrate surfaces. The red fluorescence is from Rhodamine cyclopeptide-stained F-actin. The blue fluorescence is from the DAPI-stained cell nucleus. The arrow indicates the direction of cell diffusion. As can be seen from Figure 10b, the spread of cells is oriented. Compared to the polished titanium, the spreading areas of laser-fabricated samples are significantly larger, which means the adhesion sites offered by the three types of nano-ripples are beneficial to cell attachment. Ning Liu et al. [54] uses femtosecond laser surface modification to establish a nano-ripple structure on the Fe-30Mn alloy surface. Compared to the polished sample, the nano-ripple structure surface exhibited a significant improvement in the biodegradation rate. Cell growth depends on the size of the surface topography of the material, so controlling the size of the surface morphology may be a key factor in controlling cell function. By using femtosecond lasers for surface modification, different surface properties can be prepared on the implant. Using a femtosecond laser to fabricate nano-ripples and grooves on the surface of materials is a promising way to improve the performance of the implant material.

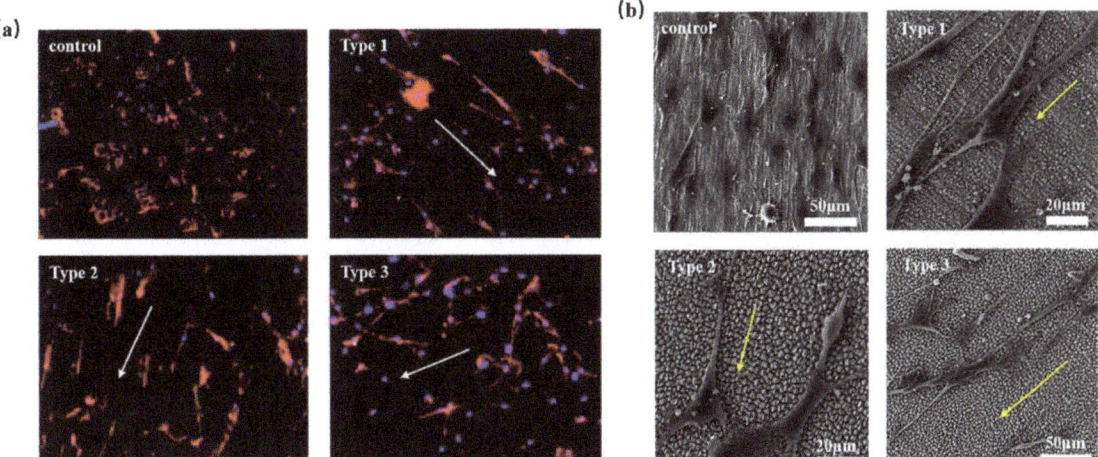

Figure 10. Ref. [103] Mesenchymal stem cells' adhesion on the polished surfaces and three types of nano-ripples after 72 h of incubations. (**a**) Fluorescence images of cytoskeletons; (**b**) SEM images of MSCs' adhesion states for the three types of nano-ripples. The both arrow indicates the direction of cell diffusion.

4.4. Reduction in Friction and Wear

LIPSS can exhibit beneficial tribological properties by reducing frictional wear. Surface topography and roughness have a significant impact on friction and wear [104]. Numerous studies [105–108] have shown that laser processing to prepare specific surface textures is an effective technique to improve surface friction performance. Jörn Bonse et al. [109] presented the latest advances in femtosecond laser surface texturing, observing the tribological properties of steel and the titanium alloy surface morphology (ripples, grooves, and spikes). Compared to the wear tracks on the surface of the polished sample, the wear tracks in the femtosecond laser processing area are almost invisible. The reason for its significant abrasion resistance is the LIPSS generated during the laser surface treatment. Figure 11 shows a sketch of the reciprocating sliding tribological test geometry (Figure 11a) along with top-view optical micrographs of the generated wear tracks on the polished Ti6Al4V alloy surface (Figure 11b) and the Spike-covered surface (Figure 11c). Additionally, top-view SEM micrographs revealing details from the wear tracks are presented (Figure 11d: initially polished, Figure 11e: LSFL, Figure 11f: Grooves, Figure 11g: Spikes). It is evident that on all laser-generated morphologies, the topmost regions have been partly worn, but the structures were not removed. The wear track and surface damage left on the polished surface is much larger than that in the laser-processed regions. The research of C. Florian et al. [110] demonstrated that femtosecond laser ablation forms a nanoscale morphology on the metal surface, resulting in a significant reduction in its coefficient of friction. Femtosecond laser treatments of metal surfaces inhibit adhesion tendencies by reducing the contact area, and the improvement in the tribological properties is due to the combined effect of LIPSSs.

4.5. Other Applications

Several other technical applications of LIPSS have been explored. Laser processing ablation obtains the desired surface features on the metal; such modified surfaces can be both beneficial and durable in phase-change heat transfer applications. Surface modification by ultra-short pulse lasers alters the heat transfer performances of the boiling system [111]. Since the LIPSSs can significantly increase the absorption rate of the surface, they will simultaneously lead to an increase in thermal radiation. Another potential application of LIPSS is related to catalytic activity in electrochemical processes [4], in which the active surface area of the electrode material is critical to the efficiency of the electrochemical reaction. LIPSSs can be applied to energy-saving components and sensors [5]. Another application of LIPSSs is in chemical analyses based on surface-enhanced Raman spectroscopy. In the future, many will explore the established and new surface functions that be created through LIPSSs, so that these materials can be better applied in mechanical engineering, healthcare, aerospace, energy, and other fields.

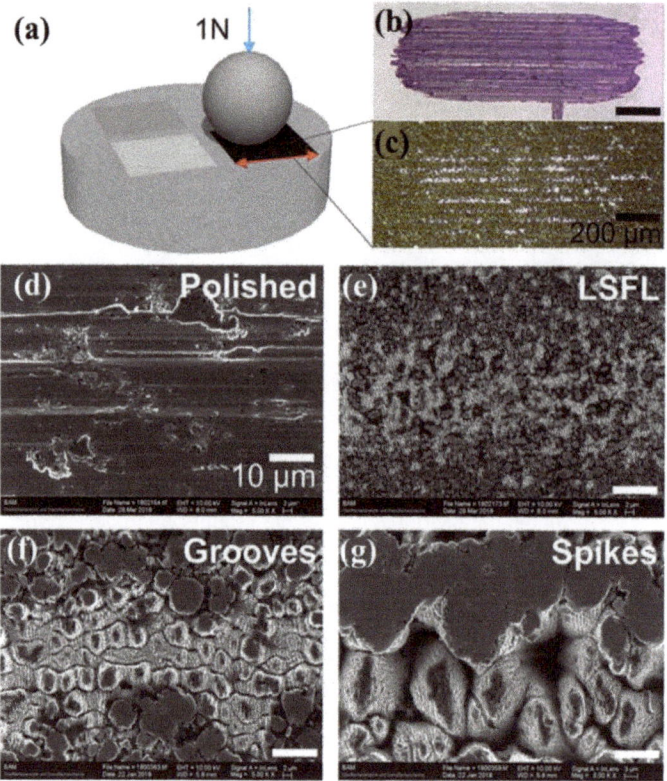

Figure 11. Ref. [110] Tribological performance of the samples after irradiation. (**a**) Sketch of the tribology setup using a steel ball of 100 Cr6 on the surface of the Ti6Al4V alloy sample. The final wear track achieved after 1000 sliding cycles is shown in (**b**) for the free surface and in (**c**) for a Spike-covered area as optical micrographs. SEM micrographs of the wear track on the different areas are shown in (**d**) for the initially polished surface, (**e**) LSFL, (**f**) Grooves, and (**g**) Spikes.

In summary, various patterns can be prepared by femtosecond lasers to improve the performance of materials. Due to the versatility of the femtosecond laser process, the correlation between surface morphology and alloy properties still has great research potential.

5. Summary and Outlook

The femtosecond laser is applied to biomedical materials as a surface modification technology. The laser-induced periodic structure generated by fs-laser action can improve the wear resistance, corrosion resistance, wetting, and biocompatibility of material surfaces. Although there have been many relevant reports, the mechanism of action of laser ablation materials to form special structures on the surface is still being explored. The correlation between surface patterns and material properties still needs to be studied continuously. In the future, numerical simulations will predictively simulate the laser processing parameters of the desired surface topography with the help of machine learning algorithms. The relevant control parameters of functionalized surfaces are identified for better industrial applications. Another trend will be continuing to explore the creation of LIPSSs and their surface capabilities so that they can be better applied in mechanical engineering, healthcare, aerospace, information, and other fields.

Author Contributions: J.L. carried out the conception of the idea of the manuscript. H.S. and M.L. collected and collated the data. H.S. wrote the original draft. J.L. reviewed and revised the original draft. F.L. and D.Y. provided guidance for the revision of the manuscript. All authors have read and agreed to the published version of the manuscript.

Funding: This research received no external funding.

Institutional Review Board Statement: Not applicable.

Informed Consent Statement: Not applicable.

Data Availability Statement: The data supporting the finding of this study are available within the article.

Conflicts of Interest: The authors declare no conflict of interest.

References

1. Townes, C.H.; Chu, S. How the Laser Happened: Adventures of a Scientist. *Phys. Today* **1999**, *52*, 59–60. [CrossRef]
2. De Spence, K.P.; Sibbett, W. 60-Fsec Pulse Generation from a Self-Mode-Locked Ti:Sapphire Laser. *Opt. Lett.* **1991**, *16*, 42–44. [CrossRef] [PubMed]
3. Ali, B.; Litvinyuk, I.V.; Rybachuk, M. Femtosecond Laser Micromachining of Diamond: Current Research Status, Applications and Challenges. *Carbon* **2021**, *179*, 209–226. [CrossRef]
4. Wang, Y.; Zhao, Y.; Qu, L. Laser Fabrication of Functional Micro-Supercapacitors. *J. Energy Chem.* **2021**, *59*, 642–665. [CrossRef]
5. Chen, M.-Q.; He, T.-Y.; Zhao, Y. Review of Femtosecond Laser Machining Technologies for Optical Fiber Microstructures Fabrication. *Opt. Laser Technol.* **2022**, *147*, 107628. [CrossRef]
6. Zhao, J.; Zhao, Y.; Peng, Y.; Lv, R.-Q.; Zhao, Q. Review of Femtosecond Laser Direct Writing Fiber-Optic Structures Based on Refractive index Modification and Their Applications. *Opt. Laser Technol.* **2022**, *146*, 107473. [CrossRef]
7. Kumar, R.; del Pino, A.P.; Sahoo, S.; Singh, R.K.; Tan, W.K.; Kar, K.K.; Matsuda, A.; Joanni, E. Laser Processing of Graphene and Related Materials for Energy Storage: State of the Art and Future Prospects. *Prog. Energy Combust. Sci.* **2022**, *91*, 100981. [CrossRef]
8. Siuzdak, K.; Haryński, Ł.; Wawrzyniak, J.; Grochowska, K. Review on Robust Laser Light Interaction with Titania—Patterning, Crystallisation and Ablation Processes. *Prog. Solid State Chem.* **2021**, *62*, 100297. [CrossRef]
9. Zhao, B.; Zheng, X.; Lei, Y.; Xie, H.; Zou, T.; Yuan, G.; Xin, W.; Yang, J. High-Efficiency-and-Quality Nanostructuring of Molybdenum Surfaces by Orthogonally Polarized Blue Femtosecond Lasers. *Appl. Surf. Sci.* **2022**, *572*, 151371. [CrossRef]
10. Nivas, J.J.; Allahyari, E.; Skoulas, E.; Bruzzese, R.; Fittipaldi, R.; Tsibidis, G.D.; Stratakis, E.; Amoruso, S. Incident Angle Influence on Ripples and Grooves Produced by Femtosecond Laser Irradiation of Silicon. *Appl. Surf. Sci.* **2021**, *570*, 151150. [CrossRef]
11. Museur, L.; Manousaki, A.; Anglos, D.; Tsibidis, G.; Kanaev, A. Pathways control in modification of solid surfaces induced by temporarily separated femtosecond laser pulses. *Appl. Surf. Sci.* **2021**, *566*, 150611. [CrossRef]
12. Dashtbozorg, B.; Penchev, P.; Romano, J.-M.; Li, X.; Sammons, R.L.; Dimov, S.; Dong, H. Development of surfaces with antibacterial durability through combineds phase plasma hardening and athermal femtosecond laser texturing. *Appl. Surf. Sci.* **2021**, *565*, 150594. [CrossRef]
13. Kotsiuba, Y.; Hevko, I.; Bellucci, S.; Gnilitskyi, I. Bitmap and vectorial hologram recording by using femtosecond laser pulses. *Sci. Rep.* **2021**, *11*, 16406. [CrossRef] [PubMed]
14. Zhang, T.; Wei, D.; Lu, E.; Wang, W.; Wang, K.; Li, X.; Zhang, L.-C.; Kato, H.; Lu, W.; Wang, L. Microstructure evolution and deformation mechanism of A+ B dual-phase Ti-Xnb-Yta-2zr alloys with high performance. *J. Mater. Sci. Technol.* **2022**, *131*, 68–81. [CrossRef]
15. Guo, L.; Naghavi, S.A.; Wang, Z.; Varma, S.N.; Han, Z.; Yao, Z.; Wang, L.; Liu, C. On the design evolution of hip implants: A review. *Mater. Des.* **2022**, *216*, 110552. [CrossRef]
16. Wang, J.C.; Liu, Y.J.; Liang, S.X.; Zhang, Y.S.; Wang, L.Q.; Sercombe, T.B.; Zhang, L.C. Comparison of microstructure and mechanical behavior of Ti-35nb manufactured by laser powder bed fusion from elemental powder mixture and prealloyed powder. *J. Mater. Sci. Technol.* **2022**, *105*, 1–16. [CrossRef]
17. Cui, Y.-W.; Chen, L.-Y.; Qin, P.; Li, R.; Zang, Q.; Peng, J.; Zhang, L.; Lu, S.; Wang, L.; Zhang, L.-C. Metastable pitting corrosion behavior of laser powder bed fusion produced Ti-6al-4v in Hank's solution. *Corros. Sci.* **2022**, *203*, 110333. [CrossRef]
18. Candel, J.J.; Amigó, V. Recent advances in laser surface treatment of titanium alloys. *J. Laser Appl.* **2011**, *23*, 022005. [CrossRef]
19. Mohazzab, B.F.; Jaleh, B.; Fattah-alhosseini, A.; Mahmoudi, F.; Momeni, A. Laser surface treatment of pure titanium: Microstructural analysis, wear properties, and corrosion behavior of titanium carbide coatings in Hank's physiological solution. *Surf. Interfaces* **2020**, *20*, 100597. [CrossRef]

20. Liu, R.; Chi, Z.; Cao, L.; Weng, Z.; Wang, L.; Li, L.; Saeed, S.; Lian, Z.; Wang, Z. Fabrication of biomimetic superhydrophobic and anti-icing Ti6al4v alloy surfaces by direct laser interference lithography and hydrothermal treatment. *Appl. Surf. Sci.* **2020**, *534*, 147576. [CrossRef]
21. Wang, L.; Xie, L.; Shen, P.; Fan, Q.; Wang, W.; Wang, K.; Lu, W.; Hua, L.; Zhang, L.-C. Surface microstructure and mechanical properties of Ti-6al-4v/Ag nanocomposite prepared by fsp. *Mater. Charact.* **2019**, *153*, 175–183. [CrossRef]
22. Wang, L.; Xie, L.; Lv, Y.; Zhang, L.C.; Chen, L.-Y.; Meng, Q.; Qu, J.; Zhang, D.; Lu, W. Microstructure evolution and superelastic behavior in Ti-35nb-2ta-3zr alloy processed by friction stir processing. *Acta Mater.* **2017**, *131*, 499–510. [CrossRef]
23. Thesleff, A.; Ortiz-Catalan, M.; Branemark, R. Low plasticity burnishing improves fretting fatigue resistance in bone-anchored implants for amputation prostheses. *Med. Eng. Phys.* **2022**, *100*, 103755. [CrossRef]
24. Li, Z.-X.; Bao, Y.-T.; Wu, L.-K.; Cao, F.-H. Oxidation and tribological properties of anodized Ti45al8.5nb alloy. *Trans. Nonferrous Met. Soc. China* **2021**, *31*, 3439–3451. [CrossRef]
25. Kazemi, M.; Ahangarani, S.; Esmailian, M.; Shanaghi, A. Investigating the corrosion performance of Ti-6al-4v biomaterial alloy with hydroxyapatite coating by artificial neural network. *Mater. Sci. Eng.* **2022**, *278*, 115644. [CrossRef]
26. Garrido, B.; Dosta, S.; Cano, I.G. Bioactive glass coatings obtained by thermal spray: Current status and future challenges. *Boletín De La Soc. Española De Cerámica Y Vidr.* **2021**, in press. [CrossRef]
27. Gabor, R.; Cvrček, L.; Doubková, M.; Nehasil, V.; Hlinka, J.; Unucka, P.; Buřil, M.; Podepřelová, A.; Seidlerová, J.; Bačáková, L. Hybrid coatings for orthopaedic implants formed by physical vapour deposition and microarc oxidation. *Mater. Des.* **2022**, *219*, 110811. [CrossRef]
28. Yang, Z.; Zhu, C.; Zheng, N.; Le, D.; Zhou, J. Superhydrophobic surface preparation and wettability transition of titanium alloy with micro/nano hierarchical texture. *Materials* **2018**, *11*, 2210. [CrossRef]
29. Gräf, S. Formation of laser-induced periodic surface structures on different materials: Fundamentals, properties and applications. *Adv. Opt. Technol.* **2020**, *9*, 11–39. [CrossRef]
30. Liang, C.; Hu, Y.; Wang, H.; Xia, D.; Li, Q.; Zhang, J.; Yang, J.; Li, B.; Li, H.; Han, D.; et al. Biomimetic cardiovascular stents for in vivo re-endothelialization. *Biomaterials* **2016**, *103*, 170–182. [CrossRef]
31. Khorkov, K.S.; Kochuev, D.A.; Dzus, M.A.; Prokoshev, V.G. Wettability surface control on stainless steel by lipss formation. *J. Phys. Conf. Ser.* **2021**, *1822*, 012010. [CrossRef]
32. Ijaola, A.O.; Bamidele, E.A.; Akisin, C.J.; Bello, I.T.; Oyatobo, A.T.; Abdulkareem, A.; Farayibi, P.K.; Asmatulu, E. Wettability transition for laser textured surfaces: A comprehensive review. *Surf. Interfaces* **2020**, *21*, 100802. [CrossRef]
33. Boltaev, G.S.; Alghabra, M.S.; Iqbal, M.; Ganeev, R.A.; Alnaser, A.S. Creation of azimuthally and radially directed laser-induced periodic structures on large tantalum surface. *J. Phys. Appl. Phys.* **2021**, *54*, 185109. [CrossRef]
34. Zhang, Y.; Cheng, K.; Cao, K.; Jiang, Q.; Chen, T.; Zhang, S.; Feng, D.; Sun, Z.; Jia, T. Periodic subwavelength ripples on a si surface induced by a single temporally shaped femtosecond laser pulse: Enhanced periodic energy deposition and reduced residual thermal effect. *J. Phys. Appl. Phys.* **2021**, *54*, 385106. [CrossRef]
35. Yang, C.-J.; Mei, X.-S.; Tian, Y.-L.; Zhang, D.-W.; Li, Y.; Liu, X.-P. Modification of wettability property of titanium by laser texturing. *Int. J. Adv. Manuf. Technol.* **2016**, *87*, 1663–1670. [CrossRef]
36. Yang, L.; Wei, J.; Ma, Z.; Song, P.; Ma, J.; Zhao, Y.; Huang, Z.; Zhang, M.; Yang, F.; Wang, X. The Fabrication of micro/nano structures by laser machining. *Nanomaterials* **2019**, *9*, 1789. [CrossRef]
37. Bonse, J.; Hohm, S.; Kirner, S.V.; Rosenfeld, A.; Kruger, J. Laser-induced periodic surface structures—A scientific evergreen. *IEEE J. Sel. Top. Quantum Electron.* **2017**, *23*, 1–15. [CrossRef]
38. Bonse, J.; Gräf, S. Maxwell meets marangoni—A review of theories on laser-induced periodic surface structures. *Laser Photonics Rev.* **2020**, *14*, 2000215. [CrossRef]
39. Bonse, J. Quo vadis lipss? Recent and future trends on laser-induced periodic surface structures. *Nanomaterials* **2020**, *10*, 1950. [CrossRef]
40. Kirner, S.V.; Wirth, T.; Sturm, H.; Krüger, J.; Bonse, J. Nanometer-resolved chemical analyses of femtosecond laser-induced periodic surface structures on titanium. *J. Appl. Phys.* **2017**, *122*, 104901. [CrossRef]
41. Bonse, J.; Munz, M.; Sturm, H. Structure Formation on the surface of indium phosphide irradiated by femtosecond laser pulses. *J. Appl. Phys.* **2005**, *97*, 013538. [CrossRef]
42. Varlamova, O. Evolution of femtosecond laser induced surface structures at low number of pulses near the ablation threshold. *J. Laser Micro/Nanoeng.* **2013**, *8*, 300–303. [CrossRef]
43. Reif, J.; Varlamova, O.; Uhlig, S.; Varlamov, S.; Bestehorn, M. On the physics of self-organized nanostructure formation upon femtosecond laser ablation. *Appl. Phys.* **2014**, *117*, 179–184. [CrossRef]
44. Liu, R.; Zhang, D.; Li, Z. Femtosecond laser induced simultaneous functional nanomaterial synthesis, in situ deposition and hierarchical lipss nanostructuring for tunable antireflectance and iridescence applications. *J. Mater. Sci. Technol.* **2021**, *89*, 179–185. [CrossRef]
45. Razi, S.; Varlamova, O.; Reif, J.; Bestehorn, M.; Varlamov, S.; Mollabashi, M.; Madanipour, K.; Ratzke, M. Birth of periodic micro/nano structures on 316l stainless steel surface following femtosecond laser irradiation; single and multi scanning study. *Opt. Laser Technol.* **2018**, *104*, 8–16. [CrossRef]
46. Zagoranskiy, I.; Lorenz, P.; Ehrhardt, M.; Zimmer, K. Guided self-organization of nanodroplets induced by nanosecond ir laser radiation of molybdenum films on sapphire. *Opt. Lasers Eng.* **2019**, *113*, 55–61. [CrossRef]

47. Bronnikov, K.; Gladkikh, S.; Okotrub, K.; Simanchuk, A.; Zhizhchenko, A.; Kuchmizhak, A.; Dostovalov, A. Regulating morphology and composition of laser-induced periodic structures on titanium films with femtosecond laser wavelength and ambient environment. *Nanomaterials* **2022**, *12*, 306. [CrossRef]
48. Zhang, H.; Colombier, J.-P.; Witte, S. Laser-induced periodic surface structures: Arbitrary angles of incidence and polarization states. *Phys. Rev.* **2020**, *101*, 245430. [CrossRef]
49. Echlin, M.P.; Titus, M.S.; Straw, M.; Gumbsch, P.; Pollock, T.M. Materials Response to glancing incidence femtosecond laser ablation. *Acta Mater.* **2017**, *124*, 37–46. [CrossRef]
50. Schnell, G.; Lund, H.; Bartling, S.; Polley, C.; Riaz, A.; Senz, V.; Springer, A.; Seitz, H. Heat Accumulation during femtosecond laser treatment at high repetition rate—A morphological, chemical and crystallographic characterization of self-organized structures on Ti6al4v. *Appl. Surf. Sci.* **2021**, *570*, 151115. [CrossRef]
51. Dou, H.-Q.; Liu, H.; Xu, S.; Chen, Y.; Miao, X.; Lü, H.; Jiang, X. Influence of laser fluences and scan speeds on the morphologies and wetting properties of titanium alloy. *Optik* **2020**, *224*, 165443. [CrossRef]
52. Wang, R.; Dong, X.; Wang, K.; Sun, X.; Fan, Z.; Duan, W.; Jun, M.B.-G. Polarization effect on hole evolution and periodic microstructures in femtosecond laser drilling of thermal barrier coated superalloys. *Appl. Surf. Sci.* **2021**, *537*, 148001. [CrossRef]
53. Gazizova, M.Y.; Smirnov, N.A.; Kudrayshov, S.I.; Shugurov, V.V. The effect of femtosecond laser treatment on the tribological properties of titanium nitride. *Iop Conf. Ser. Mater. Sci. Eng.* **2020**, *862*, 022054. [CrossRef]
54. Liu, N.; Sun, Y.; Wang, H.; Liang, C. Femtosecond laser-induced nanostructures on Fe-30 mn surfaces for biomedical applications. *Opt. Laser Technol.* **2021**, *139*, 106986. [CrossRef]
55. Bonse, J.; Krüger, J.; Höhm, S.; Rosenfeld, A. Femtosecond laser-induced periodic surface structures. *J. Laser Appl.* **2012**, *24*, 042006. [CrossRef]
56. Xu, S.-Z.; Dou, H.-Q.; Sun, K.; Ye, Y.-Y.; Li, Z.; Wang, H.-J.; Liao, W.; Liu, H.; Miao, X.-X.; Yuan, X.-D.; et al. Scan speed and fluence effects in femtosecond laser induced micro/nano-structures on the surface of fused silica. *J. Non-Cryst. Solids* **2018**, *492*, 56–62. [CrossRef]
57. Skoulas, E.; Manousaki, A.; Fotakis, C.; Stratakis, E. Biomimetic surface structuring using cylindrical vector femtosecond laser beams. *Sci. Rep.* **2017**, *7*, srep45114. [CrossRef] [PubMed]
58. Ji, X.; Jiang, L.; Li, X.; Han, W.; Liu, Y.; Wang, A.; Lu, Y. Femtosecond laser-induced cross-periodic structures on a crystalline silicon surface under low pulse number irradiation. *Appl. Surf. Sci.* **2015**, *326*, 216–221. [CrossRef]
59. Li, R.; Zhou, W.; Zhou, C.; Qi, Q.; Li, Y.; Yang, Y.; Zhang, W.; Zhang, P.; Dai, S.; Xu, T. Laser damage threshold of Ge8as23s69 films irradiated under single—And multiple-pulse femtosecond laser. *Ceram. Int.* **2022**, *48*, 8341–8348. [CrossRef]
60. Fraggelakis, F.; Stratakis, E.; Loukakos, P.A. Control of periodic surface structures on silicon by combined temporal and polarization shaping of femtosecond laser pulses. *Appl. Surf. Sci.* **2018**, *444*, 154–160. [CrossRef]
61. Han, W.; Han, Z.; Yuan, Y.; Wang, S.; Li, X.; Liu, F. Continuous control of microlens morphology on Si based on the polarization-dependent femtosecond laser induced periodic surface structures modulation. *Opt. Laser Technol.* **2019**, *119*, 105629. [CrossRef]
62. Fraggelakis, F.; Mincuzzi, G.; Lopez, J.; Manek-Hönninger, I.; Kling, R. Controlling 2d Laser nano structuring over large area with double femtosecond pulses. *Appl. Surf. Sci.* **2019**, *470*, 677–686. [CrossRef]
63. Romano, J.-M.; Garcia-Giron, A.; Penchev, P.; Dimov, S. Triangular laser-induced submicron textures for functionalising stainless steel surfaces. *Appl. Surf. Sci.* **2018**, *440*, 162–169. [CrossRef]
64. Milles, S.; Voisiat, B.; Nitschke, M.; Lasagni, A.F. Influence of roughness achieved by periodic structures on the wettability of aluminum using direct laser writing and direct laser interference patterning technology. *J. Mater. Process. Technol.* **2019**, *270*, 142–151. [CrossRef]
65. Vanithakumari, S.C.; Kumar, C.A.; Thinaharan, C.; Kishor, G.R.; George, R.P.; Kaul, R.; Bindra, K.S.; John, P. Laser patterned titanium surfaces with superior antibiofouling, superhydrophobicity, self-cleaning and durability: Role of line spacing. *Surf. Coat. Technol.* **2021**, *418*, 127257.
66. Durbach, S.; Hampp, N. Generation of 2d-arrays of anisotropically shaped nanoparticles by nanosecond laser-induced periodic surface patterning. *Appl. Surf. Sci.* **2021**, *556*, 149803. [CrossRef]
67. Mangababu, A.; Goud, R.S.P.; Byram, C.; Rathod, J.; Banerjee, D.; Soma, V.R.; Rao, S.N. Multi-functional gallium arsenide nanoparticles and nanostructures fabricated using picosecond laser ablation. *Appl. Surf. Sci.* **2022**, *589*, 152802. [CrossRef]
68. Yao, C.; Ye, Y.; Jia, B.; Li, Y.; Ding, R.; Jiang, Y.; Wang, Y.; Yuan, X. Polarization and fluence effects in femtosecond laser induced micro/nano structures on stainless steel with antireflection property. *Appl. Surf. Sci.* **2017**, *425*, 1118–1124. [CrossRef]
69. Sun, Y.; Chen, L.; Liu, N.; Wang, H.; Liang, C. Laser-modified Fe-30 mn Surfaces with promoted biodegradability and biocompatibility toward biological applications. *J. Mater. Sci.* **2021**, *56*, 13772–13784. [CrossRef]
70. Shin, S.; Hur, J.-G.; Park, J.K.; Kim, D.-H. Thermal damage free material processing using femtosecond laser pulses for fabricating fine metal masks: Influences of laser fluence and pulse repetition rate on processing quality. *Opt. Laser Technol.* **2021**, *134*, 106618. [CrossRef]
71. Xin, Z.; Ren, N.; Ren, Y.; Yue, X.; Han, Q.; Zhou, W.; Tao, Y.; Ye, Y. In-situ nitriding on the textured titanium alloy using femtosecond laser. *J. Mater. Res. Technol.* **2022**, *19*, 466–471. [CrossRef]
72. Yalishev, V.; Iqbal, M.; Kim, V.; Alnaser, A.S. Effect of processing environment on the wettability behavior of laser-processed titanium. *J. Phys. Appl. Phys.* **2021**, *55*, 045401. [CrossRef]

73. Yang, Y.Y.J.; Liang, C.; Wang, H.; Zhu, X.; Zhang, N. Surface microstructuring of Ti plates by femtosecond lasers in liquid ambiences: A new approach to improving biocompatibility. *Opt. Express* **2009**, *9*, 21124–21133. [CrossRef] [PubMed]
74. Dong, J.; Pacella, M.; Liu, Y.; Zhao, L. Surface engineering and the application of laser-based processes to stents—A review of the latest development. *Bioact. Mater.* **2022**, *10*, 159–184. [CrossRef] [PubMed]
75. Stratakis, E.; Bonse, J.; Heitz, J.; Siegel, J.; Tsibidis, G.D.; Skoulas, E.; Papadopoulos, A.; Mimidis, A.; Joel, A.C.; Comanns, P.; et al. Laser engineering of biomimetic surfaces. *Mater. Sci. Eng. Rep.* **2020**, *141*, 100562. [CrossRef]
76. Wang, L.; Yin, K.; Zhu, Z.; Deng, Q.; Huang, Q. Femtosecond Laser engraving micro/nanostructured poly (ether-ether-ketone) surface with superhydrophobic and photothermal ability. *Surf. Interfaces* **2022**, *31*, 102013. [CrossRef]
77. Liu, K.; Yang, C.; Zhang, S.; Wang, Y.; Zou, R.; Alamusi; Deng, Q.; Hu, N. Laser direct writing of a multifunctional superhydrophobic composite strain sensor with excellent corrosion resistance and anti-icing/deicing performance. *Mater. Des.* **2022**, *218*, 110689. [CrossRef]
78. Yin, K.; Du, H.; Luo, Z.; Dong, X.; Duan, J.-A. Multifunctional micro/nano-patterned ptfe near-superamphiphobic surfaces achieved by a femtosecond laser. *Surf. Coat. Technol.* **2018**, *345*, 53–60. [CrossRef]
79. Dusser, B.S.Z.; Soder, H.; Faure, N.; Colombier, J.; Jourlin, M.; Audouard, E. Controlled nanostructrures formation by ultra fast laser pulses for color marking. *Opt. Express* **2010**, *18*, 2913–2924. [CrossRef]
80. Wang, C.; Huang, H.; Qian, Y.; Zhang, Z.; Huang, W.; Yan, J. Nitrogen assisted formation of large-area ripples on Ti6al4v surface by nanosecond pulse laser irradiation. *Precis. Eng.* **2022**, *73*, 244–256. [CrossRef]
81. Zhang, D.; Liu, R.; Li, Z. Irregular lipss produced on metals by single linearly polarized femtosecond laser. *Int. J. Extrem. Manuf.* **2021**, *4*, 015102. [CrossRef]
82. Milovanović, D.S.; Gaković, B.; Radu, C.; Zamfirescu, M.; Radak, B.; Petrović, S.; Miladinović, Z.R.; Mihailescu, I.N. Femtosecond laser surface patterning of steel and titanium alloy. *Phys. Scr.* **2014**, *T162*, 014017. [CrossRef]
83. Florian, C.; Kirner, S.V.; Krüger, J.; Bonse, J. Surface functionalization by laser-induced periodic surface structures. *J. Laser Appl.* **2020**, *32*, 022063. [CrossRef]
84. Stoian, R.; Colombier, J.-P. Advances in ultrafast laser structuring of materials at the nanoscale. *Nanophotonics* **2020**, *9*, 4665–4688. [CrossRef]
85. Gnilitskyi, I.; Derrien, T.J.-Y.; Levy, Y.; Bulgakova, N.M.; Mocek, T.; Orazi, L. High-Speed manufacturing of highly regular femtosecond laser-induced periodic surface structures: Physical origin of regularity. *Sci. Rep.* **2017**, *7*, 8485. [CrossRef] [PubMed]
86. Wang, S.; Liu, K.; Yao, X.; Jiang, L. Bioinspired surfaces with superwettability: New insight on theory, design, and applications. *Chem. Rev.* **2015**, *115*, 8230–8293. [CrossRef]
87. Wang, Y.; Zhang, J.; Li, K.; Hu, J. Surface characterization and biocompatibility of isotropic microstructure prepared by uv laser. *J. Mater. Sci. Technol.* **2021**, *94*, 136–146. [CrossRef]
88. Simoes, I.G.; Dos Reis, A.C.; Da Costa Valente, M.L. Analysis of the influence of surface treatment by high-power laser irradiation on the surface properties of titanium dental implants: A systematic review. *J. Prosthet. Dent.* **2021**, *in press*. [CrossRef]
89. Yang, K.; Shi, J.; Wang, L.; Chen, Y.; Liang, C.; Yang, L.; Wang, L.-N. Bacterial anti-adhesion surface design: Surface patterning, roughness and wettability: A review. *J. Mater. Sci. Technol.* **2022**, *99*, 82–100. [CrossRef]
90. Cunha, A.; Serro, A.P.; Oliveira, V.; Almeida, A.; Vilar, R.; Durrieu, M.C. Wetting behaviour of femtosecond laser textured Ti–6al–4v surfaces. *Appl. Surf. Sci.* **2013**, *265*, 688–696. [CrossRef]
91. Vorobyev, A.Y.; Guo, C. Multifunctional surfaces produced by femtosecond laser pulses. *J. Appl. Phys.* **2015**, *117*, 033103. [CrossRef]
92. Liu, E.; Lee, H.J.; Lu, X. Superhydrophobic surfaces enabled by femtosecond fiber laser-written nanostructures. *Appl. Sci.* **2020**, *10*, 2678. [CrossRef]
93. Jalil, S.A.; Akram, M.; Bhat, J.A.; Hayes, J.J.; Singh, S.C.; ElKabbash, M.; Guo, C. Creating superhydrophobic and antibacterial surfaces on gold by femtosecond laser pulses. *Appl. Surf. Sci.* **2020**, *506*, 144952. [CrossRef] [PubMed]
94. Cunha, A.; Zouani, O.F.; Plawinski, L.; Botelho do Rego, A.M.; Almeida, A.; Vilar, R.; Durrieu, M.C. Human mesenchymal stem cell behavior on femtosecond laser-textured Ti-6al-4v surfaces. *Nanomedicine* **2015**, *10*, 725–739. [CrossRef] [PubMed]
95. Dumas, V.; Guignandon, A.; Vico, L.; Mauclair, C.; Zapata, X.; Linossier, M.T.; Bouleftour, W.; Granier, J.; Peyroche, S.; Dumas, J.-C.; et al. Femtosecond laser nano/micro patterning of titanium influences mesenchymal stem cell adhesion and commitment. *Biomed. Mater.* **2015**, *10*, 055002. [CrossRef] [PubMed]
96. Kumari, R.; Scharnweber, T.; Pfleging, W.; Besser, H.; Majumdar, J.D. Laser surface textured titanium alloy (Ti–6al–4v)—Part li—studies on bio-compatibility. *Appl. Surf. Sci.* **2015**, *357*, 750–758. [CrossRef]
97. Klos, A.; Sedao, X.; Itina, T.E.; Helfenstein-Didier, C.; Donnet, C.; Peyroche, S.; Vico, L.; Guignandon, A.; Dumas, V. Ultrafast laser processing of nanostructured patterns for the control of cell adhesion and migration on titanium alloy. *Nanomaterials* **2020**, *10*, 864. [CrossRef] [PubMed]
98. Raimbault, O.; Benayoun, S.; Anselme, K.; Mauclair, C.; Bourgade, T.; Kietzig, A.-M.; Girard-Lauriault, P.-L.; Valette, S.; Donnet, C. The effects of femtosecond laser-textured Ti-6al-4v on wettability and cell response. *Mater. Sci. Eng. Mater. Biol. Appl.* **2016**, *69*, 311–320. [CrossRef]
99. Wang, Y.; Yu, Z.; Li, K.; Hu, J. Study on the effect of surface characteristics of short-pulse laser patterned titanium alloy on cell proliferation and osteogenic differentiation. *Mater. Sci. Eng. Mater. Biol. Appl.* **2021**, *128*, 112349. [CrossRef]

100. Stanciuc, A.-M.; Flamant, Q.; Sprecher, C.M.; Alini, M.; Anglada, M.; Peroglio, M. Femtosecond laser multi-patterning of zirconia for screening of cell-surface interactions. *J. Eur. Ceram. Soc.* **2018**, *38*, 939–948. [CrossRef]
101. Kedia, S.; Bonagani, S.K.; Majumdar, A.G.; Kain, V.; Subramanian, M.; Maiti, N.; Nilaya, J.P. Nanosecond laser surface texturing of type 316l stainless steel for contact guidance of bone cells and superior corrosion resistance. *Colloid Interface Sci. Commun.* **2021**, *42*, 100419. [CrossRef]
102. Borcherding, K.; Marx, D.; Gätjen, L.; Specht, U.; Salz, D.; Thiel, K.; Wildemann, B.; Grunwald, I. Impact of laser structuring on medical-grade titanium: Surface characterization and in vitro evaluation of osteoblast attachment. *Materials* **2020**, *13*, 2000. [CrossRef] [PubMed]
103. Luo, X.; Yao, S.; Zhang, H.; Cai, M.; Liu, W.; Pan, R.; Chen, C.; Wang, X.; Wang, L.; Zhong, M. Biocompatible nano-ripples structured surfaces induced by femtosecond laser to rebel bacterial colonization and biofilm formation. *Opt. Laser Technol.* **2020**, *124*, 105973. [CrossRef]
104. Luo, J.; Sun, W.; Duan, R.; Yang, W.; Chan, K.; Ren, F.; Yang, X.-S. Laser surface treatment-introduced gradient nanostructured tizrhftanb refractory high-entropy alloy with significantly enhanced wear resistance. *J. Mater. Sci. Technol.* **2022**, *110*, 43–56. [CrossRef]
105. Bonse, J.; Koter, R.; Hartelt, M.; Spaltmann, D.; Pentzien, S.; Höhm, S.; Rosenfeld, A.; Krüger, J. Tribological performance of femtosecond laser-induced periodic surface structures on titanium and a high toughness bearing steel. *Appl. Surf. Sci.* **2015**, *336*, 21–27. [CrossRef]
106. Bonse, J.; Koter, R.; Hartelt, M.; Spaltmann, D.; Pentzien, S.; Höhm, S.; Rosenfeld, A.; Kruger, J. Femtosecond laser-induced periodic surface structures on steel and titanium alloy for tribological applications. *Appl. Phys.* **2014**, *117*, 103–110. [CrossRef]
107. Bonse, J.; Höhm, S.; Koter, R.; Hartelt, M.; Spaltmann, D.; Pentzien, S.; Rosenfeld, A.; Krüger, J. Tribological performance of sub-100-Nm femtosecond laser-induced periodic surface structures on titanium. *Appl. Surf. Sci.* **2016**, *374*, 190–196. [CrossRef]
108. Pan, X.; He, W.; Cai, Z.; Wang, X.; Liu, P.; Luo, S.; Zhou, L. Investigations on femtosecond laser-induced surface modification and periodic micropatterning with anti-friction properties on Ti6al4v titanium alloy. *Chin. J. Aeronaut.* **2022**, *35*, 521–537. [CrossRef]
109. Bonse, J.; Kirner, S.V.; Griepentrog, M.; Spaltmann, D.; Krüger, J. Femtosecond laser texturing of surfaces for tribological applications. *Materials* **2018**, *11*, 801. [CrossRef]
110. Florian, C.; Wonneberger, R.; Undisz, A.; Kirner, S.V.; Wasmuth, K.; Spaltmann, D.; Krüger, J.; Bonse, J. Chemical effects during the formation of various types of femtosecond laser-generated surface structures on titanium alloy. *Appl. Phys.* **2020**, *126*, 266. [CrossRef]
111. Kumar, G.U.; Suresh, S.; Kumar, C.S.; Back, S.; Kang, B.; Lee, H.J. A review on the role of laser textured surfaces on boiling heat transfer. *Appl. Therm. Eng.* **2020**, *174*, 115274. [CrossRef]

Article

Preparation and Degradation Characteristics of MAO/APS Composite Bio-Coating in Simulated Body Fluid

Zexin Wang, Fei Ye, Liangyu Chen *, Weigang Lv, Zhengyi Zhang, Qianhao Zang, Jinhua Peng, Lei Sun and Sheng Lu *

School of Materials Science and Engineering, Jiangsu University of Science and Technology, Zhenjiang 212003, China; xxkissbaby@126.com (Z.W.); woshiyefei123@126.com (F.Y.); wzxlwg@126.com (W.L.); bassick@126.com (Z.Z.); sdu_zangqianhao@126.com (Q.Z.); jinhua19930517@126.com (J.P.); lei.sun2.o@nio.com (L.S.)
* Correspondence: lychen@just.edu.cn (L.C.); lusheng_ktz@just.edu.cn (S.L.); Tel.: +86-511-84401171 (L.C.); +86-511-84401188 (S.L.)

Abstract: In this work, ZK60 magnesium alloy was employed as a substrate material to produce ceramic coatings, containing Ca and P, by micro-arc oxidation (MAO). Atmospheric plasma spraying (APS) was used to prepare the hydroxyapatite layer (HA) on the MAO coating to obtain a composite coating for better biological activity. The coatings were examined by various means including an X-ray diffractometer, a scanning electron microscope and an energy spectrometer. Meanwhile, an electrochemical examination, immersion test and tensile test were used to evaluate the in vitro performance of the composite coatings. The results showed that the composite coating has a better corrosion resistance. In addition, this work proposed a degradation model of the composite coating in the simulated body fluid immersion test. This model explains the degradation process of the MAO/APS coating in SBF.

Keywords: ZK60 magnesium; micro-arc oxidation; atmospheric plasma spraying; corrosion

1. Introduction

As a crucial research field in the study of the emerging clinical applications of orthopedics, some metal materials benefit from excellent machining performance, low cost, and compatibility with the human body, advantages which have drawn tremendous attention in the last several years [1–6]. Stainless steels and titanium alloys are usually the most commonly used biomedical metallic materials due to their satisfactory mechanical properties and comparative biocompatibility. However, there is a large gap in terms of elastic moduli between the natural bones and the materials mentioned above. Furthermore, patients usually suffer as a result of pain and financial burdens when using these metal materials because such materials are difficult to degrade in the human body and, therefore, a second operation is often required [7–10].

Compared with stainless steels, zirconium alloys and titanium alloys, magnesium alloys exhibit unique advantages including biodegradability, similar density and moduli to the bones, and reliable bone induction [11–15]. Owing to their comparative moduli, magnesium and its alloys are enabled to serve as bone implants and fixation materials. Unfortunately, such an extremely fast degradation rate of magnesium alloys can produce a large amount of hydrogen in the human body, leading to severe alkalosis, which significantly limits the applications of magnesium alloys as a biologically active material. Therefore, reducing the degradation rate to match the bone healing rate while considering its excellent biological activity has been considered to be the key factor for the performance of magnesium alloys [16–20]. Recently, preparing a coating with good biological activity and corrosion resistance by using micro-arc oxidation is one of the most common strategies to achieve surface modification for bio-implanted magnesium alloys [21].

Micro-arc oxidation (MAO) (also known as plasma electrolytic oxidation (PEO)) is a surface modification technology developed based on anodic oxidation. MAO uses instant high-temperature sintering in the micro-arc discharging zone to generate a coating with high hardness, high strength, insulation, wear resistance, corrosion resistance, high-temperature resistance, and other excellent properties in the valve metals [22–24]. The coating fabricated by this technology can be firmly attached to the magnesium alloy substrate with good compactness and excellent corrosion resistance. Besides, the whole MAO process is stable and reliable, with good repeatability and environmental friendliness. Compared with other surface modification technologies, the MAO coating shows high porosity and favorable corrosion resistance, which are considered by many researchers to be important characteristics in terms of achieving high-performance coatings [25–29]. During the MAO process, the surface of the metal substrate undergoes a breakdown discharge process under the instantaneous high voltage input, which results in the formation of countless discharge channels. Furthermore, phenomena such as melting, solidification and accumulation would take place in the channels that formed on the surface of the MAO coating. It should be noted that the coating is porous; hence, these pores can serve as the channels that lead to the ingression of the corrosive medium. To overcome this drawback, the researchers fabricated an additional coating with high-performance and bio-active substances on the surfaces of the MAO coatings via post-treatment, which could improve the corrosion resistance and enhance the biological activity of the magnesium implants. New research suggests that bioceramics hydroxyapatite (HA), with a Ca/P atomic ratio of 1.67, is a major inorganic component of bone tissue (such as human and animal bones and teeth). HA is not only non-toxic but also has good biocompatibility and osteoinductivity [30–34], and has gradually become an indispensable biological medium in the fabrication of the composite coatings on the surface of magnesium alloys.

However, pure HA still has shortcomings such as poor mechanical properties, high brittleness, low strength and poor fatigue resistance in the physiological environment, which limits its application in the field of clinical medicine. To overcome these shortcomings, the favorable biological activity and osteoinductive properties of the HA materials are utilized in order to directly deposit or coat them on the metal surface to form a coating that combines physical and biological properties, which has set off a wave of revolution in the area of surface modification technology [35].

Atmospheric plasma spraying (APS) is a technology that is widely used in the preparation of coatings [36–40]. Tremendous efforts have been made in the development of composite biological coatings involving the use of MAO and APS methods [41]. Daroonparvar et al. [42] fabricated an MAO coating on the surface of magnesium alloy, and then a nano-TiO_2 coating was coated on the MAO by using atmospheric plasma spraying technology. The results show that nano-TiO_2 particles can penetrate the micropores and defects produced during the MAO process after plasma spraying, effectively preventing the corrosive medium from penetrating the magnesium implants. Xizhi et al. [43] fabricated a Yb_2SiO_5 coating on the surface of magnesium alloy MAO coating by plasma spraying, achieving a good sealing effect on the MAO coating. In addition, the composite coating exhibits satisfactory performance, such as good compatibility and strong bonding force between the composite coating and the substrate. In addition, for biomedical implant materials, understanding the degradation behavior of the coating in the biological environment is the prerequisite for regulating the degradation rate. Gu et al. [44,45] found that a coating of corrosion product was deposited on the surface of the sample during the immersion process in the simulated body fluid. It is demonstrated that the corrosion product coating enables a reduction in the corrosion rate of the sample in the simulated fluid. Xiao et al. [46] showed that the degradation process of MAO coating is accompanied by the generation of degradation products and the deposition of the Ca–P product coating. Yao et al., Jie et al. and Wang et al. [47–49] analyzed the degradation behavior of the coating through the electrochemical impedance spectrum (EIS) test and proposed a chemical reaction mechanism that promotes the formation of the corrosion product coating.

In this work, a composite bio-coating was fabricated on the ZK60 magnesium surface via the combination of micro-arc oxidation and plasma spraying surface modifications, which aims to obtain favorable surface performance, good biological activity, and better corrosion resistance for the magnesium implants. First, the MAO treatment was used to fabricate a bio-coating on the ZK60 magnesium alloy. Then, the HA powder was selected to coat on the MAO coating by plasma spraying in order to obtain the HA contained MAO/APS composite bio-coating. Finally, a long-term immersion test was performed on the prepared composite bio-coating to investigate its corrosion behavior in simulated body fluid.

2. Materials and Methods

2.1. MAO/APS Composite Coating Structure Design

In this work, a micro-arc oxidation coating is first prepared on the surface of ZK60 magnesium alloy, and after that, the HA coating is prepared by plasma spraying on the surface of the MAO coating. Such a composite coating is considered to effectively combine excellent corrosion resistance and good biological activity. Thus, a biological composite coating with excellent performance is designed.

2.1.1. MAO Process

Wrought ZK60 magnesium alloy with the dimension of Φ 25 mm × 5 mm is used as the substrate material in this work. Before the MAO process, the substrates were polished using a Veiyee M-1 metallographic pre-mill machine (ZhiJin matelreader, Laizhou, China) in sequence with 600, 800, 1000, 1200 grit SiC waterproof sandpaper, and then treated by grit blasting. Finally, the roughened ZK60 substrate was placed in an ultrasonic cleaner for 10 min and dried in air for use. The chemical compositions of the substrates are shown in Table 1.

Table 1. Compositions of the wrought ZK60 magnesium alloy (wt.%).

Element	Zn	Zr	Impurities	Mg
Content	4.8–6.2	>0.45	≤0.30	balance

The MAO equipment utilizes the WHD-20 bipolar AC pulse device (Harbin Institute of Technology, Harbin, China). During the MAO test, the Mg alloy sample was used as the anode and the stainless-steel tank was used as the cathode. The magnesium alloy substrate was fixed with aluminum alloy by screw connection, and then the sample was completely immersed in the electrolyte in a hanging manner. During the test, the bio-electrolyte was cooled by circulating water and its temperature was kept below 30 °C. The specific parameters of the bio-electrolyte are shown in Table 2, where NaOH is used to maintain the pH of the bio-electrolyte at 13. The parameters of micro-arc oxidation are shown in Table 3.

Table 2. Composition of bio-electrolyte for MAO on ZK60 alloy.

Composition.	Na_2SiO_3	$Ca(AC)_2$	$(NaPO_3)_6$	NaH_2PO_4	NaOH
Concentration/(g/L)	6.0	0.5	0.8	0.5	-

Table 3. Parameters of micro-arc oxidation.

Power Control Mode	Current Density/(A/dm^2)	Frequency/Hz	Duty Cycle/%	Response Time/min
Constant current	20	500	+40/−60	15

2.1.2. Plasma Spraying Process

Before plasma spraying, the HA powder was dried at 200 °C for 3 h in a furnace. The average size of HA powder particles is 12 µm, the Ca/P ratio is about 1.67, the purity is about 99%, and the powder is spherical. The particle morphology is shown in Figure 1. The prepared micro-arc oxidation sample was placed in the fixture that was used for spraying, and the plasma jet was used to sweep the surface of sample 1 to 2 times to dry its surface, which avoids the breakdown of the coating. The detailed spraying parameters are shown in Table 4.

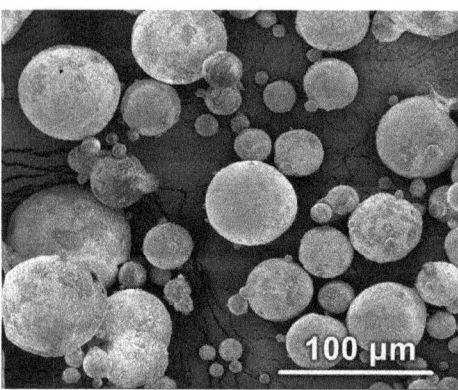

Figure 1. The SEM images of HA powder.

Table 4. Plasma spraying parameters selected for depositing HA coating.

Parameters	Spraying
Voltage (V)	60
Current (A)	500
Powder feed rate (g/s)	0.3
Spray step (mm)	3
Gun traverse rate (mm/s)	100
Main gas Ar (dm^3/s)	4.1
Secondary gas N$_2$ (dm^3/s)	1
Feed gas Ar (dm^3/s)	2.1
Distance to the substrate (mm)	110

2.2. Microstructural Characterization

A scanning electron microscope (SEM, JSM-6480A, JEOL, Tokyo, Japan), equipped with an energy dispersive X-ray spectrometer, was used to characterize the corrosion morphology of the coating and simulated body fluid at different periods and analyze the constituent elements of the corrosion sample.

The phase constituents of the coating and the corrosion products after immersion in simulated body fluid were characterized by X-ray diffraction (XRD, Shimadzu Corporation, Tokyo, Japan) with a scanning speed of 2°/min, scanning range (2θ) of 20–90°, and an accelerating voltage of 40 kV.

2.3. Tensile Tests

The bonding strength of the coating was examined by using the CMT5205 tester (SKYAN, ShenZhen, China). The test method is based on the GB/T8642-2002 standard that was specially designed for plasma-sprayed coatings. The initial surface of the MAO/APS coatings was subjected to grit blasting. Then, the tested surface was glued to the tensile fixture, and reinforced with screws. The glued samples were placed in air for 3 h, and then placed in the furnace at 100 °C for 4 h of heat treatment. Finally, the samples were cooled to room temperature for the tensile test. The loading speed of the tensile force was set to 165 N/s.

2.4. Electrochemical Tests

The AC impedance method was used to analyze the corrosion behavior of the coating in the SBF. The electrochemical impedance test in this experiment was also carried out on the CS2310 electrochemical workstation (CorrTest, Wuhan, Hubei, China), and the measured frequency range was 100 kHz~100 mHz. The reference electrode was a saturated calomel electrode, the auxiliary electrode was a platinum electrode, the sample was a working electrode, the medium used was the SBF solution, the salt bridge was a saturated KCl solution, and the temperature was 37 °C. The equivalent circuit was fitted by using the ZSimpWin software (V3.60, eDAQ, Colorado Springs, CO, USA) to characterize the corrosion resistance of the coatings.

2.5. MAO Experiment

ZK60 magnesium alloys with and without the surface treatment were selected for a 20-day simulated body fluid immersion experiment. The ion content in the simulated body fluid (Simulated Body Fluid, SBF) used was the same as that in human blood (as shown in Table 5), and the reagents and dosages used to configure 1000 mL of SBF are shown in Table 6. When configuring, the container was placed in a constant temperature water bath at 36.5 ± 0.5 °C, the reagents were completely dissolved in the order listed in the table, and finally, the pH value of the SBF was adjusted to 7.45 with Tris and HCl. The static weight-loss method was used to measure the changes of the sample weight during the immersion test. The pH value of the SBF was measured by using the Banter-920-UK precision pH meter (Want Balance Instrument, ChangZhou, China). The immersion test was conducted with three sets of samples.

Table 5. Ion concentration of SBF and blood plasma (mol/L).

Solution	Na^+	K^+	Mg^{2+}	Ca^{2+}	Cl^-	HCO_3^-	HPO_4^{2-}	SO_4^{2-}
SBF	142.0	5.0	1.5	2.5	148.8	4.2	1.0	0.5
Blood Plasma	142.0	5.0	1.5	2.5	103.0	27.0	1.0	0.5

Table 6. Reagents and dose for 1000 mL SBF.

Reagent	NaCl	$NaHCO_3$	KCl	$K_2HPO_4·3H_2O$	$MgCl_2·6H_2O$	1.0M-HCl	Na_2SO_4	$CaCl_2$
Dosage (g)	8.035	0.355	0.225	0.231	0.311	39 mL	0.072	0.278

Before immersion, each sample (including substrate) was weighed to establish its original mass (M) with a balance, then each sample was soaked in a plastic container with SBF, and the container was placed in a constant temperature water bath at 37 °C. The ratio of the exposed area of the sample (cm^2) to the volume of SBF (mL) was 1:10. The sequential pH value of SBF was recorded and the SBF was replaced by a new one every 24 h. Samples were taken out every five days, cleaned and dried, and their corroded masses were weighed (M1). The weight loss rate is:

$$R_{wt} = (M - M1)/M$$

where:

R_{wt}—weight loss rate;
M—the mass of the sample before corrosion (g);
M1—the mass of the sample after corrosion (g).

3. Results and discussion

3.1. Surface Morphology of MAO/APS Biological Composite Coating

Figure 2 shows the surface morphology of MAO coating under optimized parameters and MAO/APS composite coating. Figure 2a reveals that MAO coating after sandblasting pretreatment has an uneven distribution of micropores, with small pores in some specific areas. The reason for this situation is that the surface of the substrate after the sandblasting pretreatment is extremely uneven, with considerable tiny pits. The surface condition extends the ion transmission path and causes the reaction to require more energy supply.

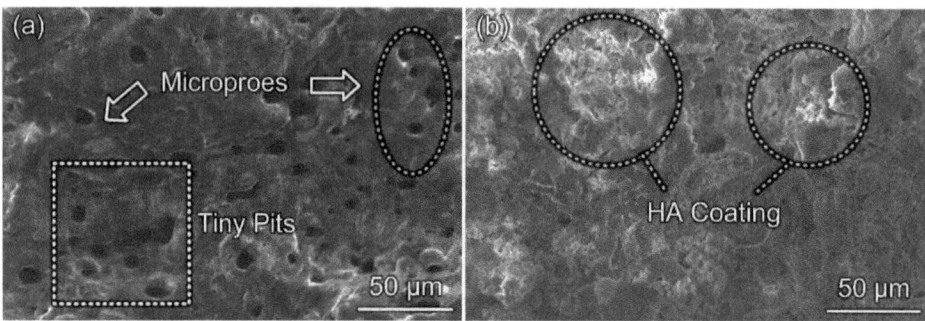

Figure 2. Morphologies of coatings: (**a**) surface morphology of MAO coating; (**b**) surface morphology MAO/APS coating.

Figure 2b shows the surface morphology of MAO/APS composited coating under an optimized spraying process. It reveals that HA particles are heated by plasma jet to form small spherical molten droplets with a diameter of about 8–10 μm and impact the MAO coating at a high speed. At this moment, the small spherical molten droplets were stacked and spread out on the MAO coating surface, and finally, the HA coating was formed after the spraying process.

At this time, the interface between HA particles became blurred, and the melting degree of the particles increased. There were no obvious unmelted particles on the coating. The particles were flattened to a high degree and finally formed the uniform HA coating.

3.2. Analysis of Bonding Strength and Fracture of MAO/APS Composite Biological Coating

The test results show that the composite coating that exhibits good bonding strength is 20.2 MPa. Du et al. [50] showed that the bonding strength of the lower coating under the MAO process is 40–50 MPa. Since the MAO coating is produced in situ, the bonding strength of the composite coating has decreased [51].

Figure 3a shows the morphology of the cross-section of the composite coating. It reveals that the thickness of the MAO coating is 36.14 μm, and the thickness of the HA spray coating is 59.59 μm. There is no obvious interface between the two coatings, which means the coating is well combined. The MAO coatings have no obvious cracks and micropores, all of which are blind holes. This greatly improves the corrosion resistance of MAO coatings [52]. Figure 3b shows the macroscopic fracture of stretched sample and reveals that the fracture position of the composite coating is mainly at the junction of the HA coating and the MAO coating. There is still a small amount of HA remaining on the MAO coating, which is scattered on the MAO coating. Figure 4a shows the microscopic morphology of the coating fracture, which is enlarged, as shown in Figure 4b. As seen from Figure 4b, it can be found that the fracture is smooth and the remaining powder is uniform. The fracture of the composite coating is a mainly brittle fracture, with almost no plastic deformation area.

Figure 3. Cross-section morphology of MAO/APS coating: (**a**) macroscopic fracture morphology of sample; (**b**) macroscopic fracture morphology of MAO/APS coating.

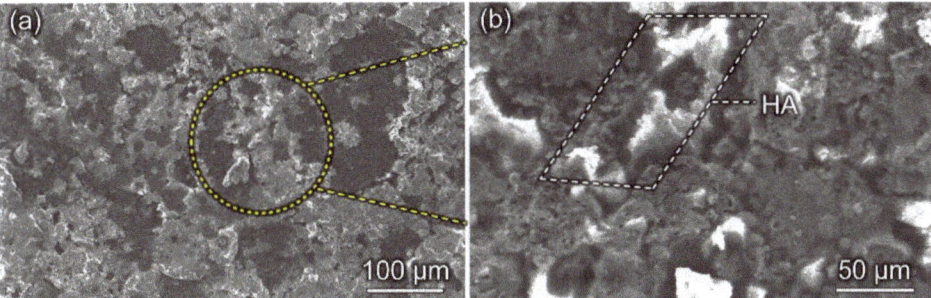

Figure 4. Tensile fracture morphology of the sample: (**a**) ×200; (**b**) ×500.

In short, the bonding method between the HA coating and the MAO substrate is mainly mechanical bonding. However, there is also a metallurgical bond in the local gap area of the MAO coating, which shows that it is not easy for the HA coating to completely fall off from the MAO substrate [53].

3.3. Phase Analysis

Figure 5 shows the XRD pattern of the MAO/APS biological composite coating. The XRD pattern shows that compared with the original HA powder diffraction peaks, the intensity of the HA coating diffraction peaks is weaker, indicating that the HA after spraying has a reduced crystallinity and the decomposition phase $Ca_3(PO_4)_2$ is produced in the coating. During the spraying, hydroxyapatite will undergo different degrees of dehydroxylation reaction to produce OHA (Oxyhdroxyapatite, $Ca_{10}(PO_4)_6(OH)_{0.5}O_{0.75}$), which contains a small number of hydroxyl groups and OA (Oxyappatite, $Ca_{10}(PO_4)_6O$), while OHA and OA can be transformed to HA when heated in a water vapor environment. In addition, the amorphous calcium phosphate (ACP) in the coating is mainly caused by OHA during the quenching process. When the temperature is higher than 1050 °C, hydroxyapatite mainly undergoes the following decomposition reactions [54]:

$$Ca_{10}(PO_4)_6(OH)_2 \rightarrow CaO + Ca_3(PO_4)_2 + H_2O \tag{1}$$

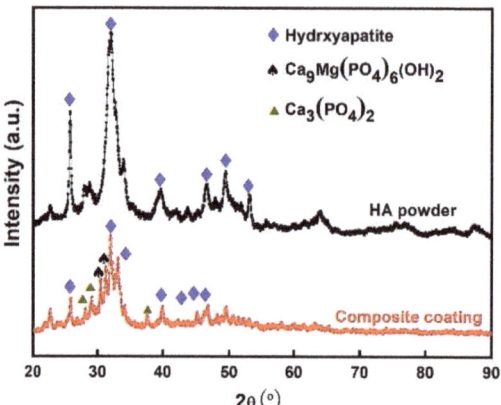

Figure 5. XRD patterns of composite coating and original HA powder.

The generation of ACP in the decomposition products has been reported and different opinions have been raised. Gross et al. [55] believed that the generation of ACP in the thermal spraying process is caused by HA powder. The dehydroxylated OHA in the coating is produced during the rapid cooling process. This statement mainly implies that three main factors are affecting the production of ACP: one is the degree of dehydroxylation of the HA particles during spraying; the other is the impact of the particles on the surface of the substrate; the third is the temperature of the substrate surface. Weng et al. [56] believe that the composition of the amorphous phase is mainly OA, and it is claimed that due to the high temperature during the spraying process, the hydroxyl group of HA particles is gradually lost, and finally, OA forms.

$Ca_9Mg(PO_4)_6(OH)_2$, namely, MgHA, is presented in the composite coating, and its formation process may be related to the crystal structure of HA itself. The HA crystal structure belongs to the hexagonal crystal system, and its lattice constants are a = b = 0.9432 nm, c = 0.9875 nm [57], as shown in Figure 6. The main component of the MAO coating is MgO, and magnesium oxide is an ionic compound. During the plasma spraying process, the instantaneous temperature is high. When the plasma jet contacts the MAO coating, free Mg^{2+} is easily generated, which easily enters the HA crystal. The Ca^{2+} vacancy formed at high temperature forms MgHA. Some cytotoxicity experiments have proved that MgHA has better biocompatibility, and animal experiments have also proved that compared with pure HA, MgHA has better osteoconductivity and bone resorption [58].

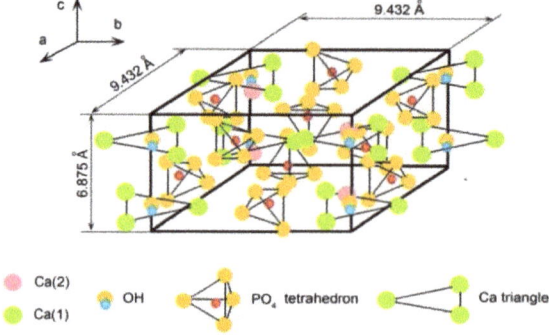

Figure 6. Diagrammatic sketch of the crystal structure of HA.

3.4. Degradation Behavior of MAO/APS Composite Coating in Simulated Body Fluid

3.4.1. pH Changes of Simulated Body Fluids and Weight Loss of Samples

The degradation rate of magnesium alloy in the human body is too fast, which will cause the generation of hydrogen. The reaction Equations are shown in (2) and (3). The hydrogen evolution reaction will lead to an increase in the pH value around the bone tissue, which will seriously affect the living environment of bone cells [59]. Therefore, detecting the pH change of the simulated body fluid during the immersion process can reflect the hydrogen evolution rate of the magnesium alloy. Figure 7 shows the pH of SBF as a function of immersion time. It can be seen from the figure that the pH values of the SBF immersed with the bare sample and the MAO sample are always higher than those of the MAO/APS sample during immersion for 20 days.

$$Mg(s) \rightarrow Mg^{2+} + 2e^- \tag{2}$$

$$2H_2O(l) + 2e^- \rightarrow 2OH^- + H_2(g) \tag{3}$$

$$Mg^{2+} + 2OH^- \rightarrow Mg(OH)_2(s) \tag{4}$$

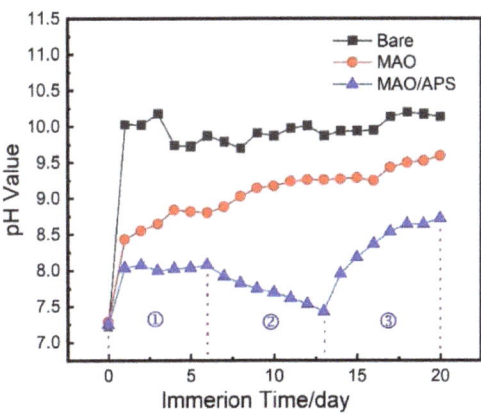

Figure 7. pH values of the SBF during immersion test.

The early stage of immersion is 0–6 days, which is tagged with 1 in Figure 7; middle-term of immersion is 7–13 days, which is tagged with 2 in Figure 7; late-term immersion is 14–20 days, which is tagged with 3 in Figure 7.

In the early stage of immersion, the pH value of the SBF immersed with the bare sample increases rapidly from 7.2 to 10.0, and the pH value is about 9.6 on the fourth day. From this point onwards, the pH value is stable. This shows that bare magnesium alloys are easily corroded in the SBF, and the main reactions that take place during this period are shown in Equations (2)–(4) [60].

Compared with the bare sample, the pH value of the SBF during the coating immersion process after the MAO treatment is slighter. Due to the low corrosion reaction rate of the MAO sample, the pH value of the SBF at the initial stage of immersion slowly increases [60].

For the composite coatings, the pH value during the immersion process is always at a relatively low level, and the change over time shows a "V" shape. The change process consists of three parts. As shown in the figure, the pH value of the SBF solution in the first stage is at a stable stage; it is basically unchanged. The second stage has a certain decrease. After 15 days, it enters the third stage, and the pH value of SBF begins to rise gradually. At the initial stage of immersion, the composite coating may only dissolve part of the amorphous phase in the simulated body fluid. Since the amorphous structure in the local area of the composite coating cannot be fully crystallized, it will dissolve faster during the immersion process [17,61,62]. Therefore, in the initial immersion process, the coating mainly locally dissolves, and the pH value of SBF is almost unchanged. The dissolution of the amorphous structure produces the Ca^{2+} and PO_4^{3-} in the SBF solution. As time goes by, calcium and phosphorus compounds may be deposited on the local area of the composite coating. During the experiment, it is found that the surface of the coating is white. The production of corrosion products consumes OH– in the SBF during this process; hence, the pH value of the SBF drops. With the prolongation of the immersion time (the third stage), the surface of the composite coating has undergone local dissolution and deposition, and the coating becomes loose. At this moment, the corrosive medium in the SBF enters the coating, causing a corrosion reaction; hence, the pH value of the system begins to gradually rise.

Figure 8 shows the weight loss of the sample during immersion in simulated body fluid. It can be seen from Figure 8 that the weight loss rates of bare samples, MAO samples and MAO/APS samples decrease in the SBF. For bare samples and MAO samples, as the immersion time increases, the substrate or coating is gradually corroded and degraded; hence, the weight loss rates of the samples gradually increase. As for the MAO/APS sample, there is the dissolution of the amorphous phase and the deposition of corrosion products during the immersion. Therefore, the weight loss rate of the MAO/APS sample gradually increases in the initial stage.

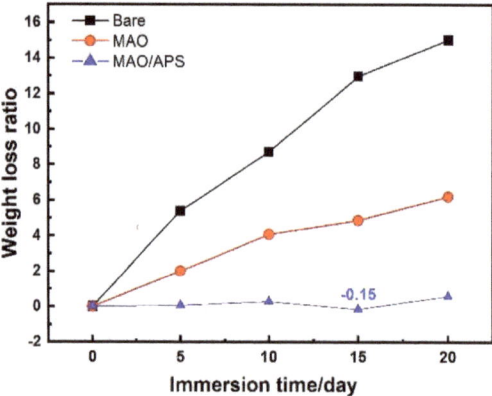

Figure 8. Weight loss rate of samples over time.

When the immersion was carried out for 15 days, the weight loss ratio of the sample showed a negative value for the first time. It could be indicated that the deposited amount of the corrosion products on the surface was greater than the dissolved amount of the coating. The degradation process of the composite coating in the immersion test becomes more complicated.

The surface coating of HA coating is beneficial in terms of inducing the accumulation of corrosion products containing Ca–P in the solution during the immersion process, the inner coating of MAO and the outer coating of HA, under the joint action of the coating. For this reason, the substrate is significantly protected from corrosion by corrosive media. Comparing the bare sample and the MAO sample, it can be seen that the corrosion resistance of the MAO/APS sample is greatly improved. During the 20-day simulated body fluid immersion process, the total weight loss rate is only 0.58%.

3.4.1.1. Changes in the Macroscopic Corrosion Morphology of the Sample

Figure 9 shows the comparison of the morphology of the bare sample, MAO sample, and MAO/APS sample after immersion in SBF solution for different numbers of days. It can be seen that during the degradation process of the bare sample in the SBF solution, uniform corrosion and local corrosion take place, and the local corrosion always takes precedence over the uniform corrosion (Figure 9a). With the extension of the immersion time, the corrosive medium gradually enters from the corrosion pit and/or from the corners to the center. On the 20th day, the corner area begins to fall off, and the surface is extremely rough.

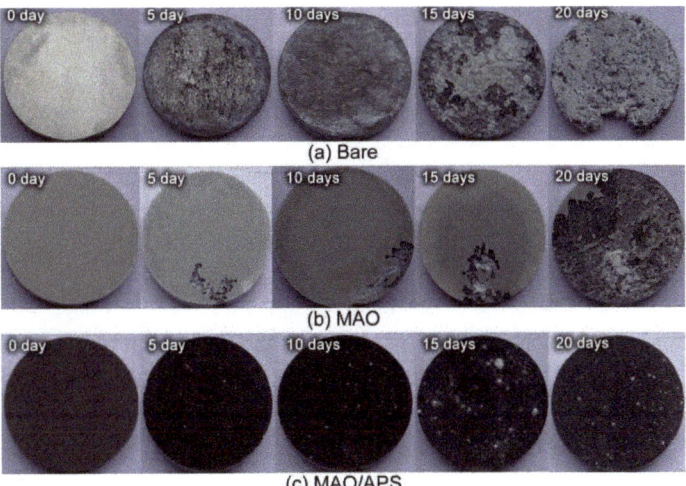

Figure 9. Macro morphology of samples after immersion: (**a**) bare; (**b**) MAO; (**c**) MAO/APS.

The degradation of the MAO sample in the SBF solution is stable (Figure 9b). In the long-term immersion process, the corrosion mainly takes place at the corners of the sample and finally leads to the peeling off of the MAO coating. On the 20th day, a small amount of the MAO coating on the surface of the sample is still present.

For the MAO/APS sample, there was no obvious change during the 20-day immersion process. On the 20th day, the sample still maintained the original integrity, and no obvious corrosion degradation on the surface is found.

Figure 10 shows the surface micro-topography of MAO/APS samples with immersion time. The surface of the MAO/APS sample before immersion is rough and uneven. At the initial stage of immersion, the surface of the sample is formed by a considerable number of ellipsoidal particles. As the immersion time is prolonged, the phenomenon of the accumulation of clumps can be observed in local areas. The cracks gradually spread, which divide the surface into a large number of irregular tiny polygonal shapes. Because the HA coating on the surface of the composite coating is not completely crystallized, it contains the amorphous structure and decomposition phases produced during the spraying. The decomposition phases that follow the immersion time are speculated to be hydroxy-carbonate-apatite (HCA) [63]. These decomposition products degrade in SBF solution. The rate is much higher than the degradation rate of HA crystals [64]. Therefore, in the initial stage of immersion, the ACP and decomposition phase in the coating preferentially dissolve. Due to the production of ACP and decomposition phases, supersaturated Ca^{2+} and PO_4^{3+} are produced in the SBF solution, and the deposition of Ca–P compounds is prone to take place with the extension of the immersion time, as shown in Figure 10c.

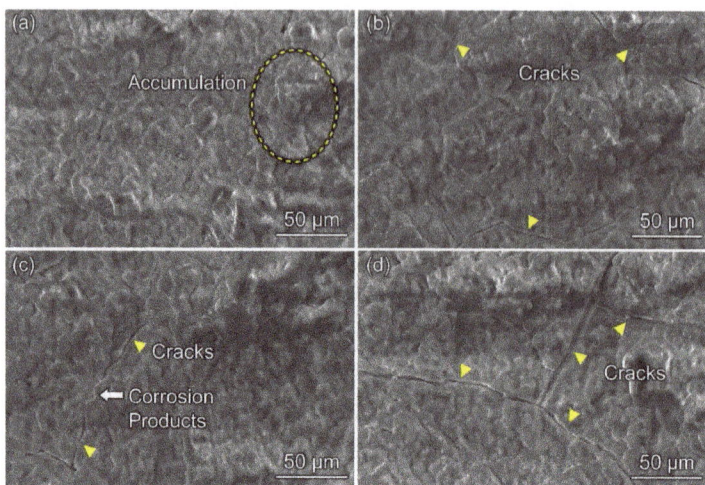

Figure 10. Morphology of samples after immersion: (**a**) 5 days; (**b**) 10 days; (**c**) 15 days; (**d**) 20 days.

In Figure 10b–d, the existence of cracks (yellow dotted line position) can be clearly observed. As the immersion time increases, the cracks gradually spread from scattered small shapes to continuous thick and deep shapes. Li et al. [65–67] reported that cracks are produced as a result of the volume shrinkage effect caused by the loss of moisture in the coating during the drying process. Apparently, such a statement cannot fully explain this phenomenon. The occurrence of the spreading of cracks may be related to the phenomenon of local corrosion [68]. Due to the inhomogeneity of the surface HA coating composition, the degradation rate of the local area is inconsistent; hence, corrosion is easily introduced at the interface position. As the immersion time increases, the corrosion gradually spreads, causing the cracks to expand. The coating is gradually divided into small pieces when the cracks are connected, which eventually causes the coating to fail.

3.4.1.2. Corrosion Products on the Surface of Coatings

Table 7 shows the changes of element content on the surface of the coating under different immersion times. It can be found that the contents of Ca and P elements increase with the immersion time, which reaches the maximum value after 15 days of immersion. The weight loss ratios of the samples can also show that the MAO/APS sample has a slight increase in weight on the 15th day, indicating that the amount of Ca-P compound deposited is greater than the local dissolution of the coating. In the later stage of immersion, Si is presented in the coating, and this element is a constituent element of the MAO coating, indicating that the composite coating, at this time, has begun to gradually fall off the HA coating.

Table 7. Variation of elements contents of the coating over immersion time (at.%).

	C	O	Mg	Ca	P	Na
5 days	20.06 ± 1.34	56.29 ± 0.61	0.80 ± 0.19	12.90 ± 0.11	9.41 ± 0.04	0.32 ± 0.10
10 days	15.36 ± 1.29	59.20 ± 0.58	1.11 ± 0.20	13.90 ± 0.12	9.82 ± 0.03	0.48 ± 0.11
15 days	15.46 ± 1.30	55.02 ± 0.60	0.80 ± 0.24	17.31 ± 0.09	10.28 ± 0.05	0.58 ± 0.12
20 days	16.25 ± 1.35	59.10 ± 0.62	0.99 ± 0.22	13.01 ± 0.11	9.32 ± 0.03	0.40 ± 0.11

It can be seen from XRD in Figure 11 that under different immersion times, the product on the surface of the coating is mainly HA. On one hand, the initial dissolution of the amorphous phase and the decomposition phase promotes the induction of HA. In the initial stage, the surface of the implant forms certain osseointegration with bone tissue, which has good biological characteristics. On the other hand, there is no obvious compound of C in the XRD pattern; hence, it is speculated that the C element may be due to the formation of CO_3^{2-} containing calcium-deficient hydroxyapatite (hydroxy-carbonate-apatite, HCA). HCA is similar to the human body's natural bone composition, so it is also called "hydroxyapatite-like". The main reason for its formation is that the CO_3^{2-} ions in the simulated body fluid replace the hydroxyl or phosphate in the HA crystal lattice, forming the A-type, B-type or AB-type substituted hydroxyapatite carbonate [69].

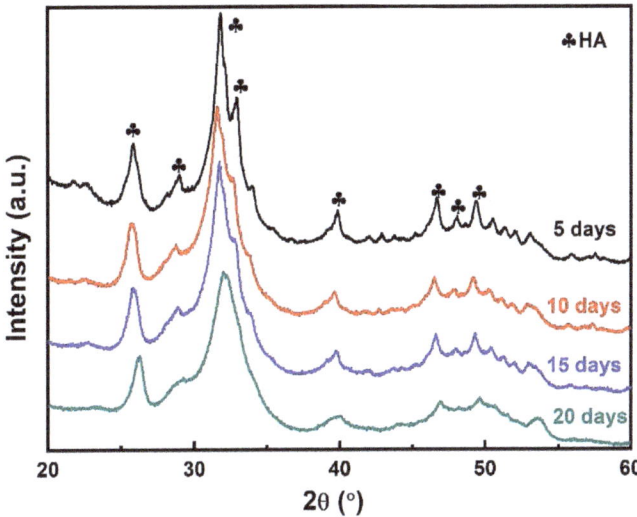

Figure 11. XRD patterns of MAO/APS coatings immersed for different numbers of days.

Figure 12 shows the Nyquist diagrams of the immersed samples; it can be found that the radius of the capacitive reactance loop and the impedance value at the low-frequency end of the MAO/APS, immersed for the tenth day, are at their maximum, and that the charge transfer resistance R_{ct}, during the corrosion reaction, is the largest. At this moment, the sample shows good corrosion resistance. The reason may be that when the sample is immersed for the tenth day, the deposition of Ca–P compound on the surface is preferentially formed at the defect, which has a good inhibitory effect on the further penetration of the corrosive medium into the coating. It can also be seen from the fitted data that the resistance R_{cp}, representing the corrosion product, exhibits a maximum value at this time. When immersed for 20 days, the R_{ct} was the smallest, which indicates that the deposition rate of Ca–P compound on the surface was significantly lower than the degradation rate of the coating. It can be seen from Figure 10 that the coating was dissolved after 15 days of immersion. Meanwhile, a certain amount of active Ca and P elements could be produced in the SBF solution during the degradation process. Hence, the biological activity of the composite coating presents a gradient change during the SBF solution immersion process, which is different from the immersion process of the MAO coating. To sum up, good corrosion resistance and biocompatibility can be received from the MAO/APS composite bio-coating.

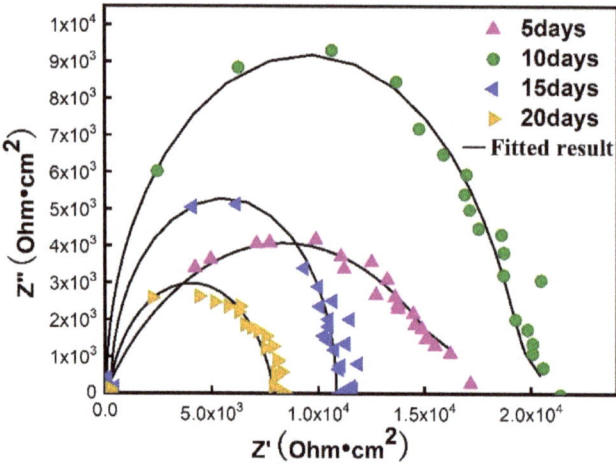

Figure 12. Nyquist plots of immersion samples in SBF.

According to the degradation model of AZ31 magnesium alloy in SBF, proposed by Gu et al. [70] and Khalajabadi et al. [71] to describe the degradation behavior of Mg/HA/MgO nanostructured materials prepared by powder metallurgy in SBF, combined with MAO/APS sample immersion process phenomena and analysis test results, the degradation model, and the equivalent circuit, are proposed. These are considered in accordance with the MAO/APS composite coating in SBF, which are mainly divided into the following three stages, and Table 8 lists the fitted result of EIS:

Table 8. EIS data of MAO/APS samples after immersion in SBF for various durations.

Immersion Time	Rs	R_{cp}	R_h	R_m	R_{ct}
5 days	19.03	-	1.1670×10^4	2861	4.934×10^5
10 days	52.20	3053	1.6200×10^4	3748	5.030×10^5
15 days	0.2342	272.1	271.9	101.1	4724
20 days	149.1	0.07597	54.79	472.4	3409

1. Early stage of immersion

During the initial immersion process of the composite membrane, a dissolution process mainly occurs. The Ca and P ions produced by the dissolution accumulate on the surface in a short time. These free Ca and P ions are easily absorbed by the damaged bone tissue to promote its growth. Therefore, part of the amorphous structure and decomposition phase existing in the HA coating is conducive, to a certain extent, to the recovery and growth of damaged bones. The equivalent circuit is shown in Figure 13a, which mainly includes three pairs of Constant Phase Element (CPE), where Rs, Rh, and Rm represent solution resistance, HA coating resistance, and MAO coating resistance, respectively.

2. Middle period of immersion

On the one hand, due to the inconsistency of the decomposition phase (HCA), amorphous structure and crystalline HA degradation rate at the initial stage of immersion, the microcracks generated at the interface between each other expand during the middle stage of immersion, as shown in Figure 13b. On the other hand, the large amount of calcium and phosphorus ions produced by the rapid degradation of the amorphous structure and the decomposition phase produce supersaturation in the local area of the composite coating, especially at the surface defects of the composite coating. These ions react with the ions in the SBF. Ca–P compounds are formed at the coating defects and deposited on the surface. The possible reactions are represented in Equations (5) and (6). It should be noted here that

HA may be crystalline hydroxyphosphorus. Greystone may also be hydroxyapatite-like, but both have good biological activity. On the other hand, as shown in Figure 13, new HA formation occurs, at this time, on the surface of the membrane, as well as the continuous expansion of cracks, and the continuous penetration of simulated body fluids into the membrane. This phenomenon is conducive to the recovery of the implant in the bone injury tissue during the formation of osseointegration. Due to the formation of the new phase of HA, the corrosion product resistance R_{cp} appears in the equivalent circuit.

$$10Ca^{2+} + 8OH^- + 6HPO_4^{2-} \rightarrow Ca_{10}(PO_4)_6(OH)_2 + 6H_2O \quad (5)$$

$$10Ca^{2+} + 8OH^- + 2OH^- \rightarrow Ca_{10}(PO_4)_6(OH)_2 \quad (6)$$

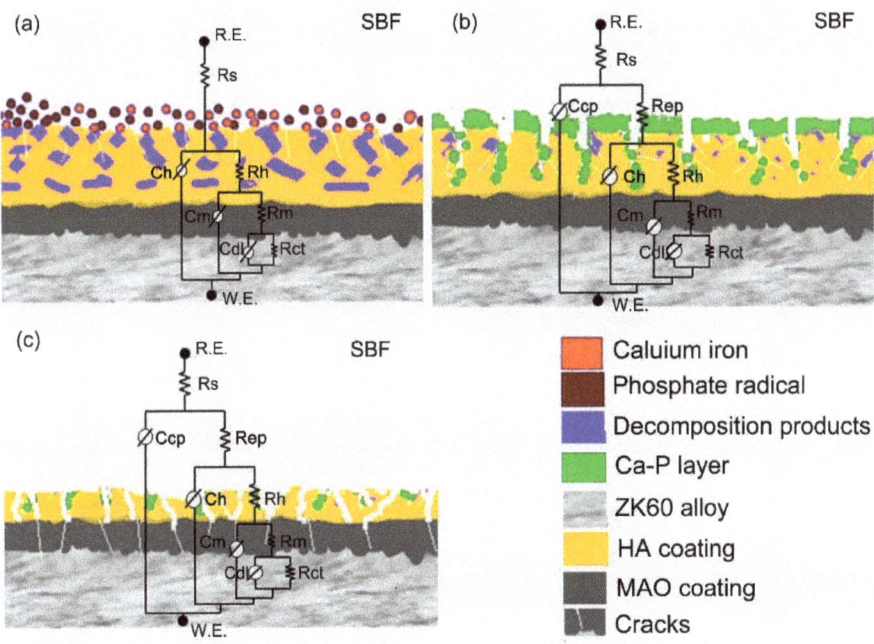

Figure 13. A physical model of the degradation process of MAO/APS samples in the SBF solution with equivalent circuits for fitting the EIS data: (**a**) initial-term of immersion; (**b**) middle-term of immersion; (**c**) late-term immersion.

3. Late time of immersion

With the progression of the immersion test, the existence of cracks and micropores causes the surface HA coating to further fail and gradually fall off. The corrosive medium further penetrates the interior and contacts the substrate. Due to the existence of the MAO coating, the service cycle of the implant during bone healing is prolonged. During the continuous immersion process, the MAO coating will gradually degrade, participate in the normal metabolism of bone growth, and be excreted from the body.

4. Conclusions

In this work, ZK60 Mg alloy was micro-arc oxidized (MAO) before undergoing the atmospheric plasma spraying process. After the examinations of microstructures, bound strength, phase constituents, immersion tests, and electrochemical measurements, some key points can be concluded as follows:

1. The surface particles of the composite coating have a higher degree of melting and flattening, and the coating is more uniform. The thickness of the MAO coating is 36.14 μm, the thickness of the HA sprayed coating is 59.59 μm, and the average bonding strength is 20.2 MPa. Fracture analysis shows that the binding mode between the composite coatings is mechanical occlusion.
2. The pH value of the composite coating sample during the SBF immersion process showed a "V" shape with time and was at the lowest level of the three, showing the lowest degradation rate, which can effectively protect the magnesium alloy substrate.
3. The decomposition phase and amorphous structure in the HA coating on the surface of the composite coating have a faster degradation rate in the SBF solution, which effectively promotes the deposition of Ca–P compounds on the surface of the composite coating. In addition, elemental analysis and XRD tests show that the Ca–P compound is a mixture of hydroxyapatite and carbonated hydroxyapatite, which is beneficial in terms of promoting the formation of the bone bond between the implant and the damaged bone tissue at the initial stage.
4. The degradation model of the composite biological coating in SBF is proposed in accordance with this experiment, and the degradation mechanism of the composite coating in the SBF, at different stages, is explained in detail. During the initial immersion process of the composite membrane, a dissolution process mainly occurs. In the middle of immersion, due to the formation of new phase of HA, corrosion product resistance, R_{cp}, appears in the equivalent circuit. Furthermore, during the later stages of immersion, the MAO coating will gradually degrade, participate in the normal metabolism of bone growth, and be excreted from the body.

Author Contributions: Z.W., L.S. and S.L. conceived and designed the experiments. F.Y., Z.Z. and W.L. performed the experiments. Z.W., Q.Z. and J.P. analyzed the data. F.Y. and L.C. wrote the manuscript. All authors have read and agreed to the published version of the manuscript.

Funding: The authors would like to acknowledge the financial support provided by the Jiangsu Province six talent peaks project (Grant No. XCL-117), the Natural Science Foundation of Jiangsu (Grant No. BK20201456), the Open Foundation of Guangxi Key Laboratory of Processing for Nonferrous Metals and Featured Materials, Guangxi University (Grant No. 2020GXYSOF01), and the Key Research and Development Program of Shaanxi (Grant No. 2020GY-251).

Institutional Review Board Statement: Not applicable.

Informed Consent Statement: Not applicable.

Data Availability Statement: The raw data supporting the conclusions of this article will be made available by the authors, without undue reservation, to any qualified researcher.

Acknowledgments: We would like to acknowledge the Instrumental Analysis Center of Jiangsu University of Science and Technology for providing technological support.

Conflicts of Interest: The authors declare no conflict of interest.

References

1. Zhang, Z.Q.; Wang, L.; Zeng, M.Q.; Zeng, R.; Mathan, B.K.; Lin, C.G. Biodegradation behavior of micro-arc oxidation coating on magnesium alloy-from a protein perspective. *Bioact. Mater.* **2020**, *5*, 398–409. [CrossRef]
2. Zhang, L.C.; Chen, L.Y.; Wang, L. Surface modification of titanium and titanium alloys: Technologies, developments, and future interests. *Adv. Eng. Mater.* **2020**, *22*, 1901258. [CrossRef]
3. Zhang, L.C.; Chen, L.Y. A review on biomedical titanium alloys: Recent progress and prospect. *Adv. Eng. Mater.* **2019**, *21*, 1801215. [CrossRef]
4. Zhang, L.C.; Attar, H.; Calin, M.; Eckert, J. Review on manufacture by selective laser melting and properties of titanium based materials for biomedical applications. *Mater. Technol.* **2016**, *31*, 66–76. [CrossRef]
5. Dai, N.; Zhang, J.; Chen, Y.; Zhang, L.C. Heat treatment degrading the corrosion resistance of selective laser melted Ti-6Al-4V alloy. *J. Electrochem. Soc.* **2017**, *164*, C428. [CrossRef]
6. Chen, L.Y.; Cui, Y.W.; Zhang, L.C. Recent development in beta titanium alloys for biomedical applications. *Metals* **2020**, *10*, 1139. [CrossRef]

7. Hornberger, H.; Virtanen, S.; Boccaccini, A.R. Biomedical coatings on magnesium alloys—A review. *Acta Biomater.* **2012**, *8*, 2442–2455. [CrossRef]
8. Tsn, S.N.; Park, I.S.; Lee, M.H. Strategies to improve the corrosion resistance of microarc oxidation (MAO) coated magnesium alloys for degradable implants: Prospects and challenges. *Prog. Mater. Sci.* **2014**, *60*, 1–71. [CrossRef]
9. Chen, J.; Lin, W.; Liang, S.; Zou, L.; Wang, C.; Wang, B.; Yan, M.; Cui, X. Effect of alloy cations on corrosion resistance of LDH/MAO coating on magnesium alloy. *Appl. Surf. Sci.* **2019**, *463*, 535–544. [CrossRef]
10. Liu, S.; Han, S.; Zhang, L.; Chen, L.Y.; Wang, L.; Zhang, L.; Tang, Y.; Liu, J.; Tang, H.; Zhang, L.C. Strengthening mechanism and micropillar analysis of high-strength NiTi–Nb eutectic-type alloy prepared by laser powder bed fusion. *Compos. Part. B Eng.* **2020**, *200*, 108358. [CrossRef]
11. Liu, Y.; Zheng, Y.; Chen, X.H.; Yang, J.A.; Pan, H.; Chen, D.; Wang, L.; Zhang, J.; Zhu, D.; Wu, S.; et al. Fundamental theory of biodegradable metals—definition, criteria, and design. *Adv. Funct. Mater.* **2019**, *29*, 1805402. [CrossRef]
12. Rabadia, C.D.; Liu, Y.J.; Zhao, C.H.; Wang, J.C.; Jawed, S.F.; Wang, L.Q.; Chen, L.Y.; Sun, H.; Zhang, L.C. Improved trade-off between strength and plasticity in titanium based metastable beta type Ti-Zr-Fe-Sn alloys. *Mater. Sci. Eng. A* **2019**, *766*, 138340. [CrossRef]
13. Zhang, L.; Chen, L.Y.; Zhao, C.; Liu, Y.; Zhang, L.C. Calculation of oxygen diffusion coefficients in oxide films formed on low-temperature annealed Zr alloys and their related corrosion behavior. *Metals* **2019**, *9*, 850. [CrossRef]
14. Chen, L.Y.; Shen, P.; Zhang, L.; Lu, S.; Chai, L.; Yang, Z.; Zhang, L.C. Corrosion behavior of non-equilibrium Zr-Sn-Nb-Fe-Cu-O alloys in high-temperature 0.01 M LiOH aqueous solution and degradation of the surface oxide films. *Corros. Sci.* **2018**, *136*, 221–230. [CrossRef]
15. Pan, H.; Kang, R.; Li, J.; Xie, H.; Zeng, Z.; Huang, Q.; Yang, C.; Ren, Y.; Qin, G. Mechanistic investigation of a low-alloy Mg–Ca-based extrusion alloy with high strength–ductility synergy. *Acta Mater.* **2020**, *186*, 278–290. [CrossRef]
16. Mao, L.; Shen, L.; Niu, J.; Zhang, J.; Ding, W.; Wu, Y.; Fan, R.; Yuan, G. Nanophasic biodegradation enhances the durability and biocompatibility of magnesium alloys for the next-generation vascular stents. *Nanoscale* **2013**, *5*, 9517–9522. [CrossRef]
17. Rúa, J.M.; Zuleta, A.A.; Ramírez, J.; Fernández-Morales, P. Micro-arc oxidation coating on porous magnesium foam and its potential biomedical applications. *Surf. Coat. Technol.* **2019**, *360*, 213–221. [CrossRef]
18. McEntire, B.J.; Bal, B.S.; Rahaman, M.N.; Chevalier, J.; Pezzotti, G. Ceramics and ceramic coatings in orthopaedics. *J. Eur. Ceram. Soc.* **2015**, *35*, 4327–4369. [CrossRef]
19. Liu, P.; Wang, J.M.; Yu, X.T.; Chen, X.B.; Li, S.Q.; Chen, D.C.; Guan, S.K.; Zeng, R.C.; Cui, L.Y. Corrosion resistance of bioinspired DNA-induced Ca–P coating on biodegradable magnesium alloy. *J. Magnes. Alloys* **2019**, *7*, 144–154. [CrossRef]
20. Durdu, S.; Aktug, S.L.; Korkmaz, K.; Yalcin, E.; Aktas, S. Fabrication, characterization and in vitro properties of silver-incorporated TiO$_2$ coatings on titanium by thermal evaporation and micro-arc oxidation. *Surf. Coat. Technol.* **2018**, *352*, 600–608. [CrossRef]
21. Li, L.H.; Kong, Y.M.; Kim, H.W.; Kim, Y.W.; Kim, H.E.; Heo, S.J.; Koak, J.Y. Improved biological performance of Ti implants due to surface modification by micro-arc oxidation. *Biomaterials* **2004**, *25*, 2867–2875. [CrossRef]
22. Xu, C.; Chen, L.Y.; Zheng, C.B.; Zhang, H.Y.; Zhao, C.H.; Wang, Z.X.; Lu, S.; Zhang, J.W.; Zhang, L.C. Improved wear and corrosion resistance of microarc oxidation coatings on Ti-6Al-4V alloy with ultrasonic assistance for potential biomedical applications. *Adv. Eng. Mater.* **2021**, *23*, 2001433. [CrossRef]
23. Asgari, M.; Aliofkhazraei, M.; Darband, G.B.; Rouhaghdam, A.S. How nanoparticles and submicron particles adsorb inside coating during plasma electrolytic oxidation of magnesium? *Surf. Coat. Technol.* **2020**, *383*, 125252. [CrossRef]
24. Li, Y.; Li, H.; Xiong, Q.; Wu, X.; Zhou, J.; Wu, J.; Wu, X.; Qin, W. Multipurpose surface functionalization on AZ31 magnesium alloys by atomic layer deposition: Tailoring the corrosion resistance and electrical performance. *Nanoscale* **2017**, *9*, 8591–8599. [CrossRef]
25. Rokosz, K.; Hryniewicz, T.; Raaen, S.; Gaiaschi, S.; Chapon, P.; Malorny, W.; Matýsek, D.; Dudek, Ł.; Pietrzak, K. *Characterization of Porous Phosphate Coatings Created on CP Titanium Grade 2 Enriched with Calcium, Magnesium, Zinc and Copper by Plasma Electrolytic Oxidation*; Preprints.org: Basel, Switzerland, 2018.
26. Muhaffel, F.; Cimenoglu, H. Development of corrosion and wear resistant micro-arc oxidation coating on a magnesium alloy. *Surf. Coat. Technol.* **2019**, *357*, 822–832. [CrossRef]
27. Dong, Y.T.; Liu, Z.Y.; Ma, G.F. The research progress on micro-arc oxidation of aluminum alloy. *IOP Conf. Ser. Mater. Sci. Eng.* **2020**, *729*, 012055. [CrossRef]
28. Dou, J.; Chen, Y.; Yu, H.; Chen, C. Research status of magnesium alloys by micro-arc oxidation: A review. *Surf. Eng.* **2017**, *33*, 731–738. [CrossRef]
29. Zhang, L.; Zhang, J.; Chen, C.F.; Gu, Y. Advances in microarc oxidation coated AZ31 Mg alloys for biomedical applications. *Corros. Sci.* **2015**, *91*, 7–28. [CrossRef]
30. Li, X.; Chen, M.; Wang, P.; Yao, Y.; Han, X.; Liang, J.; Jiang, Q.; Sun, Y.; Fan, Y.; Zhang, X. A highly interweaved HA-SS-nHAp/collagen hybrid fibering hydrogel enhances osteoinductivity and mineralization. *Nanoscale* **2020**, *12*, 12869–12882. [CrossRef]
31. Habibovic, P.; Barrère, F.; Van Blitterswijk, C.A.; de Groot, K.; Layrolle, P. Biomimetic hydroxyapatite coating on metal implants. *J. Am. Ceram. Soc.* **2002**, *85*, 517–522. [CrossRef]

32. Feng, C.; Zhang, K.; He, R.; Ding, G.; Xia, M.; Jin, X.; Xie, C. Additive manufacturing of hydroxyapatite bioceramic scaffolds: Dispersion, digital light processing, sintering, mechanical properties, and biocompatibility. *J. Adv. Ceram.* **2020**, *9*, 360–373. [CrossRef]
33. Bose, S.; Tarafder, S.; Bandyopadhyay, A. 7—Hydroxyapatite coatings for metallic implants. In *Hydroxyapatite (Hap) for Biomedical Applications*; Woodhead Publishing: Cambridge, UK, 2015; pp. 143–157.
34. Li, C.; Yao, X.; Zhang, X.; Huang, X.; Hang, R. Corrosion behavior and cytocompatibility of nanostructured hydroxyapatite hydrothermally grown on porous MgO coatings with different P contents on magnesium. *Mater. Lett.* **2020**, *264*, 127136. [CrossRef]
35. Ferraz, M.; Monteiro, F.; Manuel, C. Hydroxyapatite nanoparticles: A review of preparation methodologies. *J. Appl. Biomater. Biomech.* **2003**, *2*, 74–80.
36. Fauchais, P. Understanding plasma spraying. *J. Phys. D: Appl. Phys.* **2004**, *37*, R86–R108. [CrossRef]
37. Chen, L.Y.; Xu, T.; Wang, H.; Sang, P.; Lu, S.; Wang, Z.X.; Chen, S.; Zhang, L.C. Phase interaction induced texture in a plasma sprayed-remelted NiCrBSi coating during solidification: An electron backscatter diffraction study. *Surf. Coat. Technol.* **2019**, *358*, 467–480. [CrossRef]
38. Sang, P.; Zhao, C.; Wang, Z.X.; Wang, H.; Lu, S.; Song, D.; Xu, J.H.; Zhang, L. Particle size-dependent microstructure, hardness and electrochemical corrosion behavior of atmospheric plasma sprayed NiCrBSi coatings. *Metals* **2019**, *9*, 1342. [CrossRef]
39. Sha, J.; Liu, Y.T.; Yao, Z.J.; Lu, S.; Wang, Z.X.; Zang, Q.H.; Mao, S.H.; Zhang, L. Phase transformation-induced improvement in hardness and high-temperature wear resistance of plasma-sprayed and remelted NiCrBSi/WC coatings. *Metals* **2020**, *10*, 1688. [CrossRef]
40. Chen, L.Y.; Wang, H.; Zhao, C.; Lu, S.; Wang, Z.X.; Sha, J.; Chen, S.; Zhang, L.C. Automatic remelting and enhanced mechanical performance of a plasma sprayed NiCrBSi coating. *Surf. Coat. Technol.* **2019**, *369*, 31–43. [CrossRef]
41. Fan, X.; Liu, Y.; Xu, Z.; Wang, Y.; Zou, B.; Gu, L.; Wang, C.; Chen, X.; Khan, Z.S.; Yang, D.; et al. Preparation and characterization of 8YSZ thermal barrier coatings on rare earth-magnesium alloy. *J. Therm. Spray Technol.* **2011**, *20*, 948–957. [CrossRef]
42. Daroonparvar, M.; Yajid, M.A.M.; Yusof, N.M.; Bakhsheshi-Rad, H.R.; Hamzah, E.; Mardanikivi, T. Deposition of duplex MAO layer/nanostructured titanium dioxide composite coatings on Mg–1%Ca alloy using a combined technique of air plasma spraying and micro arc oxidation. *J. Alloys Compd.* **2015**, *649*, 591–605. [CrossRef]
43. Fan, X.; Xu, J.; Wang, Y.; Ma, H.; Zhao, S.; Zhou, X.; Cao, X. Preparation and corrosion resistance of MAO layer/Yb$_2$SiO$_5$ composite coating on Mg alloy. *Surf. Coat. Technol.* **2014**, *240*, 118–127. [CrossRef]
44. Gu, Y.; Chen, C.F.; Bandopadhyay, S.; Ning, C.; Zhang, Y.; Guo, Y. Corrosion mechanism and model of pulsed DC microarc oxidation treated AZ31 alloy in simulated body fluid. *Appl. Surf. Sci.* **2012**, *258*, 6116–6126. [CrossRef]
45. Gu, Y.; Bandyopadhyay, S.; Chen, C.F.; Ning, C.; Guo, Y. Long-term corrosion inhibition mechanism of microarc oxidation coated AZ31 Mg alloys for biomedical applications. *Mater. Des.* **2013**, *46*, 66–75. [CrossRef]
46. Lin, X.; Yang, X.; Tan, L.; Li, M.; Wang, X.; Zhang, Y.; Yang, K.; Hu, Z.; Qiu, J. In vitro degradation and biocompatibility of a strontium-containing micro-arc oxidation coating on the biodegradable ZK60 magnesium alloy. *Appl. Surf. Sci.* **2014**, *288*, 718–726. [CrossRef]
47. Pan, Y.; He, S.; Wang, D.; Huang, D.; Zheng, T.; Wang, S.; Dong, P.; Chen, C. In vitro degradation and electrochemical corrosion evaluations of microarc oxidized pure Mg, Mg-Ca and Mg-Ca-Zn alloys for biomedical applications. *Mater. Sci. Eng. C* **2015**, *47*, 85–96. [CrossRef]
48. Zhang, J.; Dai, C.; Wei, J.; Wen, Z.; Zhang, S.; Chen, C. Degradable behavior and bioactivity of micro-arc oxidized AZ91D Mg alloy with calcium phosphate/chitosan composite coating in m-SBF. *Colloids Surf. B* **2013**, *111*, 179–187. [CrossRef]
49. Wang, Y.M.; Guo, J.W.; Shao, Z.K.; Zhuang, J.P.; Jin, M.S.; Wu, C.J.; Wei, D.Q.; Zhou, Y. A metasilicate-based ceramic coating formed on magnesium alloy by microarc oxidation and its corrosion in simulated body fluid. *Surf. Coat. Technol.* **2013**, *219*, 8–14. [CrossRef]
50. Du, Q.; Wei, D.; Wang, Y.; Cheng, S.; Liu, S.; Zhou, Y.; Jia, D. The effect of applied voltages on the structure, apatite-inducing ability and antibacterial ability of micro arc oxidation coating formed on titanium surface. *Bioact. Mater.* **2018**, *3*, 426–433. [CrossRef]
51. Tang, H.; Han, Y.; Wu, T.; Tao, W.; Jian, X.; Wu, Y.; Xu, F. Synthesis and properties of hydroxyapatite-containing coating on AZ31 magnesium alloy by micro-arc oxidation. *Appl. Surf. Sci.* **2017**, *400*, 391–404. [CrossRef]
52. Xia, Q.; Li, X.; Yao, Z.; Jiang, Z. Investigations on the thermal control properties and corrosion resistance of MAO coatings prepared on Mg-5Y-7Gd-1Nd-0.5Zr alloy. *Surf. Coat. Technol.* **2021**, *409*, 126874. [CrossRef]
53. Wang, X.; Li, B.; Zhou, L.; Ma, J.; Zhang, X.; Li, H.; Liang, C.; Liu, S.; Wang, H. Influence of surface structures on biocompatibility of TiO$_2$/HA coatings prepared by MAO. *Mater. Chem. Phys.* **2018**, *215*, 339–345. [CrossRef]
54. Vilardell, A.M.; Cinca, N.; Garcia-Giralt, N.; Dosta, S.; Cano, I.G.; Nogués, X.; Guilemany, J.M. In-vitro comparison of hydroxyapatite coatings obtained by cold spray and conventional thermal spray technologies. *Mater. Sci. Eng. C* **2020**, *107*, 110306. [CrossRef]
55. Gross, K.A.; Phillips, M.R. Identification and mapping of the amorphous phase in plasma-sprayed hydroxyapatite coatings using scanning cathodoluminescence microscopy. *J. Mater. Sci. Mater. Med.* **1998**, *9*, 797–802. [CrossRef]
56. Weng, J.; Liu, X.; Zhang, X.; Ma, Z.; Ji, X.; Zyman, Z. Further studies on the plasma-sprayed amorphous phase in hydroxyapatite coatings and its deamorphization. *Biomaterials* **1993**, *14*, 578–582. [CrossRef]

57. Shen, H.Z.; Guo, N.; Zhao, L.; Shen, P. Role of ion substitution and lattice water in the densification of cold-sintered hydroxyapatite. *Scr. Mater.* **2020**, *177*, 141–145. [CrossRef]
58. Xiong, L.; Wang, P.; Hunter, M.N.; Kopittke, P.M. Bioavailability and movement of hydroxyapatite nanoparticles (HA-NPs) applied as a phosphorus fertiliser in soils. *Environ. Sci. Nano* **2018**, *5*, 2888–2898. [CrossRef]
59. Song, J.; Jin, P.; Li, M.; Liu, J.; Wu, D.; Yao, H.; Wang, J. Antibacterial properties and biocompatibility in vivo and vitro of composite coating of pure magnesium ultrasonic micro-arc oxidation phytic acid copper loaded. *J. Mater. Sci. Mater. Med.* **2019**, *30*, 49. [CrossRef]
60. Wang, Z.X.; Xu, L.; Zhang, J.W.; Ye, F.; Lv, W.G.; Xu, C.; Lu, S.; Yang, J. Preparation and degradation behavior of composite bio-coating on ZK60 magnesium alloy using combined micro-arc oxidation and electrophoresis deposition. *Front. Mater.* **2020**, *7*, 190. [CrossRef]
61. Zhang, L.C.; Xu, J.; Ma, E. Consolidation and properties of ball-milled $Ti_{50}Cu_{18}Ni_{22}Al_4Sn_6$ glassy alloy by equal channel angular extrusion. *Mater. Sci. Eng. A* **2006**, *434*, 280–288. [CrossRef]
62. Zhang, L.C.; Shen, Z.Q.; Xu, J. Mechanically milling-induced amorphization in Sn-containing Ti-based multicomponent alloy systems. *Mater. Sci. Eng. A* **2005**, *394*, 204–209. [CrossRef]
63. Landi, E.; Celotti, G.; Logroscino, G.; Tampieri, A. Carbonated hydroxyapatite as bone substitute. *J. Eur. Ceram. Soc.* **2003**, *23*, 2931–2937. [CrossRef]
64. Zheng, Z.; Zhao, M.C.; Tan, L.; Zhao, Y.C.; Xie, B.; Huang, L.; Yin, D.; Yang, K.; Atrens, A. Biodegradation behaviour of hydroxyapatite-containing self-sealing micro-arc-oxidation coating on pure Mg. *Surf. Eng.* **2021**, 1–11. [CrossRef]
65. Li, Y.; Lu, F.; Li, H.; Zhu, W.; Pan, H.; Tan, G.; Lao, Y.; Ning, C.; Ni, G. Corrosion mechanism of micro-arc oxidation treated biocompatible AZ31 magnesium alloy in simulated body fluid. *Prog. Nat. Sci. Mater. Int.* **2014**, *24*, 516–522. [CrossRef]
66. Liu, G.; Tang, S.; Li, D.; Hu, J. Self-adjustment of calcium phosphate coating on micro-arc oxidized magnesium and its influence on the corrosion behaviour in simulated body fluids. *Corros. Sci.* **2014**, *79*, 206–214. [CrossRef]
67. Agarwal, S.; Curtin, J.; Duffy, B.; Jaiswal, S. Biodegradable magnesium alloys for orthopaedic applications: A review on corrosion, biocompatibility and surface modifications. *Mater. Sci. Eng. C* **2016**, *68*, 948–963. [CrossRef] [PubMed]
68. Bocchetta, P.; Chen, L.Y.; Tardelli, J.D.; Reis, A.C.; Almeraya-Calderón, F.; Leo, P. Passive layers and corrosion resistance of biomedical Ti-6Al-4V and β-Ti alloys. *Coatings* **2021**, *11*, 487. [CrossRef]
69. Srivastava, A.; Ahmed, R.; Shah, S. Carbonic acid resistance of hydroxyapatite-containing cement. *SPE Drill. Complet.* **2020**, *35*, 088–099. [CrossRef]
70. Gu, Y.; Bandopadhyay, S.; Chen, C.F.; Guo, Y.; Ning, C. Effect of oxidation time on the corrosion behavior of micro-arc oxidation produced AZ31 magnesium alloys in simulated body fluid. *J. Alloys Compd.* **2012**, *543*, 109–117. [CrossRef]
71. Khalajabadi, S.Z.; Abdul Kadir, M.R.; Izman, S.; Kasiri-Asgarani, M. Microstructural characterization, biocorrosion evaluation and mechanical properties of nanostructured ZnO and Si/ZnO coated Mg/HA/TiO_2/MgO nanocomposites. *Surf. Coat. Technol.* **2015**, *277*, 30–43. [CrossRef]

Article

Preparation and Properties of Multilayer Ca/P Bio-Ceramic Coating by Laser Cladding

Boda Liu [1], Zixin Deng [1] and Defu Liu [1,2,*]

[1] College of Mechanical and Electrical Engineering, Central South University, Changsha 410083, China; lbdcsu@csu.edu.cn (B.L.); zixindeng66@csu.edu.cn (Z.D.)
[2] State Key Laboratory of High Performance Complex Manufacturing, Changsha 410083, China
* Correspondence: liudefu@csu.edu.cn; Tel.: +86-731-88879351

Abstract: In order to enhance the bioactivity and wear resistance of titanium (Ti) and its alloy for use as an implant surface, a multilayer Ca/P (calcium/phosphorus) bio-ceramic coating on a Ti6Al4V alloy surface was designed and prepared by a laser cladding technique, using the mixture of hydroxyapatite (HA) powder and Ti powder as a cladding precursor. The main cladding process parameters were 400 W laser power, 3 mm/s scanning speed, 2 mm spot diameter and 30% lapping rate. When the Ca/P ceramic coating was immersed in simulated body fluid (SBF), ion exchange occurred between the coating and the immersion solution, and hydroxyapatite (HA) was induced and deposited on its surface, which indicated that the Ca/P bio-ceramic coating had good bioactivity. The volume wear of Ca/P ceramic coating was reduced by 43.2% compared with that of Ti6Al4V alloy by the pin-disc wear test, which indicated that the Ca/P bio-ceramic coating had better wear resistance.

Keywords: laser cladding; Ca/P bio-ceramic coating; biocompatibility; bioactivity; wear resistance

1. Introduction

With the aging of the population and increase in joint injuries caused by traffic accidents, the demand for artificial joint replacements is growing. Titanium and its alloys, due to their excellent biocompatibility, biomechanical properties and corrosion resistance, have become the preferred materials for artificial joints [1–3]. However, these compounds are biologically inert, and have poor bone conductivity. As a result, the stem of a Ti alloy artificial joint cannot form an osseous bond with bone tissue, and long-term use in the body will cause aseptic loosening [4–6]. In addition, under the wear and corrosion of the fretting environment in vivo, Ti alloy artificial joint stems are prone to producing wear debris and metal ions, leading to the expression of bone resorption in osteoclasts and shortening of service life [7,8]. On the other hand, Ca/P bioactive materials, such as bio-glass (S520) [9], hydroxyapatite (HA) [10] and tri-calcium phosphate (TCP, $Ca_3(PO_4)_2$) [11], are regarded as attractive bone substitute materials owing to their similarity to bone apatite and biocompatibility, due to their ability to induce HA deposition in vivo to form a stable osseous bond with natural bone to achieve biological fixation [12,13]. However, the defects of these bioactive materials, such as their high degree of brittleness, low tensile strength, and poor wear resistance, limit their application in body bearing sites, such as hip and knee joints [14–16]. Therefore, combining the biological properties of bioactive materials and the mechanical properties of Ti alloys has become a research hotspot.

Some scholars have used surface modification techniques such as ion implantation [17], pulsed laser deposition [18], sol-gel [19], magnetron sputtering [20], and plasma spraying [21] to fabricate bio-ceramic coatings on the surface of Ti alloys. However, these techniques have shortcomings, such as poor interface bonding strength between substrate and coating, insufficient coating thickness, and insufficient biological activity. However, laser cladding technology [22] has some benefits such as rapid melting and solidification and good controllability, which can be used to prepare coatings with high bonding strength,

controllable thickness, and suitable physicochemical properties. Therefore, it is a very promising technique for the preparation of bio-ceramic coatings [23,24].

Liu [25] prepared a bioactive coating on a Ti6Al4V surface by laser cladding using a powder comprising $CaHPO_4 \cdot 2H_2O$, $CaCO_3$, and Ti as a precursor, composed of HA, β-TCP, etc. However, obvious large cracks appeared in the coating, and the mechanical properties needed to be improved. Yang [26] fabricated a coating on the surface of Ti6Al4V by a laser cladding process using a powder comprising HA and SiO_2. The coating contained $CaTiO_3$, $Ca_3(PO_4)_2$, Ca_2SiO_4 and other phases, and showed good biocompatibility and bioactivity. However, there were obvious pores and cracks in the interface between the coating and the substrate which limited the bearing capacity of the coating to some extent. Bajda et al. [9] applied a laser cladding technique to a prepare bioactive ceramic coating on a Ti6Al4V surface, using S520 bioactive glass powder as a precursor. The coating was approximately 100 μm thick and had a hardness range of 265–290 HV. A large amount of spherical calcium and phosphorus deposition appeared on the coating surface after immersion in simulated body fluids (SBF), which indicated that it had good biological activity. However, there were many defects, such as pores, cracks and so on. Pei [27] prepared a functional gradient carbon nanotubes/hydroxyapatite coating on the surface of a Ti substrate by laser cladding. The hardness of the coating surface was about 280.5 HV; this gradually increased with an increase in coating depth, with a maximum value of 433.5 HV, but decreased to the hardness of pure titanium (153 HV) in the transition zone. With the addition of carbon nanotubes, the hardness of the coating increased to nearly twice that of the pure hydroxyapatite coating, while exhibiting similar biological activity to a pure hydroxyapatite coating. Bioactive ceramic coatings prepared by laser cladding generally have too many defects, such as cracks and pores, due to the difference between the thermal properties of bioactive ceramic materials and Ti alloys. At the same time, the mechanical properties of the coating only remain in the hardness level, and the coatings are prone to wear. It is necessary to resolve these issues by improving the preparation process of the coating, thereby improving the performance of the coating.

In this paper, a multilayer Ca/P bio-ceramic coating on a Ti6Al4V surface prepared by laser cladding is proposed. The interface between the bio-ceramic coating and the Ti alloy should yield a good bonding strength, while the surface layer of the coating should display good bioactivity and a reasonable wear resistance. A multilayer powder layer was designed and preplaced on the Ti alloy surface as precursor, which included a transition powder layer and a bioactive powder layer. The transition powder layer was mixture of 50 wt% HA and 50 wt% Ti powder with a similar linear expansion coefficient and elastic modulus, while the bioactive powder layer was 100 wt% HA powder which was rich in calcium and phosphorus. A laser cladding technique was used to prepare the multilayer Ca/P bio-ceramic coating on the surface of the Ti alloy. Finally, the biological properties of the multilayer Ca/P bio-ceramic coating, e.g., biocompatibility and bioactivity, were investigated, and the mechanism of bioactivity was analyzed.

2. Materials and Methods

2.1. Experimental Materials

Ti6Al4V plates (Baoji Inite Medical Titanium Co. Ltd., Baoji, China), 30 mm long, 15 mm wide and 4 mm thick, were used as a substrate. The precursor powder materials used in the laser cladding were HA powder (particle size 80–85 μm, purity ≥ 99.9%, Shanghai Naiou Nano Technology Co. Ltd., Shanghai, China) and Ti powder (particle size 5–8 μm, purity ≥ 99.9%, Shanghai Naiou Nano Technology Co. Ltd., Shanghai, China). In order to reduce cracking and other problems caused by the huge difference in the thermal expansion coefficient between the coating and substrate [28], a preplaced multilayer powder was designed, which was divided into a transition powder layer and a bioactive powder layer. The transition powder layer was 50 wt% HA and 50 wt% Ti mixed powder (represented by HT in the equations below), while the bioactive powder layer was 100 wt% HA powder. The design of the multilayer powder on each sample is shown in Table 1. Table 2 lists the thermo-physical parameters of Ti6Al4V, Ti, HA, and HT. The thermo-physical parameters of HT were calculated as follows [29]:

$$M_H + M_T = 1 \tag{1}$$

$$V_T = M_T/\rho_T \tag{2}$$

$$V_H = M_H/\rho_H \tag{3}$$

$$V_H + V_T = 1 \tag{4}$$

$$Lt_{HT} = Lt_H(1 - V_T) + Lt_T V_T \tag{5}$$

$$Cp_{HT} = Cp_H(1 - M_T) + Cp_T M_T \tag{6}$$

$$\rho_{HT} = \rho_H(1 - M_T) + \rho_T M_T \tag{7}$$

where M, V, ρ, Lt, Cp represent mass fraction, volume fraction, density, linear expansion coefficient, and specific heat capacity, respectively. The subscripts H, T and HT are abbreviations of HA, Ti and HA/Ti mixed powder, respectively. Table 2 shows that the difference in the linear thermal expansion coefficient between HT and Ti6Al4V is less than that between HA and Ti6Al4V, and therefore, that the preplaced multilayer powder facilitates the binding of the coating to the Ti alloy substrate.

Table 1. Table of preplaced multilayer powder.

Layer	Mass Fraction/wt%		Weight/g
	HA	Ti	
The Transition Powder Layer	50	50	0.1
The Active Powder Layer	100	0	0.2

Table 2. Thermo-physical parameters of Ti6Al4V, Ti, HA, and HT at room temperature.

Material	Linear Thermal Expansion Coefficient 1/°C	Melting Point °C	Specific Heat Capacity J/(kg·°C)	Density kg/m^{-3}
Ti6Al4V	9.41×10^{-6}	1646 ± 42	520	4430
Ti	8.8×10^{-6}	1688	528	4500
HA	13.3×10^{-6}	1923	766	3156
HT	11.446×10^{-6}	1923	647	3828

2.2. Laser-Cladding Setup and Process

The laser cladding setup adopted in this paper is shown in Figure 1. The process required a laser system, a motion-control system, a computer-control system and auxiliary devices in the laser cladding system. The most important components were the RFL-500 medium power fiber laser (Wuhan Raycus Fiber Laser Technology Co., Ltd., Wuhan, China) and the BT-230 laser head (Raytools AG, Oberburg, Switzerland), which were connected through a QBH standard connector. The motion platform and laser cladding parameters were controlled by a computer. The two-dimensional movement of the sample was achieved by the motion platform, which made it possible to produce multitrack cladding. Auxiliary devices such as an argon protection device and a water-cooling device made the process more stable.

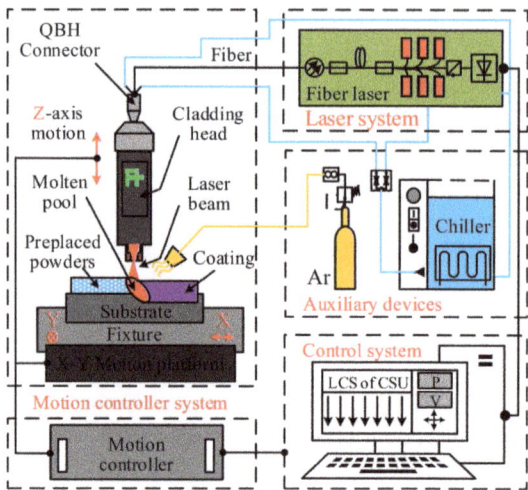

Figure 1. Schematic diagram of laser cladding set-up.

The surfaces of the Ti alloy substrates were polished with 100#, 240# or 600# SiC polishing film to remove the oxide layer, and then cleaned using an ultrasonic cleaner in ethanol or deionized water. The process of preplacing the multilayer powder on the substrate was as follows. First, 0.1 g HT powder was mixed with the sodium silicate binder, stirred evenly, and preplaced on the surface of the substrate to form a transition powder layer. It was then left to dry naturally for 15 min. Second, 0.2 g HA powder was mixed with the sodium silicate binder and preplaced on the surface of the transition powder layer to form an active powder layer. The total thickness of the multilayer powder was about 0.8 mm. Third, the sample with the preplaced multilayer powder was placed on a heating platform (60 °C) and dried for 30 min.

The sample with the multilayer powder was placed in the laser cladding system, as shown in Figure 1, to fabricate the bioactive coating using the laser cladding process. Based on a large number of previous experimental tests, a set of optimized laser cladding process parameters was applied, as follows: laser power of 400 W, scanning speed of 3 mm/s, spot diameter of 2 mm, lap rate of 30%, and argon flow rate of 10 L/min. Figure 2 shows the laser cladding process for the preparation of the multilayer Ca/P bio-ceramic coating.

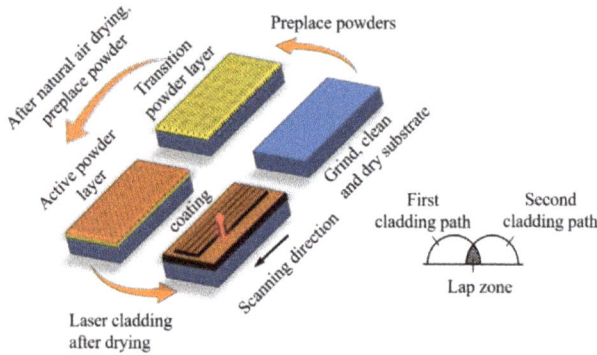

Figure 2. Flow diagram of the laser cladding process.

2.3. Microscopic Analysis of the Coating

2.3.1. Phase Test of the Coating

After cladding, the sample was placed in deionized water and cleaned with an ultrasonic cleaner. The surface phases of the coating were analyzed using Advance D8 X-ray diffraction (Bruker, Berne, Switzerland). X-ray scanning was performed using Cu/Kα radiation, a tube voltage of 40 kV, a tube current of 40 mA, a scanning speed of 2°/min, and a repetition accuracy of 0.001°; the test was performed in the range of 20° to 75°.

2.3.2. Microstructure Test of Coating

The sample was cut to a size of 15 mm × 6 mm × 4 mm along the direction perpendicular to the laser scanning direction. The sample section was polished and corroded with an etching agent (HF:HNO_3:H_2O = 2:5:93) for 15 s. A MIRA3 field emission scanning electron microscope (SEM, Tescan, Brno, Czech Republic) and an energy spectrum analyzer (EDS, Oxford Inc., Oxford, UK) were used to observe the surface morphology and test the elemental composition of the coating.

2.4. Biological Test of Coating

2.4.1. Biocompatibility Test

The biocompatibility of the ceramic coating was evaluated by a cell culture test in vitro. The cells used were MG-63 human osteosarcoma cells (Cellular Biology Institute, Shanghai, China). The medium used was a mixture of 10% fetal bovine serum (GeminiBio Foundation™, West Sacramento, CA, USA), 1% penicillin/streptomycin (Regen Biotechnology Co. Ltd., Beijing, China) and 89% DMEM low-glucose medium (CellGro-Mediatech Inc., Manassas, VA, USA). The test was conducted in a thermostatic incubator (Thermo Fisher Scientific, Waltham, MA, USA) at a temperature of 37 °C and a carbon dioxide concentration of 5%.

In this test, the Ti6Al4V substrate was used as the control, and the test sample size was 10 mm × 10 mm × 4 mm. First, the samples were sterilized in a high temperature sterilizer (121 °C, 25 min). Then, the sterilized samples were placed into 24-well plates, and each was inoculated with 9000 cells per 1.5 mL. The constant temperature incubation periods were 1 day, 3 days, or 5 days (The culture medium was replaced every 48 h). The number of samples was six per culture cycle. The cell morphology and cell diffusion were observed using a MIRA3 TESCAN field emission scanning electron microscope. Based on the MTT cell count method, the cell proliferation of the samples was determined using a Spark 10 M enzyme linked immunoassay (TESCAN, Brno, Czech Republic).

2.4.2. Bioactivity Test

The bioactivity of the Ca/P ceramic coating was evaluated by a simulated body fluid (SBF) immersion test. Table 3 presents the recipe of the SBF solution used in the test [30].

Table 3. Preparation reagent and dosage of SBF per liter.

Order	Reagent	Dosage/g
1	NaCl	8.035
2	$NaHCO_3$	0.355
3	KCl	0.225
4 [a]	K_2HPO_4	0.231
5	$MgCl_2·6H_2O$	0.311
6	1.0 mol/L HCl	39 mL
7	$CaCl_2$	0.292
8	Na_2SO_4	0.072
9	Tris	6.118
10 [b]	1.0 mol/L HCl	0–5 mL

[a] K_2HPO_4 alternatives $K_2HPO_4·3H_2O$. [b] HCl acts as a pH regulator to maintain a solution pH of 7.40.

The test sample size was 10 mm × 10 mm × 4 mm, and the Ti6Al4V substrate was also used as the control. The volume of SBF solution required for the immersion of the sample was calculated using the following Equation (8) [31]:

$$S/V = 0.05 \text{ cm}-1 \qquad (8)$$

where S is the coating immersion area and V is the volume of SBF solution.

The duration of the SBF immersion tests were 6, 12, 24, and 48 h. An ICAP7400 inductively coupled plasma emission spectrometer (Thermo Fisher Scientific, Waltham, MA, USA) was used to detect the concentrations of Ca and P in the solution for each immersion period. The number of replicates for the ICAP test was three, and the average value of the three tests was considered the final value. After 48 h of immersion, the morphologies of the coating surface were observed by MIRA3 field emission scanning electron microscopy (Tescan, Brno, Czech Republic); the elemental composition of deposition zone on the coating surface was analyzed by energy dispersive spectroscopy (EDS, Oxford Inc., Oxford, UK), and the phases of the deposition zone on the coating surface were analyzed by micro-area diffraction (Rigaku Rapid IIR, Akishima City, Tokyo, Japan).

2.5. Mechanical Properties Test of the Coating

2.5.1. Microhardness Test

A HVS-1000Z automatic digital Vickers hardness tester was used to measure the microhardness of the coating section. The load was 1.96 N and the load retention time was 20 s. The hardness of different areas along the depth direction of the coating section was measured. The hardness in the same depth direction was tested three times, the average of which was taken as the final value.

2.5.2. Wear Resistance Test

A TRB³ pin-disc friction and wear tester (Anton Paar, Graz, Austria) was used to test the friction and wear properties of the coatings. The grinding pair used in the test was an Al_2O_3 ceramic ball (diameter: 6 mm; hardness: 1650 $HV_{0.2}$). The wear resistance test parameters are listed in Table 4. At least three wear tests were performed for each test condition. The wear volume of the coating could be obtained by observing the abrasion contour with a VHX-5000 Ultra-depth microscope, and then applying Equation (9).

$$Vs = 2\pi r \cdot A \qquad (9)$$

where Vs represents the wear volume, A is the wear area of the wear contour curve (mm²), and r is the rotation radius of the grinding ball (mm).

Table 4. Experimental parameters of the wear test.

Parameter	Value	Unit
Load	5	N
Temperature	36.5 ± 1	°C
Wear time	30	mins
Rotation radius	3	mm
Rotation speed	400	r/min

3. Results and Discussion

3.1. Microstructure of the Coating

3.1.1. Phases of the Coating Surface

Figure 3 shows the X-ray diffraction pattern of the top surface of the multilayer coating fabricated by the laser cladding process. The corresponding diffraction peaks, crystal planes and phase mass fractions (RIR method) are listed in Table 5. It is seen that the coating was mainly composed of $Ca_2P_2O_7$, CaO, and $CaTiO_3$. Among these compounds,

$Ca_2P_2O_7$ can achieve osseous binding with bone tissue after implantation in the human body, and its binding strength is proportional to implantation time; additionally, it shows good biocompatibility and bioactivity [32]. CaO is one of the components of bioactive glass. Bioactive glass has good bioactivity, biocompatibility, and degradability, and is widely used in dentistry, orthopedics, and as a drug carrier [33]. In addition, $CaTiO_3$ has high hardness and superior mechanical properties, and is also used as an intermediate material to improve the adhesion between bioactive substances and metals [34,35].

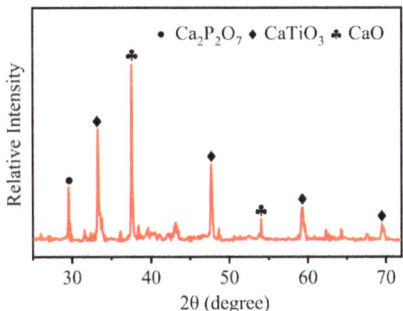

Figure 3. X-ray diffraction pattern of Ca/P ceramic coating top surface.

Table 5. The phases of the coating surface and their corresponding diffraction peaks, crystal planes, and phase mass fractions (RIR method).

Phase	Pdf Card	Diffraction Peaks/Plane		Mass Fraction (%)
$Ca_2P_2O_7$	09-0346	29.5°/[0 0 8]		7.5
CaO	82-1691	37.4°/[2 0 0]	64.2°/[3 1 1]	65.6
		54.0°/[2 2 0]	67.4°/[2 2 2]	
$CaTiO_3$	65-3287	33.3°/[1 1 0]	59.5°/[2 1 1]	26.9
		47.8°/[2 0 0]	69.9°/[2 2 0]	

The formation process of new phases of CaO, $Ca_2P_2O_7$, and $CaTiO_3$ is as follows. Under irradiation via a high-energy laser beam, a molten pool is formed on the surface of the preplaced powder layer. The HA powder then begins to decompose, releasing water vapor and forming $Ca_3(PO_4)_2$ and CaO. Through the process of heat conduction and convection, the HA powder in the transition layer also decomposes and melts into the molten pool, while elemental Ti in the transition layer and the titanium alloy substrate also enter the molten pool. Ti is a rather reactive element, and reacts with HA and its decomposition products, generating $CaTiO_3$, CaO, $Ca_2P_2O_7$, and other compounds. The specific reaction equations are as shown as (10)~(16) below [35–37].

The formation mechanism of CaO, $Ca_2P_2O_7$, and $CaTiO_3$ can be qualitatively analyzed by calculating the Gibbs free energy (ΔG^θ) of the reactions listed below. According to the thermodynamic manual of inorganic materials [38], the Gibbs free energy (ΔG^θ) of reactions (10)–(16) can be obtained as shown in Figure 4. The Gibbs free energies (10)–(16) are all negative after the temperature reaches 1100 K, which proves that the above reactions can occur spontaneously when the temperature reaches that point. Among them, reactions (10), (11), (15) occur most readily, all of which yield $Ca_3(PO_4)_2$. It was shown that HA decomposes easily and reacts at high temperature, which is the main reason for the low HA content in the cladding coating. In addition, as one of the decomposition products of

HA, $Ca_3(PO_4)_2$ will continue to react with Ti to form $CaTiO_3$ and CaO. Therefore, the final compositions of coating surface are mainly CaO, $CaTiO_3$, $Ca_2P_2O_7$.

$$Ca_{10}(PO_4)_6(OH)_2 \xrightarrow{1529.1K} 3Ca_3(PO_4)_2 + CaO + H_2O \quad (10)$$

$$2Ca_{10}(PO_4)_6(OH)_2 + 5Ti \rightarrow 5CaTiO_3 + 5CaO + 2Ca_3(PO_4)_2 + 2Ca_2P_2O_7 + H_2O \quad (11)$$

$$2Ca_3(PO4)_2 + 2Ti \rightarrow 2CaTiO_3 + 4CaO + P_4O_6 \uparrow \quad (12)$$

$$2Ca_2P_2O_7 + 2Ti \rightarrow 2CaTiO_3 + 2CaO + P_4O_6 \uparrow \quad (13)$$

$$P_4O_6 + 3Ti \rightarrow 3TiO_2 + 4P \quad (14)$$

$$Ca_{10}(PO_4)_6(OH)_2 + TiO_2 \rightarrow CaTiO_3 + 3Ca_3(PO_4)_2 + H_2O \quad (15)$$

$$CaO + TiO_2 \rightarrow CaTiO_3 \quad (16)$$

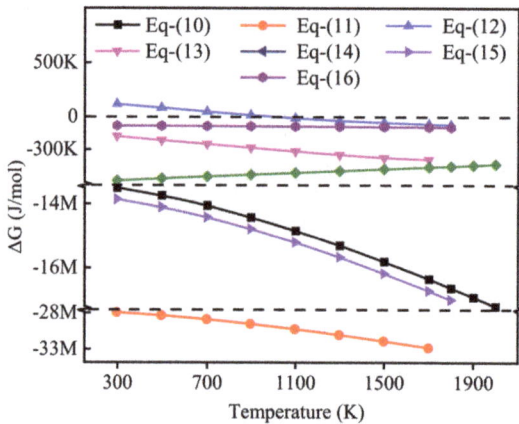

Figure 4. Gibbs free energy of Equations (10)–(16).

3.1.2. Phases of the Coating Section

Figure 5a shows the cross-section morphology of the Ca/P ceramic coating prepared by laser cladding. The coating consisted of two layers corresponding to the preplaced multilayer powder. Figure 5b shows the microscopic morphology of the bioactive layer (BL), which is mainly composed of dendritic granular grains. According to XRD (Figure 3) and EDS energy spectrum analyses, the phase composition of the granular grains was $CaTiO_3$. Figure 5c shows the microscopic morphology of the transition layer (TL), which was mainly composed of rod-like grains surrounded by a Ti matrix. Figure 6 shows the X-ray diffraction pattern of the transition layer in the coating. The corresponding diffraction peaks, crystal planes and phase mass fractions (RIR method) are listed in Table 6. According to the EDS energy spectrum analysis, the rod-like grains were composed of Ti and P. According to the X-ray diffraction of the transition layer, their corresponding diffraction peaks and crystal planes, and the Ti-P binary phase diagram (as shown in Figure 7), it can be seen that the rod-like crystals were eutectoid products of Ti_3P and Ti.

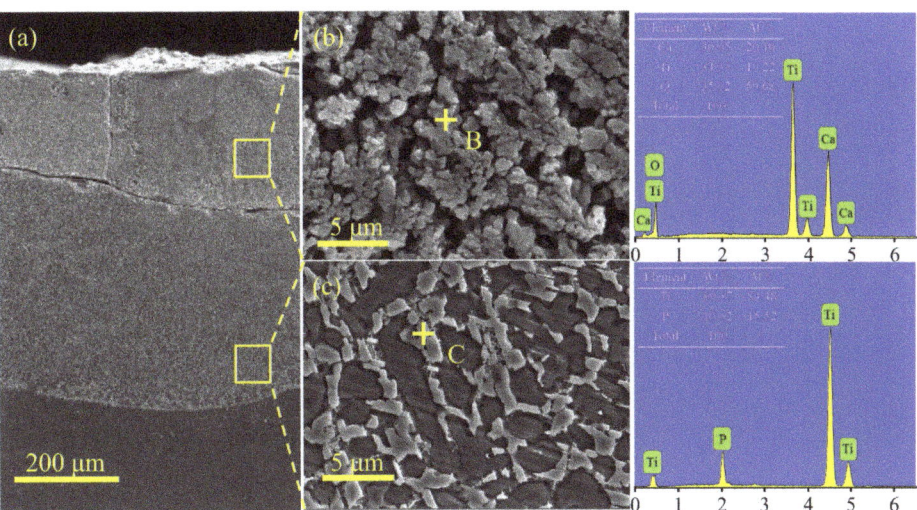

Figure 5. (**a**) Microstructure morphology of Ca/P ceramic coating section (**b**) bioactive layer (**c**) transition layer.

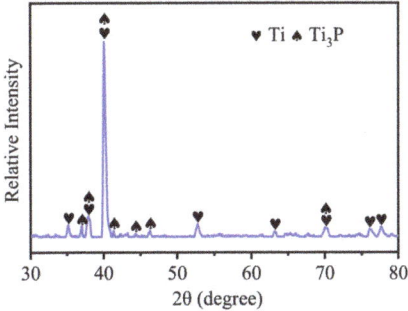

Figure 6. X-ray diffraction pattern of the phase of the transition layer.

Table 6. The phases of the transition layer in the coating and their corresponding diffraction peaks, crystal planes, and phase mass fractions (RIR method).

Phase	Pdf Card	Diffraction Peaks/Plane		Mass Fraction(%)
Ti	89-5009	35.1°/[1 0 0] 38.4°/[0 0 2] 40.2°/[1 0 1] 53.0°/[1 0 2]	63.1°/[1 1 0] 70.7°/[1 0 3] 76.3°/[1 1 2] 77.5°/[2 0 1]	50.6
Ti3P	89-2416	37.1°/[3 2 1] 38.2°/[1 1 2] 40.4°/[4 0 1] 41.5°/[1 4 1]	44.4°/[2 2 2] 46.4°/[3 1 2] 70°6/[2 6 2]	49.4

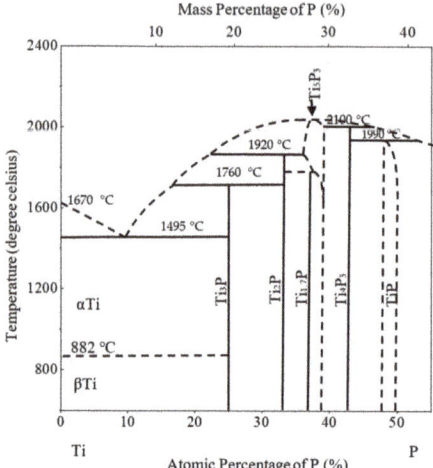

Figure 7. Ti-P binary phase diagram.

3.2. Biological Properties of the Coating

3.2.1. Biocompatibility

Figure 8 shows the cell morphology of MG-63 cells cultured on the substrate (control group) and the coating surface for 1 day, 3 days, and 5 days. The biocompatibility of the coatings could be evaluated by comparing the morphology and number of cells on the surface of the substrate and coating. After 1 day of culture, the cells began to spread and adhere to the surface of the substrate and coating, and showed a spindle shape. After 3 days of culture, the number of cells on the surface of both the substrate and coating increased significantly, and the cells began to appear pseudopodia, which may have promoted cell adhesion and migration [39]. After 5 days of culture, the number of cells on the substrate and coating was further increased, and the cells were tiled on the surfaces.

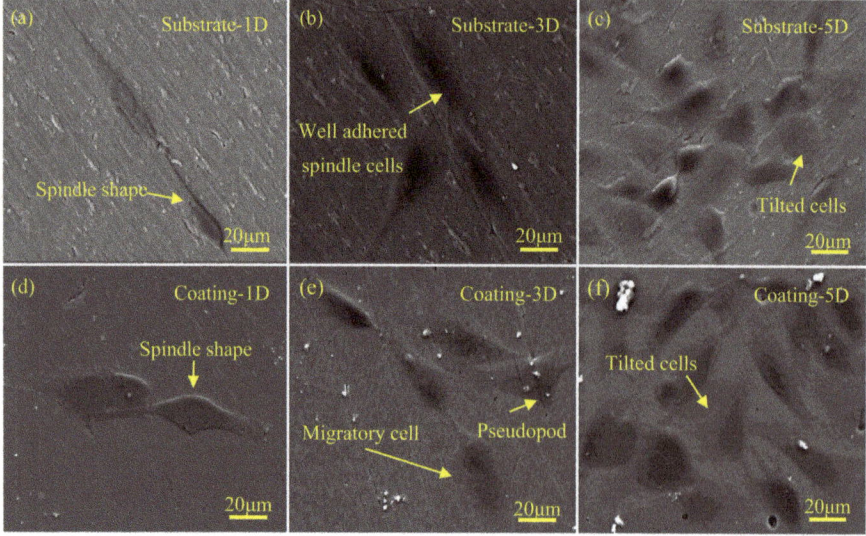

Figure 8. Morphology of MG-63 cells cultured on substrate (**a–c**) and coating (**d–f**) after 1, 3 and 5 days.

Figure 9 shows the quantitative statistics of the active MG-63 cells after 1, 3, and 5 days of culture on the surface of the substrate and coating. From a statistical point of view, there was no significant difference in the number of cells between the Ca/P ceramic coating surface and the substrate surface, indicating that the prepared Ca/P ceramic coating also had good biocompatibility.

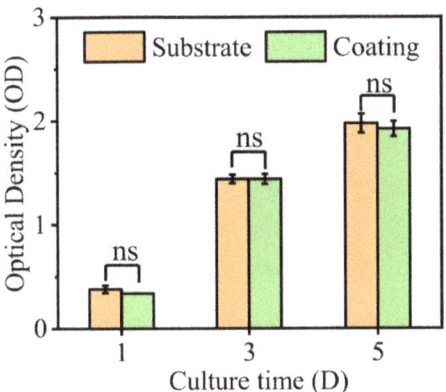

Figure 9. Quantitative statistics of active MG-63 cells after 1, 3, and 5 days of culture on the surface of the substrate and coating (ns stands for no significant difference).

3.2.2. Bioactivity

Figure 10 shows the micromorphology of the deposition on the Ti6Al4V surface (control group) and Ca/P ceramic coating surface after immersion in simulated body fluid (SBF) for 48 h. The bioactivity of the coatings was evaluated by comparing their ability to induce deposition of apatite layers in simulated body fluids. Figure 10a shows that there was only a little bit of granular deposition on the Ti alloy surface. Spectrum A shows that particle A comprised a granular deposition composed of Ca, P, O, and other elements, in which the atomic ratio of Ca to P was 1.41. Spectrum B shows that particle B was composed of Ca, C, and O, while no P was detected. Except for the deposition of a small number of particles, there was no obvious change on the Ti6Al4V surface before and after immersion. Figure 10b shows the micromorphology of the deposition layer on the surface of the Ca/P ceramic coating. The deposition layer was composed of a large number of spherical particles, which is a characteristic morphology of apatite [26,40]. Spectrum C and D show that the deposition layer was composed of Ca, P, and O, in which the atomic ratios of Ca and P were 1.7 and 1.36, respectively. To further prove the existence of apatite, the phases of the coating surface were detected before and after immersion, as shown in Figure 11. The CaO phase in the coating disappeared after immersion, and the diffraction peak intensity of $Ca_2P_2O_7$ decreased. In addition, the ceramic coating surface did not have an HA phase before immersion, while the deposition layer of ceramic coating did after 48 h of immersion, indicating that the coating could induce the deposition of apatite.

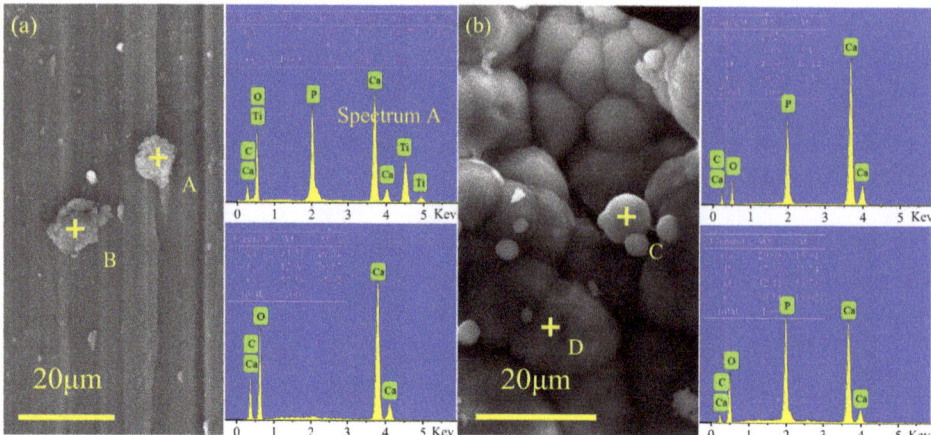

Figure 10. Micromorphology of surface deposits of substrate (**a**) and Ca/P ceramic coating (**b**) after 48 h of immersion in SBF solution.

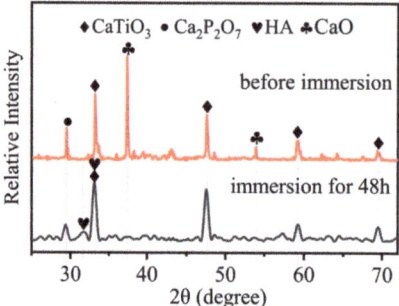

Figure 11. X-ray diffraction pattern of Ca/P ceramic coating before and after SBF immersion.

Figure 12a,b show the concentration fluctuations of Ca and P in the immersion solution and the variation rate of these elements throughout the immersion period, respectively. Each point in Figure 12a is the average value of the Ca or P concentrations obtained from three ICAP tests. The average value was used to obtain the variation rates of Ca and P, as shown in Figure 12b. As shown in Figure 12a, during the immersion period of 0–12 h, the concentration of Ca in the coating immersion solution was higher than that of the SBF standard solution, while the concentration of P was lower than that of the SBF standard solution, indicating that the dissolution of the coating phase and the precipitation of solution element occurred simultaneously at this stage, and that the dissolution rate of Ca was higher than the precipitation rate. In contrast, the dissolution rate of element P was less than the precipitation rate. During the immersion period of 12–24 h, the deposition rate of the solution was higher than the dissolution rate of the coating, which was reflected by the decrease of the concentrations of Ca and P in the solution. During the immersion period of 24–48 h, the deposition rate of the solution was less than the dissolution rate of the coating, as reflected by the increase in the concentrations of Ca in the solution. There are two possible reasons for the increase of Ca ion concentration at 48 h. On the one hand, apatite deposition is a dynamic process that occurs simultaneously along with dissolution and precipitation [41]. On the other hand, in the process of apatite-induced deposition, some intermediate products are produced, including ACP amorphous apatite, ACPP amorphous calcium pyrophosphate, etc. These intermediate products are precursors of HA, which may release Ca ions during the process of conversion to HA. The hydrolytic conversion of amorphous calcium phosphate into apatite accompanied the sustained release of calcium and orthophosphate ions [42,43]. However, during the immersion period of the Ti6Al4V alloy, the concentrations of Ca and P in the solution continued to decrease slowly, and the Ti6Al4V alloy did not show any ion exchange with the SBF solution.

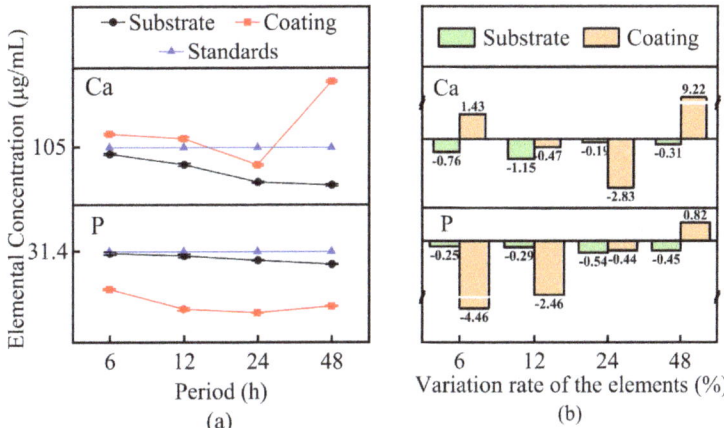

Figure 12. Concentrations of Ca and P in the immersion solution (**a**), the variation rate of the elements (**b**) during the immersion period of the coating and Ti6Al4V alloy.

Figure 12b shows the variation rates DC and DP of Ca and P ions in the solution during the immersion process of the Ti6Al4V substrate and coating:

$$DP_n = P_{n+1} - P_n/P_n \qquad (17)$$

$$DC_n = C_{n+1} - C_n/C_n \qquad (18)$$

where DP_n is the variation rate of the concentration of P ions in the period from t_n to t_{n+1}, and DC_n is the variation rate of the concentration of Ca ions in the period from t_n to t_{n+1}. Time t_1, t_2, t_3 and t_4 correspond to 6 h, 12 h, 24 h and 48 h, respectively. It can be seen from Figure 12b that ion exchange occurred between the coating and SBF solution during the immersion period, while the Ti6Al4V alloy did not show ion exchange with the SBF solution. In addition, the variation rates of the Ca and P concentrations in the coating immersion solution were much higher than in the Ti6Al4V alloy immersion solution, that is, the deposition rate of elements in the coating immersion solution was higher than in the Ti6Al4V alloy immersion solution. This is consistent with the SEM observation that less deposition had occurred on the surface of the Ti6Al4V alloy, while significant apatite deposition had occurred the coating surface.

Therefore, according to the ICAP, SEM, and XRD test results, it can be concluded that the Ca/P ceramic coating has the ability to induce hydroxyapatite, and has good bioactivity, while the titanium alloy substrate has no biological activity. This is attributed to the fact that only one deposition mode occurred during the immersion period of the titanium alloy, i.e., component nucleation, while two deposition modes occurred during the immersion period of the coating, i.e., component nucleation and structure nucleation.

During the immersion process of the titanium alloy, due to the molecular movement of saturated Ca^{2+} and PO_4^{3-} in SBF solution, the solution locally reached the nucleation site of the components, and then the component nucleation deposits; however, the deposition was slow, so there was far less deposition on the surface of titanium alloy. However, during the immersion process of the coating, active substances CaO and $Ca_2P_2O_7$ formed nucleation sites containing -OH on the surface of the coating through hydrolysis, promoting the adsorption of Ca^{2+} and PO_4^{3-} in succession, and inducing the nucleation and growth of apatite, that is, structural nucleation deposition. At the same time, the Ca^{2+}, PO_4^{3-}, and OH^- dissolved in the coating reached a supersaturated state in the solution, accelerating the movement of molecules and allowing the solution to more rapidly reach the nucleation sites of the components, so that apatite component nucleated out, as shown in Figure 13. Thus, after immersion for 48 h, an apatite deposition layer formed on the surface of the coating. The dissolution and precipitation equations are as follows [40,44,45]:

$$CaO + H_2O \rightarrow Ca^{2+} + OH \qquad (19)$$

$$Ca_2P_2O_7 + H_2O \rightarrow Ca^{2+} + PO_4^3 \qquad (20)$$

$$OH^- + Ca^{2+} + PO_4^{3-} \rightarrow Ca_{10}(PO_4)_6(OH)_2 \qquad (21)$$

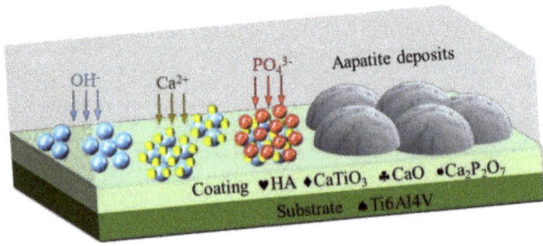

Figure 13. Schematic diagram of coating induced deposition of apatite.

3.3. Mechanical Properties of the Coating

3.3.1. Microhardness

Figure 14 shows the microhardness distribution of a ceramic coating section in the direction of depth. In the direction of depth, the sample can be divided into three regions: a coating zone (BL, TL), heat affected zone (HZ) and substrate zone (TS). In the coating area, the bioactive layer (BL) thickness was about 0.2 mm, the average microhardness was 440 $HV_{0.2}$, the transition layer (TL) thickness was about 0.4 mm, and the average microhardness is 889.75 $HV_{0.2}$. The thickness of the heat-affected zone with an average microhardness of 655.67 $HV_{0.2}$ was about 0.3 mm, and the microhardness showed a decreasing trend within this range. The average microhardness of the substrate is 340 $HV_{0.2}$. The results show that, compared with the substrate, the microhardness of the bioactive layer and the transition layer had increased by 24.1% and 161.7%, respectively. This was due to the formation of the hard phase $CaTiO_3$ in the bioactive layer and the eutectoid products Ti_3P and Ti in the transition layer. This indicates that the coating not only ensured good biocompatibility and bioactivity, but also achieved a significant improvement in hardness compared with the Ti6Al4V substrate and the related coatings, as reported by Bajda et al. [9] and Pei.

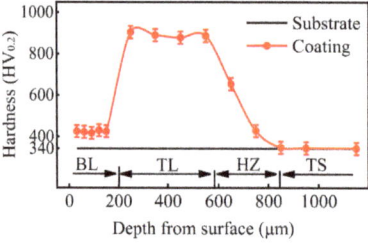

Figure 14. Microhardness distribution diagram of coating section along depth direction.

3.3.2. Wear Resistance

The wear contour curves of the Ti6Al4V alloy (control group) and coating are shown in Figure 15a, and the wear volumes taken from three repeated tests are shown in Figure 15b. The friction and wear properties of the coating can be evaluated by analyzing the wear contour and volume of both the coating and Ti6Al4V alloy. The average wear-section width and depth of the Ti6Al4V alloy were 1570 μm and 44 μm, respectively, and that of the coating were 1190 μm and 35 μm, respectively. The wear volume of the Ti6Al4V alloy was 0.829 mm^3, while that of the coating was 0.471mm^3. The variation of the wear volume conformed to the Holm-Archard wear law [46], that is, wear volume is inversely proportional to hardness. The results showed that the wear volume of the ceramic coating was reduced by 43.2% relative to the Ti6Al4V alloy, and therefore, that the ceramic coating had better wear resistance.

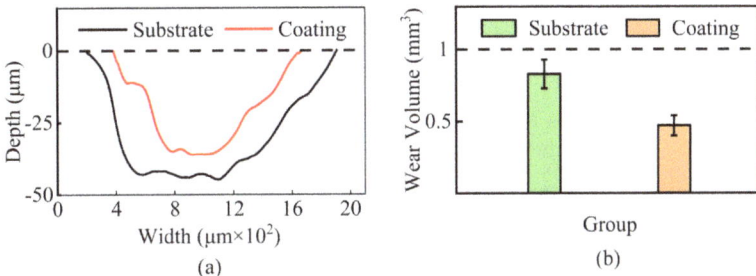

Figure 15. The wear contour curves (**a**) and wear volume (**b**) of Ti6Al4V alloy and coating.

4. Conclusions

A multilayer Ca/P bioactive ceramic coating was prepared on the surface of a Ti6Al4V alloy by the application of a laser cladding technique. The biocompatibility and bioactivity of the coating were evaluated by in vitro cell culture and simulated body fluid immersion tests, respectively. The wear resistance of the Ca/P ceramic coating was evaluated by microhardness and wear tests. The main conclusions are as follows:

1. The multilayer Ca/P bio-ceramic coating was mainly composed of CaO, CaTiO$_3$, Ca$_2$P$_2$O$_7$, Ti$_3$P, and other phases.
2. The multilayer Ca/P bio-ceramic coating exhibited biocompatibility equal to that of Ti6Al4V alloy, which is widely used in the field of medical implants.
3. The multilayer Ca/P bio-ceramic coating had good bioactivity in vitro, and could induce and deposit hydroxyapatite on its surface when immersed in SBF solution. Specifically, the coating showed obvious ion exchange during the immersion period, whereas the titanium alloy substrate did not.
4. The multilayer Ca/P bio-ceramic coating showed better microhardness and wear resistance than the Ti alloy substrate. Compared with the substrate (340HV$_{0.2}$), the microhardness of the bioactive layer (440HV$_{0.2}$) and the transition layer (889.75 HV$_{0.2}$) increased by 24.1% and 161.7%, respectively. Additionally, the wear volume of the coating was 0.471 mm^3, i.e., 43.2% less than that of Ti6Al4V alloy (0.829 mm^3).

Author Contributions: Conceptualization, B.L. and D.L.; investigation, B.L. and Z.D.; resources, B.L.; data curation, Z.D. and D.L.; writing—original draft preparation, B.L.; writing—review and editing, D.L.; visualization, B.L. and Z.D.; supervision, D.L.; project administration, D.L.; funding acquisition, D.L. All authors have read and agreed to the published version of the manuscript.

Funding: This research was funded by (1) National Natural Science Foundation of China, Grant No. 51775559; (2) The Project of State Key Laboratory of High Performance Complex Manufacturing,

Grant No. ZZYJKT2019-06, (3) The Natural Science Foundation of Hunan province of China, Grant No. 2020JJ4716.

Institutional Review Board Statement: Not applicable.

Informed Consent Statement: Not applicable.

Data Availability Statement: Data is concerned within the article.

Conflicts of Interest: The authors declare no conflict of interest.

References

1. Luo, Y.; Yang, L.; Tian, M. *Application of Biomedical-Grade Titanium Alloys in Trabecular Bone and Artificial Joints*; John Wiley & Sons: Hoboken, NJ, USA, 2013; pp. 181–216.
2. Avila, J.; Bose, S.; Bandyopadhyay, A. Additive manufacturing of titanium and titanium alloys for biomedical applications. In *Titanium in Medical and Dental Applications*; Elsevier: Amsterdam, The Netherlands, 2018; pp. 325–343.
3. Alipal, J.; Pu'Ad, N.M.; Nayan, N.; Sahari, N.; Abdullah, H.; Idris, M.; Lee, T. An updated review on surface functionalisation of titanium and its alloys for implants applications. *Mater. Today Proc.* **2021**, *42*, 270–282. [CrossRef]
4. Gallo, J.; Goodman, S.; Konttinen, Y.; Wimmer, M.; Holinka, M. Osteolysis around total knee arthroplasty: A review of pathogenetic mechanisms. *Acta Biomater.* **2013**, *9*, 8046–8058. [CrossRef]
5. Hinman, A.D.; Prentice, H.A.; Paxton, E.W.; Kelly, M.P. Modular tibial stem use and risk of revision for aseptic loosening in cemented primary total knee arthroplasty. *J. Arthroplast.* **2021**, *36*, 1577–1583. [CrossRef]
6. Koks, S.; Wood, D.J.; Reimann, E.; Awiszus, F.; Lohmann, C.H.; Bertrand, J.; Prans, E.; Maasalu, K.; Martson, A. The genetic variations associated with time to aseptic loosening after total joint arthroplasty. *J. Arthroplast.* **2020**, *35*, 981–988. [CrossRef]
7. Wiliams, D.F. Biomechanical considerations in the loosening of hip replacement prosthesis. In *Current Perspectives on Implantable Devices*; JAI Press Inc.: Stamford, CT, USA, 1989; pp. 1–45.
8. Geetha, M. Ti based biomaterials, the ultimate choice for orthopaedic implants—A review. *Prog. Mater. Sci.* **2009**, *54*, 397–425. [CrossRef]
9. Bajda, S.; Liu, Y.; Tosi, R.; Cholewa-Kowalska, K.; Krzyzanowski, M.; Dziadek, M.; Kopyscianski, M.; Dymek, S.; Polyakov, A.V.; Semenova, I.P.; et al. Laser cladding of bioactive glass coating on pure titanium substrate with highly refined grain structure. *J. Mech. Behav. Biomed. Mater.* **2021**, *119*, 104519. [CrossRef] [PubMed]
10. Mansoor, P.; Dasharath, S. Synthesis and characterization of wollastonite ($CaSiO_3$)/titanium oxide (TiO_2) and hydroxyapatite (HA) ceramic composites for bio-medical applications fabricated by spark plasma sintering technology. *Mater. Today Proc.* **2021**, *45*, 332–337. [CrossRef]
11. Roy, M.; Krishna, B.V.; Bandyopadhyay, A.; Bose, S. Laser processing of bioactive tricalcium phosphate coating on titanium for load-bearing implants. *Acta Biomater.* **2008**, *4*, 324–333. [CrossRef]
12. Zhou, X.; Siman, R.; Lu, L.; Mohanty, P. Argon atmospheric plasma sprayed hydroxyapatite/Ti composite coating for biomedical applications. *Surf. Coat. Technol.* **2012**, *207*, 343–349. [CrossRef]
13. Liu, D.; Savino, K.; Yates, M.Z. Coating of hydroxyapatite films on metal substrates by seeded hydrothermal deposition. *Surf. Coat. Technol.* **2011**, *205*, 3975–3986. [CrossRef]
14. Chakraborty, R.; Seesala, V.; Sengupta, S.; Dhara, S.; Saha, P.; Das, K.; Das, S. Comparison of osteoconduction, cytocompatibility and corrosion protection performance of hydroxyapatite-calcium hydrogen phosphate composite coating synthesized in-situ through pulsed electro-deposition with varying amount of phase and crystallinity. *Surf. Interfaces* **2018**, *10*, 1–10. [CrossRef]
15. Tlotleng, M.; Akinlabi, E.; Shukla, M.; Pityana, S. Microstructures, hardness and bioactivity of hydroxyapatite coatings deposited by direct laser melting process. *Mater. Sci. Eng. C* **2014**, *43*, 189–198. [CrossRef] [PubMed]
16. Greish, Y.E.; Al Shamsi, A.S.; Polychronopoulou, K.; Ayesh, A. Structural evaluation, preliminary in vitro stability and electrochemical behavior of apatite coatings on Ti6Al4V substrates. *Ceram. Int.* **2016**, *42*, 18204–18214. [CrossRef]
17. Ji, X.; Zhao, M.; Dong, L.; Han, X.; Li, D. Influence of Ag/Ca ratio on the osteoblast growth and antibacterial activity of TiN coatings on Ti-6Al-4V by Ag and Ca ion implantation. *Surf. Coat. Technol.* **2020**, *403*, 126415. [CrossRef]
18. Wang, D.; Chen, C.; Yang, X.; Ming, X.; Zhang, W. Effect of bioglass addition on the properties of HA/BG composite films fabricated by pulsed laser deposition. *Ceram. Int.* **2018**, *44*, 14528–14533. [CrossRef]
19. Azari, R.; Rezaie, H.R.; Khavandi, A. Investigation of functionally graded HA-TiO_2 coating on Ti–6Al–4V substrate fabricated by sol-gel method. *Ceram. Int.* **2019**, *45*, 17545–17555. [CrossRef]
20. Lenis, J.A. Structure, morphology, adhesion and in vitro biological evaluation of antibacterial multi-layer HA-Ag/SiO_2/TiN/Ti coatings obtained by RF magnetron sputtering for biomedical applications. *Mater. Sci. Eng. C* **2020**, *116*, 111268. [CrossRef] [PubMed]
21. Bansal, P.; Singh, G.; Sidhu, H.S. Improvement of surface properties and corrosion resistance of Ti13Nb13Zr titanium alloy by plasma-sprayed HA/ZnO coatings for biomedical applications. *Mater. Chem. Phys.* **2021**, *257*, 123738. [CrossRef]
22. Zhu, L.; Xue, P.; Lan, Q.; Meng, G.; Ren, Y.; Yang, Z.; Xu, P.; Liu, Z. Recent research and development status of laser cladding: A review. *Opt. Laser Technol.* **2021**, *138*, 106915. [CrossRef]

23. Li, H.C. Effect of Na$_2$O and ZnO on the microstructure and properties of laser cladding derived CaO-SiO$_2$ ceramic coatings on titanium alloys. *J Colloid Interface Sci.* **2021**, *592*, 498–508. [CrossRef] [PubMed]
24. Behera, R.R.; Hasan, A.; Sankar, M.R.; Pandey, L.M. Laser cladding with HA and functionally graded TiO$_2$-HA precursors on Ti–6Al–4V alloy for enhancing bioactivity and cyto-compatibility. *Surf. Coat. Technol.* **2018**, *352*, 420–436. [CrossRef]
25. Zhang, S.; Liu, Q.; Li, L.; Bai, Y.; Yang, B. The controllable lanthanum ion release from Ca-P coating fabricated by laser cladding and its effect on osteoclast precursors. *Mater. Sci. Eng. C* **2018**, *93*, 1027–1035. [CrossRef]
26. Yang, Y. Osteoblast interaction with laser cladded HA and SiO$_2$-HA coatings on Ti–6Al–4V. *Mater. Sci. Eng. C* **2011**, *31*, 1643–1652. [CrossRef]
27. Pei, X.; Wang, J.; Wan, Q.; Kang, L.; Xiao, M.; Bao, H. Functionally graded carbon nanotubes/hydroxyapatite composite coating by laser cladding. *Surf. Coat. Technol.* **2011**, *205*, 4380–4387. [CrossRef]
28. Tsui, Y.; Doyle, C.; Clyne, T. Plasma sprayed hydroxyapatite coatings on titanium substrates Part 1: Mechanical properties and residual stress levels. *Biomaterials* **1998**, *19*, 2015–2029. [CrossRef]
29. Vasquez, F.A.; Ramos-Grez, J.A.; Walczak, M. Multiphysics simulation of laser–material interaction during laser powder depositon. *Int. J. Adv. Manuf. Technol.* **2011**, *59*, 1037–1045. [CrossRef]
30. Oyane, K. Preparation and assessment of revised simulated body fluids. *J. Biomed. Mater. Res. Part A* **2003**, *65*, 188–195. [CrossRef]
31. Müller, L.; Müller, F.A. Preparation of SBF with different HCO$_3-$ content and its influence on the composition of biomimetic apatites. *Acta Biomater.* **2006**, *2*, 181–189. [CrossRef]
32. Kasuga, T. Bioactive calcium pyrophosphate glasses and glass-ceramics. *Acta Biomater.* **2005**, *1*, 55–64. [CrossRef]
33. Henao, J.; Poblano-Salas, C.; Monsalve, M.; Corona-Castuera, J.; Barceinas-Sanchez, O. Bio-active glass coatings manufactured by thermal spray: A status report. *J. Mater. Res. Technol.* **2019**, *8*, 4965–4984. [CrossRef]
34. Chen, C.-Y.; Ozasa, K.; Katsumata, K.-I.; Maeda, M.; Okada, K.; Matsushita, N. CaTiO$_3$ nanobricks prepared from anodized TiO$_2$ nanotubes. *Electrochem. Commun.* **2012**, *22*, 101–104. [CrossRef]
35. Yadi, M.; Esfahani, H.; Sheikhi, M.; Mohammadi, M. CaTiO$_3$/α-TCP coatings on CP-Ti prepared via electrospinning and pulsed laser treatment for in-vitro bone tissue engineering. *Surf. Coat. Technol.* **2020**, *401*, 126256. [CrossRef]
36. Roy, M.; Balla, V.; Bandyopadhyay, A.; Bose, S. Compositionally graded hydroxyapatite/tricalcium phosphate coating on Ti by laser and induction plasma. *Acta Biomater.* **2011**, *7*, 866–873. [CrossRef] [PubMed]
37. Lusquiños, F.; De Carlos, A.; Pou, J.; Arias, J.L.; Boutinguiza, M.; León, B.; Pérez-Amor, M.; Driessens, F.C.M.; Hing, K.; Gibson, I.; et al. Calcium phosphate coatings obtained by Nd: YAG laser cladding: Physicochemical and biologic properties. *J. Biomed. Mater. Res. Part A* **2003**, *64*, 630–637. [CrossRef]
38. Ye, D. *Manual of Practical Inorganic Thermodynamics Data*; Metallurgical Industry Press: Beijing, China, 2002.
39. Xie, K.; Yang, Y.; Jiang, H. Controlling cellular volume via mechanical and physical properties of substrate. *Biophys. J.* **2018**, *114*, 675–687. [CrossRef] [PubMed]
40. Paital, S.R.; Dahotre, N.B. Wettability and kinetics of hydroxyapatite precipitation on a laser-textured Ca-P bioceramic coating. *Acta Biomater.* **2009**, *5*, 2763–2772. [CrossRef]
41. Lu, X.; Leng, Y. Theoretical analysis of calcium phosphate precipitation in simulated body fluid. *Biomaterials* **2005**, *26*, 1097–1108. [CrossRef] [PubMed]
42. Dridi, A.; Riahi, K.Z.; Somrani, S. Mechanism of apatite formation on a poorly crystallized calcium phosphate in a simulated body fluid (SBF) at 37 °C. *J. Phys. Chem. Solids* **2021**, *156*, 110122. [CrossRef]
43. Edén, M. Structure and formation of amorphous calcium phosphate and its role as surface layer of nanocrystalline apatite: Implications for bone mineralization. *Materialia* **2021**, *17*, 101107. [CrossRef]
44. Bakher, Z. Solubility study at high phosphorus pentoxide concentration in ternary system CaCO3 + P2O5 + H2O at 25, 35 and 70 °C. *Fluid Phase Equilib.* **2018**, *478*, 90–99. [CrossRef]
45. Bakhsheshi-Rad, H.; Hamzah, E.; Ismail, A.F.; Kasiri-Asgarani, M.; Daroonparvar, M.; Parham, S.; Iqbal, N.; Medraj, M. Novel bi-layered nanostructured SiO2/Ag-FHAp coating on biodegradable magnesium alloy for biomedical applications. *Ceram. Int.* **2016**, *42*, 11941–11950. [CrossRef]
46. Yamamoto, S. Physical meaning of the wear volume equation for nitrogenated diamond-like carbon based on energy considerations. *Wear* **2016**, *368–369*, 156–161. [CrossRef]

Article

Characterization and In Vitro Studies of Low Reflective Magnetite (Fe₃O₄) Thin Film on Stainless Steel 420A Developed by Chemical Method

Reghuraj Aruvathottil Rajan [1,*], Kaiprappady Kunchu Saju [1] and Ritwik Aravindakshan [2]

[1] Mechanical Engineering Division, School of Engineering, Cochin University of Science and Technology, Kalamassery, Cochin 682022, Kerala, India; kksaju@cusat.ac.in
[2] Mechanical Engineering Division, Toc H Institute of Science and Technology, Cochin 682314, Kerala, India; ritwik@tistcochin.edu.in
* Correspondence: reghuraj@alameen.edu.in

Abstract: Stainless steel has been the most demanded material for surgical utensil manufacture due to superior mechanical properties, sufficient wear, and corrosion resistance. Surgical grade 420A stainless steel is extensively used for producing sophisticated surgical instruments. Since these instruments are used under bright light conditions prevalent in operation theatres, the reflection from the material is significant which causes considerable strain to the eye of the surgeon. Surgical instruments with lower reflectance will be more efficient under these conditions. A low reflective thin-film coating has often been suggested to alleviate this inadmissible difficulty. This paper reports the development of an optimum parametric low reflective magnetite coating on the surface of SS 420A with a black color using chemical hot alkaline conversion coating technique and its bioactivity studies. Coating process parameters such as coating time, bath temperature, and chemical composition of bath are optimized using Taguchi optimization techniques. X-ray photoelectron spectroscopy (XPS) analysis was used to identify the composition of elements and the chemical condition of the developed coating. Surface morphological studies were accomplished with a scanning electron microscope (SEM). When coupled with an energy-dispersive X-ray analysis (EDAX), compositional information can also be collected simultaneously. Invitro cytotoxicity tests, corrosion behavior, the effect of sterilization temperature on adhesion property, and average percentage reflectance (R) of the developed coating have also been evaluated. These results suggest adopting the procedure for producing low reflective conversion coatings on minimally invasive surgical instruments produced from medical grade 420A stainless steel.

Keywords: magnetite; conversion coating; optimization; invitro cytotoxicity; corrosion

Citation: Aruvathottil Rajan, R.; Saju, K.K.; Aravindakshan, R. Characterization and In Vitro Studies of Low Reflective Magnetite (Fe₃O₄) Thin Film on Stainless Steel 420A Developed by Chemical Method. *Coatings* 2021, *11*, 1145. https://doi.org/10.3390/coatings11091145

Academic Editor: Liqiang Wang

Received: 1 August 2021
Accepted: 14 September 2021
Published: 21 September 2021

Publisher's Note: MDPI stays neutral with regard to jurisdictional claims in published maps and institutional affiliations.

Copyright: © 2021 by the authors. Licensee MDPI, Basel, Switzerland. This article is an open access article distributed under the terms and conditions of the Creative Commons Attribution (CC BY) license (https://creativecommons.org/licenses/by/4.0/).

1. Introduction

Martensitic stainless steel of grade 410, 420A, 420B, and 420C has been widely used for the manufacturing of cutting and non-cutting surgical instruments [1,2]. In addition to biocompatibility, martensitic stainless steel has high strength, hardness, stiffness, rigidity, resilient, and non-corrosive nature [3]. Titanium (Ti) and its alloys are also used in various biomedical applications. Conventional as well as modern surface modification technologies have emerged in the recent decades. The mechanical, chemical, and biological properties are improved by adopting various surface modification methods. Conventional methods like sand blasting, alkali treatment, and plasma treatment have minor enhancements in the surface properties due to the intricate geometry of the work piece. To overcome these kinds of limitations, many modern surface modification methodologies such as laser surface modification, physical vapor deposition (PVD), and plasma spray-PVD show better performance in surface properties [4]. Key-hole or laparoscopic surgeries have been the focus of the emerging medical advancements. A detailed review of different laparoscopic surgical staplers and mechanical components such as gears, links, pivots, and sliders used in performing their required functions is available. This aids in the identification of individual mechanisms used for certain tasks. It will make it easier to grasp the interrelationships between each sub-component that is required or will be used to complete difficult tasks [5]. Minimally invasive surgical instruments are generally made of 420A grade stainless steel. High intensity bright lights are used during minimally invasive or key-hole surgeries to improve the efficacy of surgeon by enhancing proper visualization of the operating field. But the reflection of incident light from the top layers of instruments hinders surgeon's visual field and leads to lack of precision in key-hole surgeries [6]. Low reflective top layers are key to reducing the disturbances caused due to reflection from bright surgical surfaces [7]. Conversion coating techniques are a widely used method for blackening ferrous materials. Hot aqueous alkaline process is a chemical conversion coating process. In such a process, the substrate surface is converted to black color due to chemical reaction between metal and molten salts, salt solution having an alkaline nature or with air at elevated temperatures [8]. The oxide layers developed on the substrate surface does not change the chemical state of the underlying surface.

In recent years various techniques have been adopted for blackening materials for different applications. Solar absorber made of AISI 316L has been deposited with a black coating using electrochemical deposition technique to reduce reflection [9]. Grey cast iron used for manufacturing electric stove was blackened by using a mixture of sodium hydroxide and sodium nitrate. The major component of the black oxide layer was magnetite and this oxide layer protects the hot plates of the electric stove [10]. The chemical conversion coating process has also been utilized to develop black formation on copper and its alloys. These coatings enhance bonding strength between copper and various polymers [11,12]. The anodizing method has been adopted to modify AZ31 Mg alloys to enhance corrosion resistance and biocompatibility, especially used for biomedical devices [13]. Electroless nickel coating technique has also been adopted to produce coating on the substrate surface to improve various properties of the base material [14]. A black film of titanium aluminum nitride and titanium carbon nitride has been produced on the stainless steel by physical vapor deposition which reduced the undesired light reflection, especially for minimally invasive surgical instruments [3,6]. Diamond-like carbon (DLC) coating has been developed on surgical instruments to enhance the intracorporeal instrument surface and durability of devices by using ionized evaporation method [15]. Despite the fact that many methods for coating metals with black oxide have been established, the scope of reducing reflection from surgical grade 420A stainless steel, which is extensively used for surgical instruments, has not been studied at length.

This paper reports the development of a black iron oxide (magnetite, Fe_3O_4) coating on the SS 420A surface using a hot aqueous alkaline treatment technique having a black color. The coating parameters such as coating duration, chemical composition of bath, temperature of salt solution, pickling effect, and pH value influence the anti-reflection characteristics of the coating developed [11,16–19]. Taguchi optimization technique was used to obtain optimal parameter low-reflective coating on the substrate surface. The XPS analysis, SEM analysis invitro toxicity, corrosion behavior, and sterilization effect of the developed coating were also studied extensively as a part of this research work.

2. Methodology

2.1. Sample Preparation and Coating Procedure

Surgical grade 420A stainless steel sample with the size of 25 mm × 25 mm × 3 mm has been produced from the ingot. The impurity from the surface was removed by blasting process. Glass beads with 60–80 screen size and 0.250–0.180 mm grain size were employed for an efficient blasting process. Further, it was degreased using an alkaline solution. The sample was pickled by using 20% dilute hydrochloric acid and cleaned in running water. For each experimental trial, laboratory-grade sodium hydroxide (NaOH), sodium nitrate ($NaNO_3$), and sodium dichromate ($Na_2Cr_2O_7$)—(Nice Chemicals Ltd., Kerala, India) were dissolved in distilled water to prepare alkaline salt solution. The prepared alkaline mixture is poured into a stainless steel bath. The temperature of the bath can be controlled by using a regulator circuit together with a heating coil capable of heating 200 °C. Further, the samples were immersed inside the alkaline bath for the coating process. The Equations (1)–(3) describe the chemical reactions between stainless steel 420A and as prepared alkaline salt solution which causes the formation of magnetite. Further, the samples were cleaned in de-ionized water and allowed to dry in the air.

$$4Fe + NaNO_3 + 5NaOH \rightarrow 3Na_2FeO_2 + FeO + H_2O + NH_3 \tag{1}$$

$$7Fe + 5Na_2FeO_2 + Na_2Cr_2O_7 + 2NaNO_3 + 5NaOH + 13H_2O \\ \rightarrow 3Fe_3O_4 + Fe_2O_3 + FeO + 2Cr(OH)_3 + 19\,NaOH + 2NH_3 \tag{2}$$

$$7Fe + 5Na_2FeO_2 + Na_2Cr_2O_7 + 2NaNO_3 + 13H_2O \\ \rightarrow 3Fe_3O_4 + Fe_2O_3 + FeO + 2Cr(OH)_3 + 14NaOH + 2NH_3 \tag{3}$$

The ranges of coating parameters of the hot alkaline conversion treatment process were fixed by conducting numerous trials. The development of a homogeneous coating on the SS 420A surface is reliant on different coating variables, such as coating time, bath temperature, and chemical composition of salt solution. The alkaline salt solution is prepared by varying the weight percentage (wt.%) of sodium dichromate in the composition. Three different wt.% of sodium dichromate were considered. Wt.% of 250, 300, and 350 g/L were fixed after conducting the number of trial experiments. It was seen that no uniform black color coating was formed at less than 250 g/L and the coating changed into golden yellow color when we used more than 350 g/L. The coating time was adjusted between 50 and 60 min and the bath temperature was regulated between 115 and 125 °C. Experimental trials showed that lower bath temperature and duration in alkaline solution produced non-uniform black traces on the substrate surface. The duration and bath temperature above 60 min. and 125 °C lead to a change in color, which is undesirable. The wt.% of sodium hydroxide, sodium nitrate, and the pH value of the bath are maintained at constant values of 400 g/L, 320 g/L, and 8 respectively. The images of non-coated and coated stainless steel 420A substrate is as shown in Figure 1. The average percentage reflection (R) is a measurement of the amount of light reflected in the visible region from the substrate surface. A Varian Cary 5000 UV-VIS-NIR spectrophotometer (spectral range of 175–3300 nm, Agilent Technologies, Santa Clara, CA, USA) was used to measure the reflection of incident light (visible range) from the coated surface.

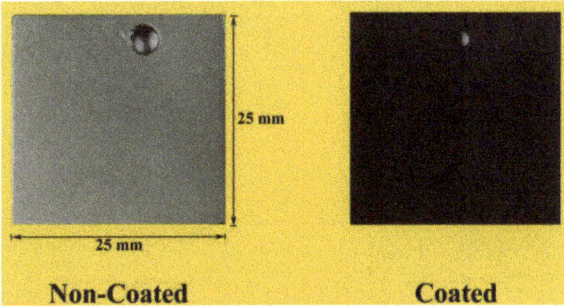

Figure 1. Non-coated and coated stainless steel 420A substrate.

2.2. Taguchi Design of Experiments

To determine the optimal parameter combination for minimizing the average percentage reflection (R), Taguchi orthogonal array (OA) was utilized. The coating time (D), bath temperature (T), and weight % of sodium dichromate (C) are selected as control elements and their levels are as shown in Table 1. The signal-to-noise (S/N) ratio was performed using Minitab-17 software. It is mainly categorized according to the required output characteristics, that are lower-the-better, higher-the-better, or nominal-the-better [20]. Average percentage reflectance (R) should follow lower-the-better criteria, since lower R value is appreciable to obtain a low-reflective coating on SS 420A substrate surface. Higher-the-better condition is adopted to obtain the optimum combination of control factors from the main effect plot [21]. Since a high S/N ratio value indicates the optimum coating parameters in order to avoid the experimental noises as much as possible. Analysis of variance (ANOVA) and regression analysis were also conducted. Contour plots were used to investigate the relationship of output characteristics and two different control variables by analyzing discrete contours [22]. A conventional L9 OA for a 3 parameters 3 levels are as given in Table 2.

Table 1. Selected parameters and experimental levels.

Input Parameter	Symbol	Levels		
		A	B	C
Coating Time (min)	D	50	55	60
Bath temp (°C)	T	115 ± 2	120 ± 2	125 ± 2
Wt.% of sodium dichromate (g/L)	C	250	300	350

Table 2. The L9 OA, experimental results for avg. percentage reflectance.

Sample Code	OA			Avg. Percentage Reflectance	S/N Ratio
	D	T	C	R (%)	R (dB)
I	1	1	1	9.75	−19.7801
II	1	2	2	9.64	−19.6815
III	1	3	3	7.93	−17.9855
IV	2	1	2	9.01	−19.0945
V	2	2	3	8.21	−18.2869
VI	2	3	1	8.27	−18.3501
VII	3	1	3	7.9	−17.9525
VIII	3	2	1	9.04	−19.1234
IX	3	3	2	7.93	−17.9855

2.3. X-ray Photoelectron Spectroscopy (XPS)

XPS was used to identify the compositional and chemical states of the coating produced on the SS 420A substrate. Spectrum were plotted using a XPS, PHI 5000 Versa Probe II, (ULVAC-PHI Inc, Hagisono, Japan) having test parameters with beam spot of 200 μm, power of 15 KV monochromatic X-ray stimulation source Al-Kα (hv=1486.6 eV). Survey scans were recorded with an X-ray source power of 50 W and pass energy of 187.85 eV. High-resolution spectra of the major elements were recorded at 46.95 eV pass energy. The morphological studies and elemental composition of the coating were also analyzed using a scanning electron microscope (JEOL Model JSM-6390LV, JEOL, Tokyo, Japan) equipped with an energy dispersive X-ray analysis (EDAX, Oxford XMX N, Wiesbaden, Germany).

2.4. Cytotoxity Assessment

Direct extract method according to ISO 10993-5 standard was used for invitro cytotoxicity assessment [23,24]. The L-929 cell line was used for the study. The coated and non-coated substrates were disinfected by autoclaving before conducting the assessment. The cultures were incubated in a minimum essential medium (MEM) supplemented with fetal bovine serum (FBS) for 24–26 h at 37 °C. For extract preparation, the disinfected specimens were submerged in the culture medium for 72 ± 2 h and kept at 50 ± 2 °C. Positive and negative controls were prepared by diluting phenol solution and incubating ultra-high molecular weight polyethylene (UHMWPE) for 72 ± 2 h at 50 ± 2 °C with culture medium. The prepared extract was adjusted to 100, 50, and 25 percentages respectively by mixing the very similar cell culture. To determine the cytotoxic effects, various concentrations of test samples as well as positive and negative controls were seeded on the L-929 cell line and kept at 37 ± 1 °C for 24–26 h. The cellular reaction was investigated by analyzing the incubated cells cultured with positive, negative, and test controls using a microscope.

2.5. Corrosion Analysis

Quantitative assessment of corrosion was investigated using potentiodynamic polarization (PDP) tests according to ASTM standard G5-94 [25]. A standard three-electrode with 250 mL capacity cell was used for corrosion analysis. The electrochemical workstation was fitted with saturated calomel electrode (SCE) as the reference electrode, and platinum was employed as auxiliary electrode respectively. Coated, non-coated (control) SS 420A samples were used as working electrode during corrosion test. The electrolyte used for performing corrosion analysis was simulated body fluid (SBF) having a pH of 7.4 to create a natural tissue environment. The methodology adopted to create SBF was Kokubo method [26,27]. The SBF electrolyte was regulated at a temperature of 37 ± 1 °C.

A 1-cm^2 area of the working electrode was kept in contact with the prepared electrolyte. The specimens were allowed to remain in contact with the solution for duration of 3600 s until a constant open circuit potential (OCP) was achieved. The OCP was increased from an initial value of −1000 mV to 1000 mV. The sweep rate was set at 0.0005 mV/s. After conducting the potentiodynamic test on both the coated and non-coated SS 420A, Tafel extrapolation was conducted on polarization curves to calculate the corrosion current density (I_{corr}). The other corrosion parameters were measured to verify the corrosion resistance properties of non-coated and coated samples.

The ASTM B117 standard was followed for performing the salt spray test [28]. The reliability of the test is extremely influenced by the specimen type, evaluation criteria selected, and operating variables. The solution was formed by dissolving 5 ± 1 parts by mass of NaCl in 95 parts of water. The specific gravity and pH value was kept between 1.0268–1.0413 and 6.5–7.2 at 35 °C. During the testing, samples were exposed to a temperature of 35 ± 2 °C in the salt spray chamber for 24 h.

2.6. Repeated Sterilization and Morphology Studies

A repeated sterilization test was performed as per ISO 17665-1 standard for assessing the effect of sterilization temperature on the adhesion property of the black oxide coating [29]. In the conventional method, the temperature of the steam varies from 121–134 °C for a period of 15–21 min to ensure adequate sterilization. In rapid cycle, the samples were held at 134 °C for 3 min in an autoclave. To complete the sterilization process, samples were later allowed to cool. The entire sterilization procedure requires 45 min to one hour. Moreover, surface morphological studies of optimal parameter combination black oxide (Fe_3O_4)-coated SS 420A sample were accomplished with a scanning electron microscope (JEOL Model JSM-6390LV, JEOL, Tokyo, Japan).

3. Results and Discussion

3.1. Process Parameter Optimization

Table 2 shows the observed responses for different combinations of trials as well as the computed S/N ratios corresponding to avg. percentage reflectance (R). The obtained avg. S/N ratios for each level of input parameters are tabulated in Table 3. The ranks are assigned based on the statistical delta values. The parameter corresponding to the highest rank has a major influence on the coating process. From Figure 2, the highest mean S/N ratio for avg. percentage reflectance correspond to 60 min. coating time, 125 °C bath temperature, and 350 g/L wt.% of sodium dichromate. Therefore, the predicted optimal parameter combination for getting low avg. reflectance using Taguchi method were found as D = 60 min., T = 125 °C, and C = 350 g/L and it is represented as D3-T3-C3.

Table 3. Avg. S/N ratios for reflectance.

Level	D	T	C
1	−19.15	−18.94	−19.08
2	−18.58	−19.03	−18.92
3	−18.35	−18.11	−18.07
Delta	0.8	0.92	1.01
Rank	3	2	1

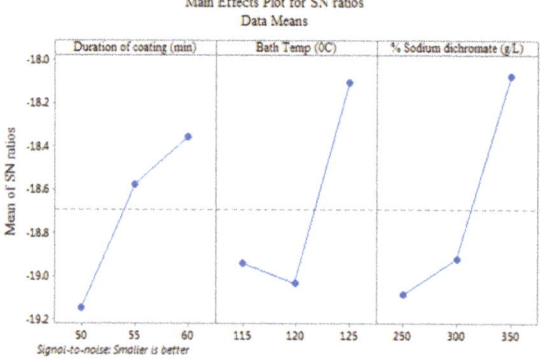

Figure 2. Mean effect plot for reflectance(R).

Table 4 shows the conformation test results for verifying Taguchi's estimated avg. reflectance parameter combination. When comparing S/N ratios of estimated and optimum process parameters, it was found that they were almost identical. On comparing the S/N ratio value of the preliminary parameter combination, an improvement of 2.217 dB was noticed for the Taguchi estimated parameter combination. Additionally, the %R value decreased by 29.21% with respect to the preliminary parameter combination. Hence, the estimated ideal combination was adopted to produce an excellent low-reflective coating on stainless steel 420A with the least R value.

Table 4. Confirmation test for Taguchi predicted optimum parameter combination.

Output Characteristics	Preliminary Parameter Combination	Optimal Combination of Parameter	
		Predicted Values	Experimental Values
Level	D1-T1-C1	D3-T3-C3	D3-T3-C3
Avg. reflectance (%)	9.75	D3-T3-C3	7.546
S/N ratio (dB)	−19.7801	−17.1491	−17.5631
S/N ratio enhancements (dB)	2.217	-	-
Percentage decrement in R	29.21%	-	-

Analysis of variance (ANOVA) results were obtained corresponding to the output characteristics and summarized in Table 5. The p values corresponding to input parameters clearly depict the considerable influence of parameters during magnetite formation on SS 420A. The contributions of each parameter in terms of percentage were calculated and wt.% of sodium dichromate has remarkable influence in the formation of magnetite.

Equation (4) shows the mathematical model was formulated between output characteristics and input variables using regression analysis. The corresponding R-sq value is 85.41%, which denotes the effectiveness of the developed linear regression equation in predicting R within the given input range. The confirmation tests were also conducted to verify the validity of the developed mathematical model. Taguchi predicated optimum combination was considered as input values and the obtained results are summarized in Table 6. From the table, it is observed that, the R value predicted by the formulated mathematical model is nearer to the experimental results, which is desirable.

$$\text{Avg. percentage reflectance (R)} = 26.26 - 0.0817 \times (D) - 0.0843 \times (T) - 0.01007 \times (C) \qquad (4)$$

Table 5. ANOVA results for average reflectance value (R).

Parameters	DF	SS	MS	F	P	% Contribution
Coating time (D)	2	1.08179	0.54088	224.33	0.004	24.55
Bath temp (T)	2	1.56349	0.78174	324.23	0.003	35.49
wt.% of $Na_2Cr_2O_7$ (C)	2	1.75582	0.87991	364.11	0.003	39.85
Error	2	0.00482	0.00241	-	-	-
Total	8	4.40589	-	-	-	-

Table 6. Validation of developed mathematical model.

Coating Characteristics	Optimal Parameter Combination	Predicted Value	Experimental Value
Reflectance (R)	D3-T3-C3 (60 min, 125 °C, 350 g/L)	7.296%	7.546%

Figure 3A–C shows the contour plots that illustrate the desirable average percentage reflectance and corresponding control factors. It is seen that higher bath temperature, longer coating duration, and larger wt.% of sodium dichromate lead to generate coating with low reflectance.

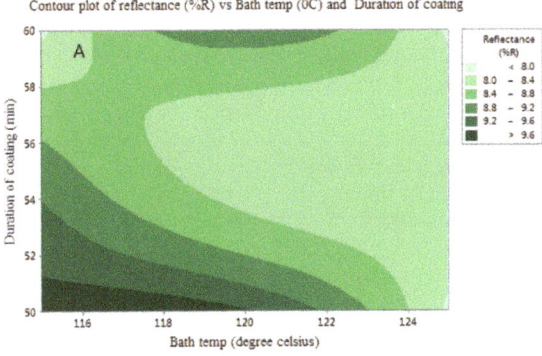

Figure 3. Cont.

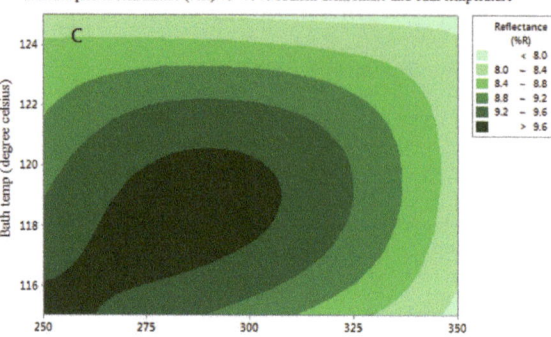

Figure 3. (**A**) Effect of bath temperature and duration of coating on reflectance. (**B**) Effect of wt.% of sodium dichromate and duration of coating on reflectance. (**C**) Effect of wt.% of sodium dichromate and bath temperature on reflectance.

3.2. XPS Analysis of Coating

The surface survey according to XPS indicated the presence of C 1s (85.3%), O 1s (14.1%), Fe 2p3 (0.4%), and Cr 2p3 (0.17%) as in Figure 4. Figure 5A,B represents the high resolution spectrum of C 1s and O 1s. The Fe 2p3/2 spectrum was fitted with major and minor peaks of Fe^{3+} and Fe^{2+} corresponding to the binding energy of 712.4 eV and 709.9 eV as shown in Figure 5C and it is well in accordance with the values of magnetite found. The details about FeO and γ-Fe_2O_3 can be provided easily by separating the corresponding satellite structure [30]. Therefore, the developed black oxide coating is magnetite (Fe_3O_4) [31,32]. The high-resolution spectrum of Cr 2p3/2 displays metallic species of Cr and formation of Cr_2O_3 corresponding to 573.6 eV and 576.1 eV respectively as in Figure 5D. The multiplet splitting at Cr 2p3/2 peak is only because of the presence of chromic oxide (Cr^{3+}) since only trivalent chromium out of the possible oxides of Cr shows pronounced multiplet splitting [33]. To ensure the presence of chromium, the elemental composition of the coating is also analyzed by SEM-EDAX. From Figure 6, the presence of chromium is ensured in the developed coating.

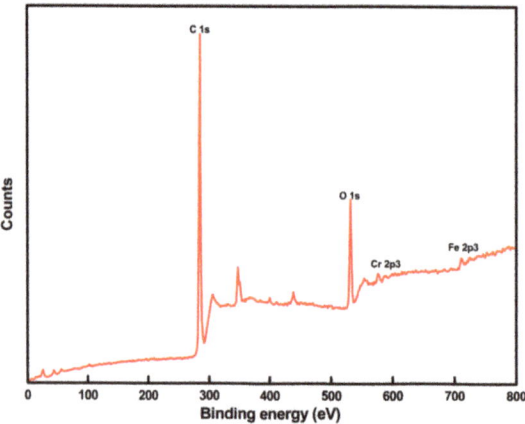

Figure 4. The XPS survey spectra recorded from a thin film magnetite on SS 420A.

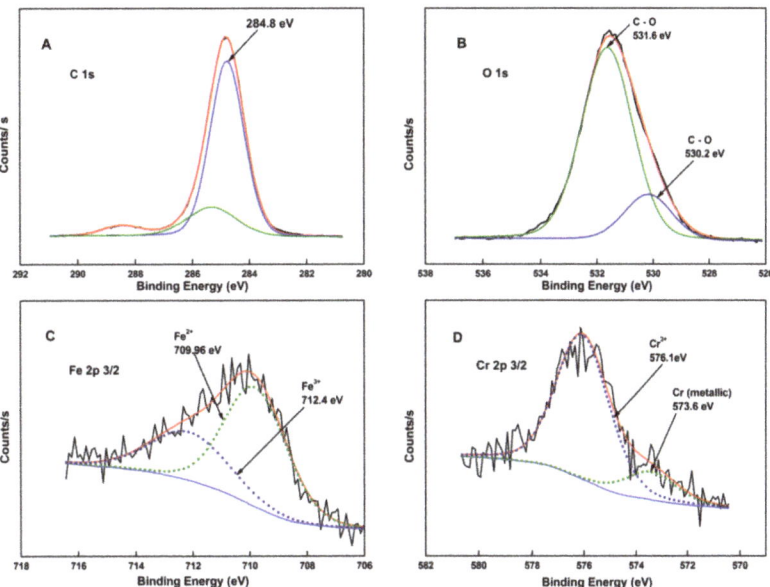

Figure 5. The XPS high-resolution spectrum from the fractured surface of coated sample (**A**) C 1s, (**B**) O 1s, (**C**) Fe 2p3/2, (**D**) Cr 2p3/2.

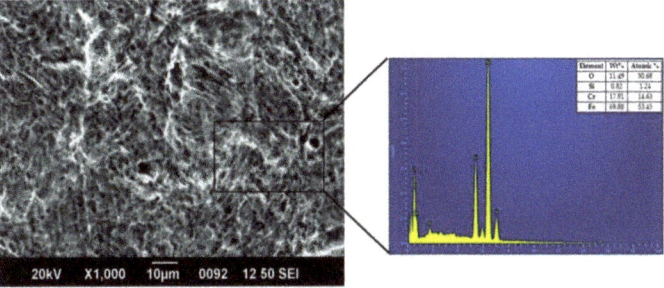

Figure 6. Microstructure and EDAX spectra of Fe_3O_4-coated SS 420A.

3.3. In Vitro Cytotoxicity Study

Cellular response of L-929 cell line cultured with extracts of different samples was graded and evaluated. The cytotoxicity was evaluated by comparing the cell response grade of test samples with negative and positive controls. Negative and positive controls gave grades 0 and 4 as expected. The three diluted extracts, 100%, 50%, and 25% of non-coated (control) and coated SS 420A samples have given a grade of 2, which is acceptable by ISO 10993-5 [23,24]. Figure 7A–F are the microscopic images of the L-929 cell line exposed for a duration of 24 h with 100%, 50% and 25% proportions of extracts of non-coated and coated SS 420A. From the figure, it has been shown that up to 50% of cells are round, devoid of cytoplasmic fine particles, showing no cellular lysis, with vacant spaces between cells. These findings suggest that the developed low reflective chemically formed magnetite on stainless steel 420A is non-toxic.

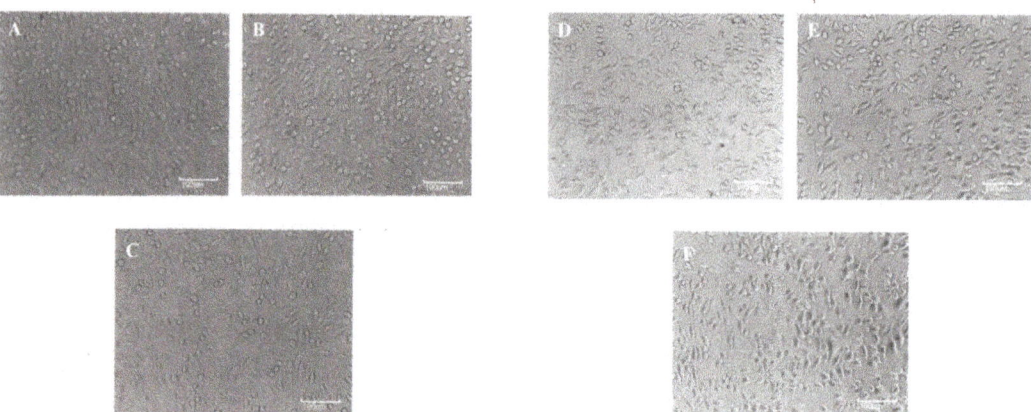

Figure 7. Microscopic image of L-929 cells exposed to different proportions (100%, 50% and 25%) of extract media of (**A–C**) non-coated SS 420A (**D–F**) coated SS 420.

3.4. Potentiodynamic Polarization and Salt Spray Studies

The corrosion behavior of control and coated SS 420A was evaluated using an electrochemical workstation together with simulated body fluid (SBF) environment and repeated at least three times. The polarization curves for different samples are given in Figure 8. The measured electrochemical parameters from PDP curves are tabulated in Table 7. From the table, it was noticed that corrosion current density (I_{corr}) having a strict linear relationship with corrosion rate (CR). Thus, the lower the I_{corr} value the greater the resistance to corrosion [34]. Hence the magnetite over-layered SS 420A sample has slight improvement in resistance to accelerated corrosion. Moreover, the higher E_{corr} value of magnetite layered SS 420A, indicates chemical inertness and the lowest corrosion tendency. The measured R_p value also indicates that the coated substrate shows slight improvement against applied accelerated corrosion compared to the non-coated SS 420A. The corrosion resistance of coated samples was also verified by the salt spray test. Samples were visually inspected and no rust was formed after 24 h test and it shows the developed coating can withstand corrosion.

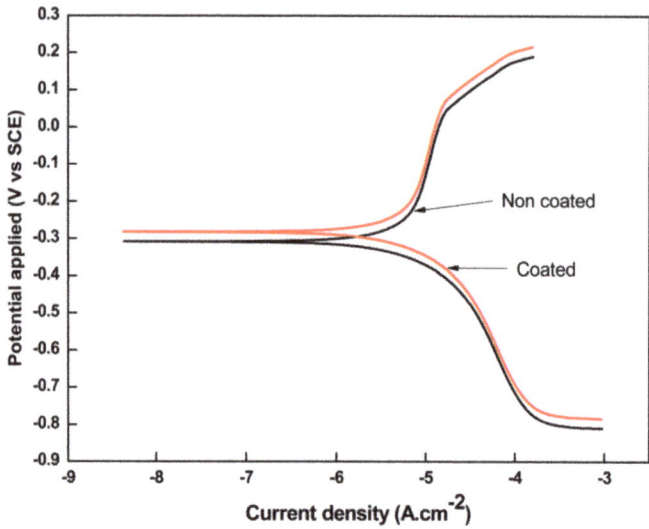

Figure 8. Potentiodynamic polarization curves of non-coated (control) and coated SS 420A.

Table 7. Polarization parameters of coated and non-coated SS 420A in SBF.

Sample	Corrosion Parameters					
	E_{corr} (mv)	I_{corr} (mA/cm^2)	β_a (mV·decade^{-1})	β_c (mV·decade^{-1})	R_p (Ω Cm^2)	Corrosion Rate (mm/y)
SS 420A Coated Sample	−283.57	0.0239	−910.88	392.74	8855.7	0.310
SS 420A Non coated (Control Sample)	−308.51	0.0351	−863.36	379.16	8029.1	0.413

3.5. Effect of Repeated Sterilization on Coating

The SEM images (20 kV and 1000× magnification) of coated samples before and after conducting 100 cycles of repeated sterilization as per ISO standards are shown in Figure 9. No damages like discoloration peel off were observed after the repeated autoclaving process. From Figure 9B, minor stress relief cracks were observed. This is due to sudden cooling of the coating and the substrate from treatment temperature to room temperature [35]. If the developed coating detaches due to the sterilization temperature from the instrument during the surgery, it can lead to serious damages to surrounding tissues. The average percentage reflectance (R) value was measured to analyze the effect of repeated sterilization on coatings. The average percentage reflectance graph of coated SS420A before and after steam sterilization is given in Figure 10A,B. The measured average percentage reflectance value of coated substrate before steam sterilization is 7.546%. From Figure 10B, the coating exhibits excellent antireflective property with a minimum R value in the visible region of 8.32% after undergoing100 cycles of repeated sterilization on uniform black magnetite forming SS 420A. There is smaller variation in the R value of coated sample before and after repeated sterilization procedure. Even though the reflection of light comparatively is much lesser than not coated SS 420A. This indicates, repeated sterilization does not adversely affect the antireflective property of the developed coating.

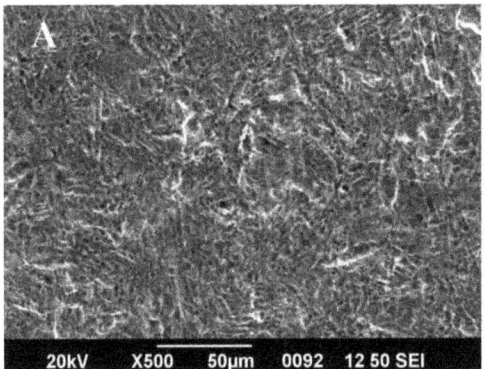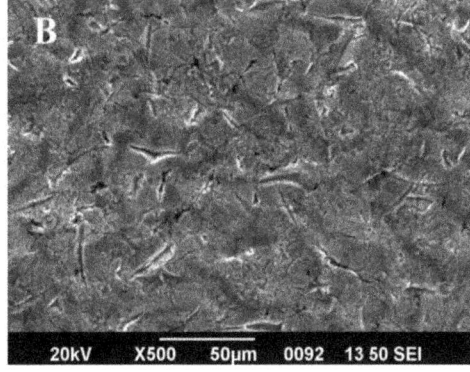

Figure 9. SEM images of coated SS 420A (**A**) before sterilization (**B**) after 100 cycles of sterilization.

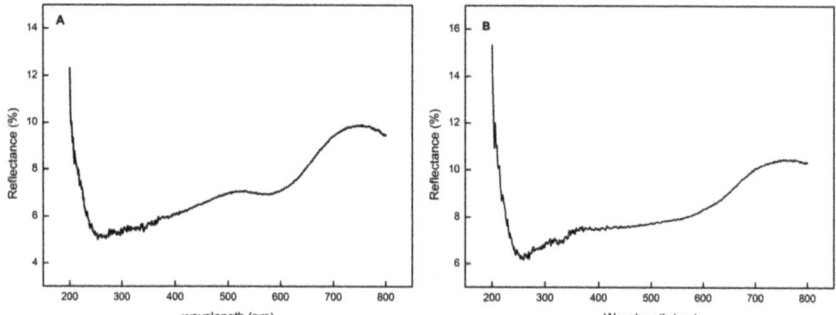

Figure 10. The average percentage of reflectance in the visible region (**A**) before steam sterilization (**B**) after 100 cycles of steam sterilization.

4. Conclusions

A low reflective black iron oxide (magnetite) coating was produced on stainless steel 420A using chemical hot aqueous alkaline conversion coating technique. The coating parameter values of alkaline conversion coating process were optimized using the Taguchi optimization technique. The optimum coating parameter combination for minimum average percentage reflection (R) value is obtained as D3-T3-C3 (D = 60 min., T = 125 °C and C = 350 g/L). The R value of uniform black coated SS 420A is reduced to 7.546% from a value of 56.14% of non-coated samples. It is seen that the chemical composition of bath and bath temperature has greatest impact in magnetite formation on SS 420A. This coating reduces the undesired difficulties of surgeons due to scattering and reflection of incident light in the visible spectrum from the exteriors of surgical equipment manufactured by SS 420A. The developed coating was characterized using XPS analysis and the formation of magnetite (Fe_3O_4) was identified. An invitro cytotoxicity test was performed using the L-929 cell line and confirmed the non-toxic behavior of the developed coating. Corrosion studies showed that magnetite layered SS 420A has better corrosion resistance properties than the non-coated substrate. A repeated sterilization test was conducted to measure the ability of the developed coating to withstand the sterilization process. The average percentage reflectance (R) value slightly increased to 8.32% after 100 cycles of autoclaving, which is desirable. These results suggest that the developed coating may be used for producing a biocompatible low-reflective black coating for surgical equipment composed of medical grade 420A stainless steel.

Author Contributions: R.A.R., contributed to the experimental and analytical work of the research; K.K.S., contributed to the validation and evaluation of the findings reported in the paper; R.A., contributed to evaluation of the findings. All authors have read and agreed to the published version of manuscript.

Funding: We did not receive any specific grant from funding agencies in the public, commercial, or not-for-profit sectors.

Institutional Review Board Statement: Not applicable.

Informed Consent Statement: Not applicable.

Data Availability and Statement: The authors declare that data supporting the findings of this study are available within the article.

Acknowledgments: We would like to thank DST-SAIF, Cochin, NIIST Trivandrum, ICAR-CIFT, Cochin and Jayon Implants, Palakkad, for providing all necessary support towards this research work.

Conflicts of Interest: The authors declare that they have no conflict of interests.

Abbreviations

Nomenclature

SS	Stainless steel
XPS	X-ray photoelectron spectroscopy
R	Average percentage reflection
AISI	American iron and steel institute
DLC	Diamond-Like Carbon
CVD	Chemical vapor deposition
PVD	Physical vapor deposition
wt.%	Weight percentage
D	Duration of coating
T	Bath temperature
C	wt% of sodium dichromate
S/N	Signal-to-noise
OA	Orthogonal Array
DF	Degrees of freedom
SS	Sum of squares
MS	Mean square
F	Fisher's ratio
P	Probability value
MEM	Minimum essential medium
FBS	Fetal bovine serum
UHMWPE	Ultra-high molecular weight polyethylene
SCE	Saturated calomel electrode
PDP	Potentiodynamic polarization
SBF	Simulated body fluid
OCP	Open circuit potential
I_{corr}	Corrosion current
E_{corr}	Corrosion potential
R_p	Polarization resistance
β_a	Anodic Tafel slope
β_c	Cathodic Tafel slope
CR	Corrosion rate

References

1. Festas, A.; Ramos, A.; Davim, J.P. Medical devices biomaterials—A review. *Proc. Inst. Mech. Eng. Part L J. Mater. Des. Appl.* **2019**, *234*, 218–228. [CrossRef]
2. Chauhan, L.R.; Singh, M.; Bajpai, J.K.; Misra, K.; Agarwal, A. Development of chemical conversion coating for blackening of a grade of stainless steel useful for fabrication of optical devices. *J. Surf. Sci. Technol.* **2017**, *32*, 99. [CrossRef]
3. Ali, S.; Rani, A.M.A.; Baig, Z.; Ahmed, S.W.; Hussain, G.; Subramaniam, K.; Hastuty, S.; Rao, T.V. Biocompatibility and corrosion resistance of metallic biomaterials. *Corros. Rev.* **2020**, *38*, 381–402. [CrossRef]
4. Zhang, L.-C.; Chen, L.-Y.; Wang, L. Surface modification of titanium and titanium alloys: Technologies, developments, and future interests. *Adv. Eng. Mater.* **2020**, 1–16. [CrossRef]
5. Lim, J.J.; Erdman, A.G. A review of mechanism used in laparoscopic surgical instruments. *Mech. Mach. Theory* **2003**, *38*, 1133–1147. [CrossRef]
6. Hollstein, F.; Louda, P. Bio-compatible low reflective coatings for surgical tools using reactive d.c.-magnetron sputtering and arc evaporation—a comparison regarding steam sterilization resistance and nickel diffusion. *Surf. Coat. Technol.* **1999**, *120–121*, 672–681. [CrossRef]
7. Reghuraj, A.; Saju, K. Black oxide conversion coating on metals: A review of coating techniques and adaptation for SAE 420A surgical grade stainless steel. *Mater. Today Proc.* **2017**, *4*, 9534–9541. [CrossRef]
8. Ooi, S.W.; Yan, P.; Vegter, R.H. Black oxide coating and its effectiveness on prevention of hydrogen uptake. *Mater. Sci. Technol.* **2018**, *35*, 12–25. [CrossRef]
9. Lira-Cantú, M.; Sabio, A.M.; Brustenga, A.; Gomez-Romero, P. Electrochemical deposition of black nickel solar absorber coatings on stainless steel AISI316L for thermal solar cells. *Sol. Energy Mater. Sol. Cells* **2005**, *87*, 685–694. [CrossRef]
10. Nagode, A.; Klančnik, G.; Schwarczova, H.; Kosec, B.; Gojić, M.; Kosec, L. Analyses of defects on the surface of hot plates for an electric stove. *Eng. Fail. Anal.* **2012**, *23*, 82–89. [CrossRef]

11. Lebbai, M.; Kim, J.-K.; Szeto, W.K.; Yuen, M.M.F.; Tong, P. Optimization of black oxide coating thickness as an adhesion promoter for copper substrate in plastic integrated-circuit packages. *J. Electron. Mater.* **2003**, *32*, 558–563. [CrossRef]
12. Kim, J.-K.; Woo, R.S.; Hung, P.Y.; Lebbai, M. Adhesion performance of black oxide coated copper substrates: Effects of moisture sensitivity test. *Surf. Coat. Technol.* **2006**, *201*, 320–328. [CrossRef]
13. Zaffora, A.; Di Franco, F.; Virtù, D.; Pavia, F.C.; Ghersi, G.; Virtanen, S.; Santamaria, M. Tuning of the Mg alloy AZ31 anodizing process for biodegradable implants. *ACS Appl. Mater. Interfaces* **2021**, *13*, 12866–12876. [CrossRef]
14. Sarkar, S.; Mukherjee, A.; Baranwal, R.K.; De, J.; Biswas, C.; Majumdar, G. Prediction and parametric optimization of surface roughness of electroless Ni-Co-P coating using Box-Behnken design. *J. Mech. Behav. Mater.* **2019**, *28*, 153–161. [CrossRef]
15. Kaneko, M.; Hiratsuka, M.; Alanazi, A.; Nakamori, H.; Namiki, K.; Hirakuri, K. Surface reformation of medical devices with DLC coating. *Materials* **2021**, *14*, 376. [CrossRef]
16. Sharma, A.; Rani, R.U.; Mayanna, S. Thermal studies on electrodeposited black oxide coating on magnesium alloys. *Thermochim. Acta* **2001**, *376*, 67–75. [CrossRef]
17. Tepe, B.; Gunay, B. Evaluation of pre-treatment processes for HRS (hot rolled steel) in powder coating. *Prog. Org. Coat.* **2008**, *62*, 134–144. [CrossRef]
18. Li, G.; Niu, L.; Lian, J.; Jiang, Z. A black phosphate coating for C1008 steel. *Surf. Coat. Technol.* **2004**, *176*, 215–221. [CrossRef]
19. Zhao, M.; Wu, S.; Luo, J.; Fukuda, Y.; Nakae, H. A chromium-free conversion coating of magnesium alloy by a phosphate–permanganate solution. *Surf. Coat. Technol.* **2006**, *200*, 5407–5412. [CrossRef]
20. Mohsin, I.; He, K.; Li, Z.; Zhang, F.; Du, R. Optimization of the polishing efficiency and torque by using taguchi method and ANOVA in robotic polishing. *Appl. Sci.* **2020**, *10*, 824. [CrossRef]
21. Zalnezhad, E.; Sarhan, A.A.D.; Hamdi, M. Optimizing the PVD TiN thin film coating's parameters on aerospace AL7075-T6 alloy for higher coating hardness and adhesion with better tribological properties of the coating surface. *Int. J. Adv. Manuf. Technol.* **2012**, *64*, 281–290. [CrossRef]
22. Kumar, S.; Singh, R.; Jaiswal, R.; Kumar, A. Optimization of process parameters of electron beam welded Fe49Co2V alloys. *Int. J. Eng. Trans. B Appl.* **2020**, *33*, 870–876. [CrossRef]
23. Liu, X.; Rodeheaver, D.P.; White, J.C.; Wright, A.M.; Walker, L.M.; Zhang, F.; Shannon, S. A comparison of in vitro cytotoxicity assays in medical device regulatory studies. *Regul. Toxicol. Pharmacol.* **2018**, *97*, 24–32. [CrossRef]
24. Vidal, M.N.P.; Granjeiro, J.M. Cytotoxicity tests for evaluating medical devices: An alert for the development of biotechnology health products. *J. Biomed. Sci. Eng.* **2017**, *10*, 431–443. [CrossRef]
25. Farzam, M.; Baghery, P.; Dezfully, H.R.M. Corrosion study of steel API 5A, 5L and AISI 1080, 1020 in drill-mud environment of Iranian hydrocarbon fields. *Int. Sch. Res. Not.* **2011**, *2011*, 681535. [CrossRef]
26. Zaludin, M.A.F.; Jamal, Z.A.Z.; Derman, M.N.; Kasmuin, M.Z. Fabrication of calcium phosphate coating on pure magnesium substrate via simple chemical conversion coating: Surface properties and corrosion performance evaluations. *J. Mater. Res. Technol.* **2019**, *8*, 981–987. [CrossRef]
27. Kokubo, T.; Takadama, H. How useful is SBF in predicting in vivo bone bioactivity? *Biomaterials* **2006**, *27*, 2907–2915. [CrossRef] [PubMed]
28. Mitchell, J.; Crow, N.; Nieto, A. Effect of surface roughness on pitting corrosion of AZ31 Mg Alloy. *Metals* **2020**, *10*, 651. [CrossRef]
29. Vannozzi, L.; Catalano, E.; Telkhozhayeva, M.; Teblum, E.; Yarmolenko, A.; Avraham, E.S.; Konar, R.; Nessim, G.D.; Ricotti, L. Graphene oxide and reduced graphene oxide nanoflakes coated with glycol chitosan, propylene glycol alginate, and polydopamine: Characterization and cytotoxicity in human chondrocytes. *Nanomaterials* **2021**, *11*, 2105. [CrossRef] [PubMed]
30. Ramazanov, S.; Sobola, D.; Orudzhev, F.; Knápek, A.; Polčák, J.; Potoček, M.; Kaspar, P.; Dallaev, R. Surface modification and enhancement of ferromagnetism in BiFeO$_3$ nanofilms deposited on HOPG. *Nanomaterials* **2020**, *10*, 1990. [CrossRef] [PubMed]
31. Yamashita, T.; Hayes, P. Analysis of XPS spectra of Fe^{2+} and Fe^{3+} ions in oxide materials. *Appl. Surf. Sci.* **2008**, *254*, 2441–2449. [CrossRef]
32. Grosvenor, A.; Kobe, B.A.; Biesinger, M.C.; McIntyre, N.S. Investigation of multiplet splitting of Fe 2p XPS spectra and bonding in iron compounds. *Surf. Interface Anal.* **2004**, *36*, 1564–1574. [CrossRef]
33. Liu, R.; Conradie, J.; Erasmus, E. Comparison of X-ray photoelectron spectroscopy multiplet splitting of Cr 2p peaks from chromium tris(β-diketonates) with chemical effects. *J. Electron Spectrosc. Relat. Phenom.* **2016**, *206*, 46–51. [CrossRef]
34. Faghani, G.; Rabiee, S.M.; Nourouzi, S.; Elmkhah, H. Corrosion behavior of TiN/CrN nanoscale multi-layered coating in Ringer's solution. *Int. J. Eng.* **2020**, *33*, 329–336. [CrossRef]
35. Hanoz, D.; Settimi, A.G.; Dabalà, M. Characterization of black coating on Fe360 steel obtained with immersion in aqueous solutions. *Surf. Interfaces* **2021**, *26*, 101317. [CrossRef]

Article

Microstructural and Mechanical Properties Characterization of Graphene Oxide-Reinforced Ti-Matrix Composites

Zhaomei Wan, Jiuxiao Li *, Dongye Yang and Shuluo Hou

School of Materials Engineering, Shanghai University of Engineering Science, Shanghai 201620, China; wanzhaomei123@163.com (Z.W.); ydy_hit@163.com (D.Y.); hsl111hsl@163.com (S.H.)
* Correspondence: lijiuxiao@126.com

Abstract: The 0.1–0.7 wt.% graphene oxide (GO)-reinforced Ti-matrix composites (TMCs) were prepared by the hot-pressed sintering method. The effects of GO content on the mechanical properties of TMCs were investigated. The microstructure of TMCs was analyzed. The results show that the microstructure of Ti and TMCs is equiaxed α. The average grain size of TMCs decreases with GO increasing. GO can react with Ti to form TiC at high temperatures. Meanwhile, GO is also presented in the matrix. The hardness of TMCs is higher than that of pure Ti. The maximum hardness is 320 HV, which is 43% higher than that of pure Ti. The yield strength of Ti-0.5 wt.% GO sintered at 1373 K is 1324 MPa, 77% more than pure Ti. The strengthening mechanism of TMCs is the fine-grained strengthening and the reinforcement that bear the stress from the matrix. The friction coefficient of Ti-0.3 wt.% GO sintered at 1373 K comes up to 0.50, which is reduced by 0.2 compared with pure Ti.

Keywords: Ti-matrix composites (TMCs); graphene oxide (GO); mechanical properties; reinforcement

Citation: Wan, Z.; Li, J.; Yang, D.; Hou, S. Microstructural and Mechanical Properties Characterization of Graphene Oxide-Reinforced Ti-Matrix Composites. Coatings 2022, 12, 120. https://doi.org/10.3390/coatings12020120

Academic Editor: Anton Ficai

Received: 6 December 2021
Accepted: 16 January 2022
Published: 21 January 2022

Publisher's Note: MDPI stays neutral with regard to jurisdictional claims in published maps and institutional affiliations.

Copyright: © 2022 by the authors. Licensee MDPI, Basel, Switzerland. This article is an open access article distributed under the terms and conditions of the Creative Commons Attribution (CC BY) license (https://creativecommons.org/licenses/by/4.0/).

1. Introduction

Ti and Ti alloys, with their excellent mechanical properties, corrosion resistance and relatively low density, are essential materials for structural applications in aerospace, defense, automotive, etc. [1–5]. In the past, a large amount of research has been carried out to improve the properties of Ti alloys by adding reinforcing phases, such as graphite, carbon nanotubes, TiC and TiB, etc. [6–10].

Compared with the common materials, graphene has excellent mechanical properties, thermal and electrical conductivity [11–14]. As a reinforcing phase, graphene was widely used to improve the properties of metal-matrix composites [15–19]. Shin et al. [20] reported that few-layer graphene-reinforced Al-matrix composites by powder metallurgic method. The compressive strength of Al-matrix composites with 0.7 vol.% graphene was twice that of pure Al. Chen et al. [21] Added different graphene contents into the Cu matrix, the results showed that the yield strength of Cu-matrix composites with 0.6 vol.% graphene was about twice as pure Cu, and the friction coefficient of the composite was about 0.25, which was about 40% of that of pure Cu. There are hydroxyl, carboxyl and oxidizing functional groups on the surface of graphene oxide (GO), which can improve the dispersion of GO in the matrix [22]. Shuai et al. [23] prepared GO-reinforced magnesium alloy by laser melting method, and the research results showed that the compression yield strength of the magnesium alloy with 1wt.% graphene was increased by 30%. Mu et al. [24] prepared graphene nanosheets discontinuous reinforced Ti-matrix composites (TMCs) by powder metallurgy. The test results showed that the ultimate strength of TMCs containing 0.1 wt.% graphene nanoplates (GNPS) was 54.2% higher than that of the Ti matrix. Cao et al. [25] synthesized GNPs-reinforced TMCs with 1.2 vol.% GNPs. Compared with the monolithic titanium alloy, the composite with 1.2 vol.% GNPs exhibits significantly improved elastic modulus and strength. The sliding wear test shows that the wear volume loss of composite with 1.2 vol.% GNPs decreased by 18% than pure Ti. Dong et al. [26] fabricated oxide nanosheets (GONs)-reinforced TMCs composites. It was revealed that yield strength and ultimate tensile strength of TMCs with the 0.6 wt.% GONs were increased by 7.44% and 9.65% as compared to those of pure Ti. Cao et al. [27] reported the tensile strength at room temperature after R&A can reach 1206 MPa for 0.3 wt.% GNP-reinforced TMCs, which increased by 46% compared with pure Ti.

These reported TMCs demonstrated that the interface between graphene and Ti matrix owns effective load-transfer ability due to the strong Ti-C ionic bond and TiC reaction products effect to enhance mechanical properties. However, graphene is not easy to be uniformly dispersed in matrix attributed to graphemic nano-characters and strong Vander Waals forces between graphene, limiting the improvement of mechanical properties. In this paper, powder metallurgy (PM) is a favorable method in MMCs due to its low cost, flexibility, and ease of control. Irregular Ti powder has a much higher apparent volume than spherical powders, which is beneficial for the uniform distribution of GO. GO at a low content level usually exhibit excellent strengthening effect in TMCs, which can be attributed to its various oxygen functional groups (hydroxyl, carboxyl acid, and epoxy) on the surface and sheet edges, improving dispersibility in solvents. The interface wettability between in-situ TiC and titanium matrix is excellent, which can significantly improve the interface bonding and matching, and improve the interface strength. In this work, GO-reinforced TMCs were prepared and microstructures were observed. In addition, the mechanical properties and strengthening mechanisms of GO-reinforced TMCs are discussed.

2. Experimental Procedure

The particle size of pure Ti powder is about 150 um. The dispersion of GO was mixed with pure Ti by ultrasonic stirring method and then GO-reinforced TMCs were sintered in a vacuum hot-pressed sintering furnace. The mass percentages of GO are 0.1 wt.%, 0.3 wt.%, 0.5 wt.% and 0.7 wt.% respectively. The sample was sintered for 1 h (1273 K, 1373 K, pressure of 30 MPa, vacuum of 1×10^{-3} Pa) in a vacuum hot-pressed sintering furnace. The details of TMCs reinforced with 0–0.7 wt.% GO are shown in Table 1.

Table 1. Materials of different GO content at different sintering temperatures.

Materials	GO Content (wt.%)	Sintering Temperature (K)
Ti1	0	1273
TMC1	0.1	1273
TMC2	0.3	1273
TMC3	0.5	1273
TMC4	0.7	1273
Ti2	0	1373
TMC5	0.1	1373
TMC6	0.3	1373
TMC7	0.5	1373
TMC8	0.7	1373

The microstructure was analyzed by optical microscope (OM, Olympus-BX53, Olympus, Tokyo, Japan) and scanning electron microscope (SEM, NOVA, NanoSEM 230, FEI, Hillsboro, OR, USA). The phase analysis of the sample was carried out by X-ray diffraction (XRD, D-max 2550 V, Japan Science Corporation, Tokyo, Japan) with a Cu-Kα radiation source (5–100°, 5°/min). The accelerating voltage and tube current are 30 KV and 25 mA. The average grain size was measured by the average interception method. The hardness of the sample was measured by hardness tester (MH-VK, Shanghai Taiming Optical Instrument Co., Ltd., Shanghai, China) at 200 N for 15 s. Nine different points were taken for each sample to be averaged. Room compression test was conducted on Zwick–Z020 (Zwick, Ulm, Germany). The compression rate is 0.6 mm/min. The size of compression samples is ⌀4 mm × 6 mm. A tribological test was conducted on a friction and wear testing machine (HT-1000, Zhongke Kaihua Technology Development Co., Ltd., Lanzhou, China). The method of dry friction of the ball plate was adopted. GCr15 ball with a diameter of 6 mm was used as friction pair. The sliding time was 1200 s under a regular load of 3 N. Friction and wear experiments were repeated three times to take the average value. The wear width, depth and volume were measured by surface profiler (MT-500, producerZhongke Kaihua Technology Development Co., Ltd., Lanzhou, China).

3. Result Discussion

3.1. Phase Analysis

Figure 1 shows the XRD pattern of TMCs. Figure 1a,b show samples sintered at 1273 K and 1373 K. The peak of α-Ti is detected by XRD. The C peak is at 26.5° (002). Figure 1c is the locally enlarged view of 25–28° in Figure 1b. The peak of TiC appeared at 41.08 (200). Figure 1d is the locally enlarged view of 40.4–42° in Figure 1b. The peak of TiO is at 29.4° (101) and 30.8° (103). It is evident that the intensity of TiC increases with the increase of the GO content. The result indicates that GO reacts with Ti at high temperatures. The Gibbs free energy of TiC from the reaction of GO with Ti can be calculated by the following formula [28].

$$Ti + C = TiC \tag{1}$$

$$\Delta G = -184571.8 + 41.382t - 5.042\, Tlnt + 2.425 \times 10 - 3T2 - 9.79 \times 105/T \tag{2}$$
$$(T < 1939\ K).$$

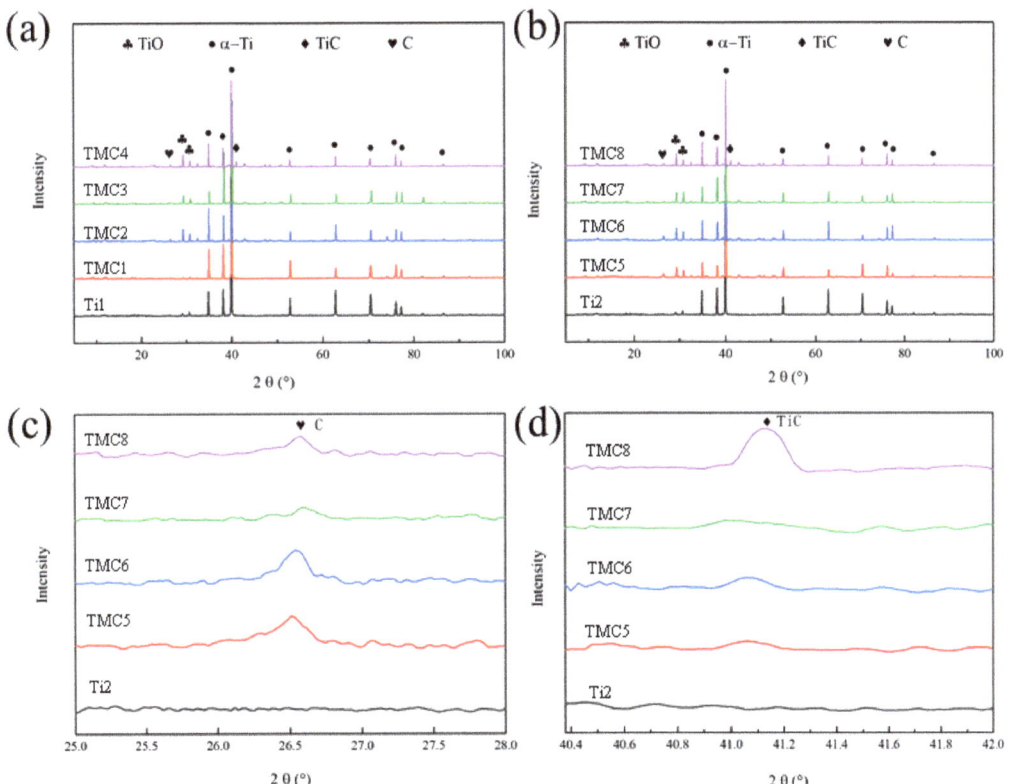

Figure 1. X-ray diffraction of Ti and TMCs sintered at: (**a**) 1273 K; (**b**) 1373 K; (**c**) 25–28° in (**b**); (**d**) 40.4–42.0° in (**b**).

The calculated Gibbs free energy of TMCs sintered at 1273 K and 1373 K are about −157.75 kJ/mol and −174 kJ/mol. It shows that the hard second phase TiC can be formed spontaneously via a reaction between GO and Ti during hot-pressed sintering.

3.2. Microstructure

The SEM micrographs of TMCs are outlined in Figure 2. The grain boundary is relatively apparent. The microstructure of Ti and TMCs is equiaxed α. There are no well-defined pores in the material. With the increase of GO content, the GO aggregation can be displayed in Figure 2e,i,j. With the increase of sintering temperature, Ti atoms in the composites can migrate effectively, thus achieving better densification. The average grain size of TMCs decreases with GO increasing. The average grain sizes of Ti and TMCs are listed in Table 2. The average grain size of TMC8 is about 38.3 μm, which decreases by 52% compared with that of Ti2. In the process of grain growth, GO and TiC can hinder the movement of the grain boundary, and the grain growth is restricted. So, the grain size of TMCS is smaller than pure Ti.

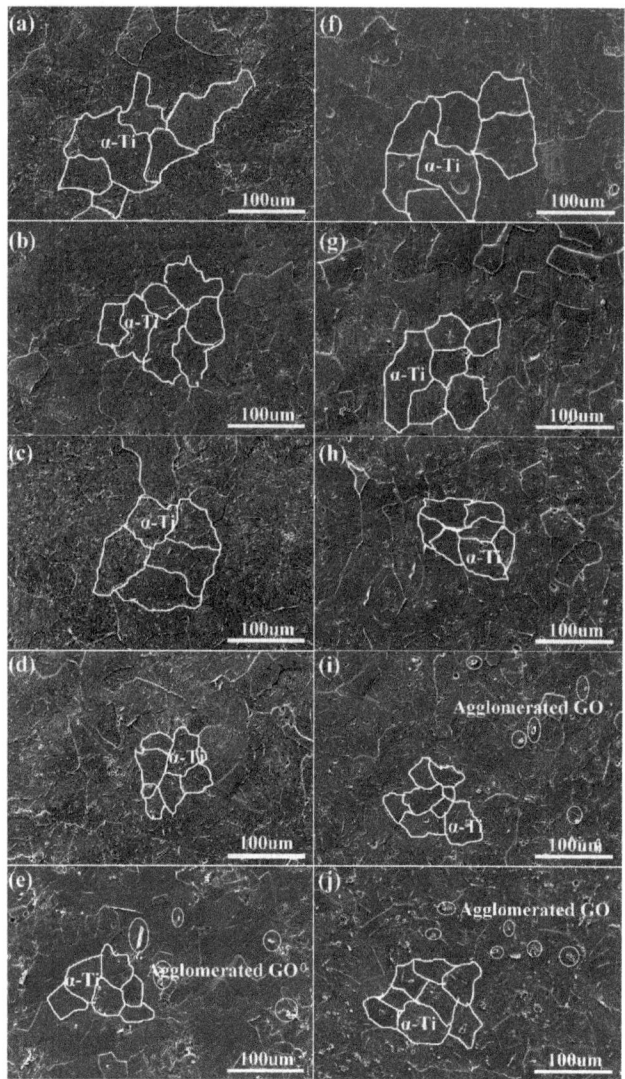

Figure 2. The SEM diagram of TMCs. (**a**) Ti1; (**b**) TMC1; (**c**) TMC2; (**d**) TMC3; (**e**) TMC4; (**f**) Ti2; (**g**) TMC5; (**h**) TMC6; (**i**) TMC7; (**j**) TMC8.

Table 2. Average grain Size of Ti and TMCs.

Materials	Average Grain Size (μm)	Materials	Average Grain Size (μm)
Ti1	78.5 ± 0.4	Ti2	80.3 ± 0.6
TMC1	70.6 ± 0.8	TMC5	76.5 ± 0.8
TMC2	57.2 ± 0.7	TMC6	63.4 ± 0.7
TMC3	43.5 ± 0.9	TMC7	50.6 ± 0.8
TMC4	36.7 ± 0.6	TMC8	38.3 ± 0.5

The EDS of TMC8 is shown in Figure 3. Location 1 is the enrichment area of O and C, and location 2 is the enrichment area of C. Ti is less at 1 and 2 locations. Figure 4 is the EDS of TMC8, which contains both strip-like second phase and granular matter. Figure 4a is the line scanning of striped second phase matter. The content of C and O is higher than that in other areas, and the content of Ti is lower than that in other areas. According to the XRD analysis results, it is inferred that it was undamaged flake-like GO. The result is consistent with Dong's work [29]. Figure 4b is the analysis results of granular matter. The C appears, and the content of Ti is reduced compared with other areas. Combined with the analysis of XRD results, it is judged that this is TiC generated by the reaction of GO and Ti. Similar results have been reported [30,31].

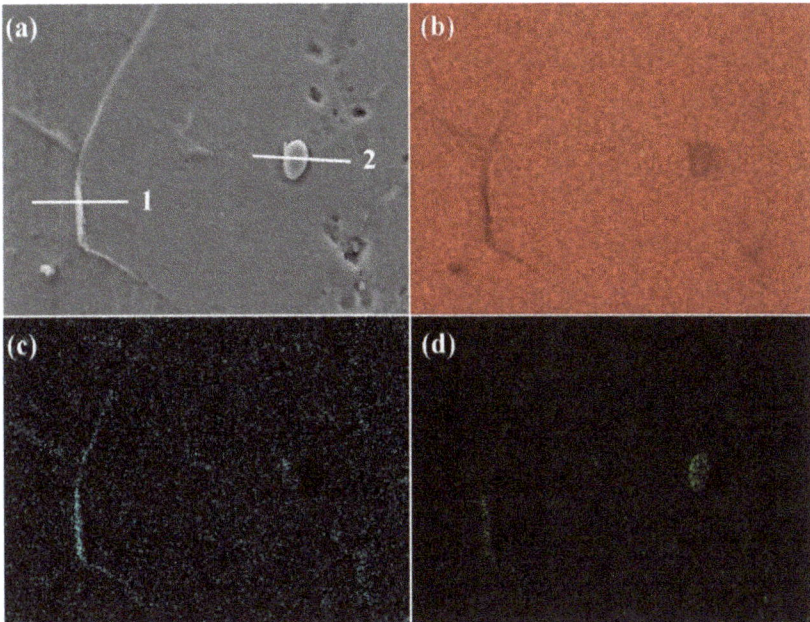

Figure 3. EDS of TMC8. (a) Surface scanning of TMC8; (b) Ti distribution in (a); (c) O distribution in (a); (d) C distribution in (a).

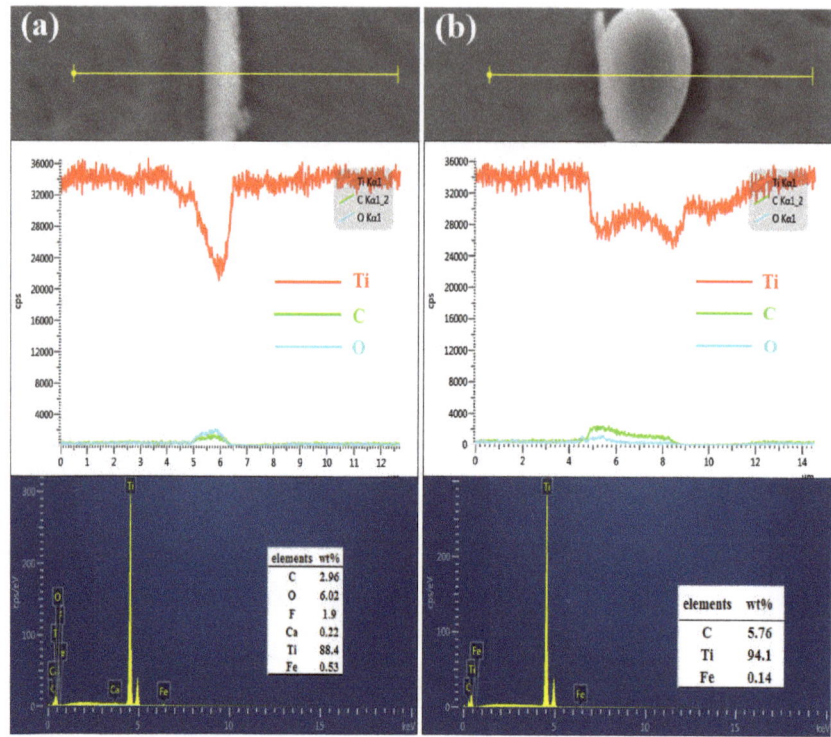

Figure 4. EDS of TMC8. (**a**) Line scanning surface of striped second-phase matter; (**b**) line scanning surface of granular matter.

3.3. Properties Analysis

3.3.1. Hardness

The hardness of Ti and TMCs is shown in Figure 5 and Table 3. The hardness of the TMCs becomes higher and higher with the increase of GO content. The hardness of TMC4 and TMC8 are 300 HV and 320 HV, which is separately increased by 42% and 43% compared with that of pure Ti. As is mentioned above, GO reacts with pure Ti to form TiC at high temperature. The hard second phase TiC can resist local plastic deformation of TMCs and improve the hardness of TMCs. The higher the temperature, the greater the activation energy and the more TiC, which raises the hardness of TMCs sintered at 1373 K.

Table 3. Hardness of Ti and TMCs.

Materials	Hardness/HV	Materials	Hardness/HV
Ti1	210 ± 3	Ti2	223 ± 5
TMC1	239 ± 6	TMC5	241 ± 6
TMC2	245 ± 4	TMC6	250 ± 4
TMC3	268 ± 2	TMC7	280 ± 3
TMC4	300 ± 5	TMC8	320 ± 4

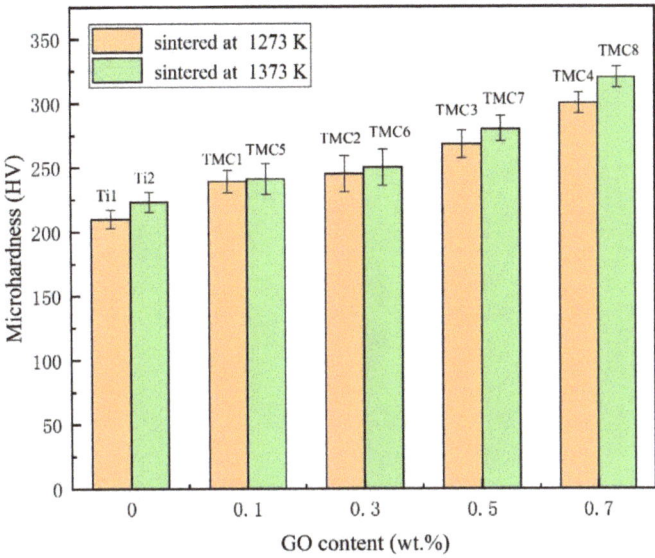

Figure 5. Hardness of Ti and TMCs.

3.3.2. Compression Properties

Figure 6a,b show the compressive stress–strain curves of Ti and TMCs sintered at 1273 K and 1373 K. The detail data are listed in Table 4. The yield strength of TMC4 is 1024 MPa, which is increased by 42% compared with that of Ti1. The yield strength of TMC7 is 1324 MPa, which is increased by 77% compared with that of Ti2. The yield strength detail of Ti and TMCs is listed in Table 3. The yield strength increases with the increase of GO content sintered at 1273 K. But, the yield strength of TMC8 is lower than TMC7. This is due to the agglomeration of GO. Similar results have been reported by Liu [32].

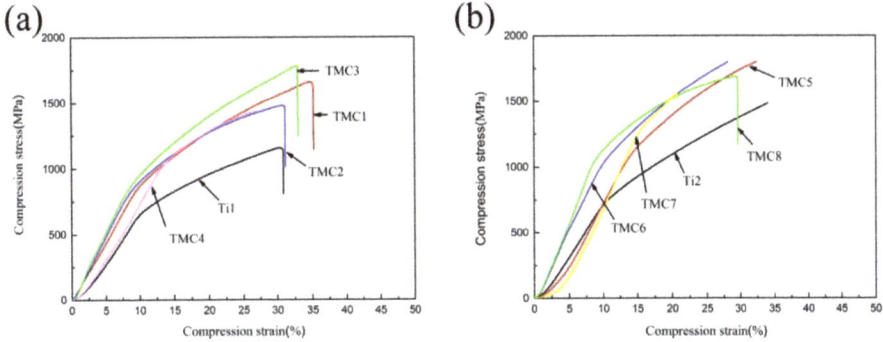

Figure 6. Stress–strain curve of Ti and TMCs sintered at (**a**) 1273 K and (**b**) 1373 K.

Table 4. Compressive properties of Ti and TMCs.

Materials	Yield Strength/MPa	Materials	Yield Strength/MPa
Ti1	721 ± 3	Ti2	748 ± 5
TMC1	859 ± 5	TMC5	1025 ± 7
TMC2	906 ± 6	TMC6	1135 ± 8
TMC3	977 ± 4	TMC7	1324 ± 6
TMC4	1024 ± 7	TMC8	1146 ± 9

Figure 7 shows the OM diagram near the compression fracture of TMC7. Elongated α-Ti grains and fractured reinforcement are observed in Figure 7. It can be inferred that the reinforcement bears the load during the compression process, thereby improving the strength of the TMCs.

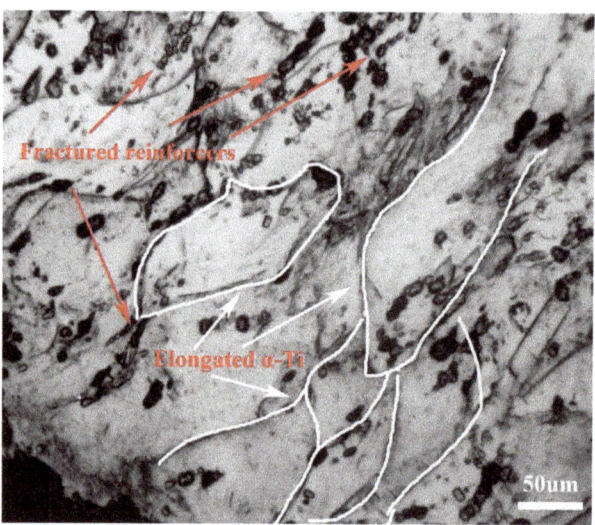

Figure 7. OM diagram near the compression fracture of TMC7.

There are two reasons for strengthening TMCs.

1. The reinforcement GO and TiC bear part of the load during the deformation process. The shear hysteresis model is usually used to evaluate the strength of TMCs. The yield strength (σ_c) can be expressed as [20,33]

$$\sigma_c = \sigma_r V_r (1 - \frac{l_c}{2l}) + \sigma_m (1 - V_r) \quad (3)$$

σ_m is the yield strength of Ti matrix, σ_r is the fracture strength of reinforcement, V_r is the volume fraction of reinforcement, l is the length of reinforcement, and l_c is the critical length of the reinforcement.

$$l_c = \sigma_r \frac{Al}{\tau_m S} \quad (4)$$

τ_m is the shear strength of Ti matrix. A is the cross-sectional areas of reinforcement, S is Ti-reinforcement interfacial areas. A = wt, S = (w + t) l, w and t is the width and thickness of reinforcement.

2. The change of yield stress can be calculated using the Hall–Petch formula [22]:

$$\sigma_y = \sigma_0 + kD^{-0.5} \quad (5)$$

where σ_y is the yield stress; σ_0 is the friction force to be overcome by dislocation motion; k is the constant related to the material, k = 0.68 MPa·m$^{0.5}$; D is the grain size. The grains of TMCs are smaller than that of pure Ti. The yield stress increases with the average grain sizes decreasing. The yield stress of TMCs are bigger than that of Ti.

3.3.3. Tribological Properties

The Friction coefficient of Ti and TMCs under dry conditions is shown in Figure 8 and listed in Table 4. Figure 8a shows friction coefficients of Ti and TMCs sintered at 1273 K.

There is a pre-grinding period at the beginning of friction, and the friction coefficient is rising sharply. The friction coefficient is stable in the period of 300–900 s. Such as Table 4, in Table 5, average friction coefficient of Ti1, TMC1, TMC2, TMC3 and TMC4 are 0.82, 0.70, 0.64, 0.55 and 0.67, respectively. Compared with Ti1, the friction coefficient of the TMCs is reduced. It indicates that GO may play a lubrication role. The friction coefficient of TMCs sintered at 1373 K is shown in Figure 8b. During 300–900 s, average friction coefficients of Ti2, TMC5, TMC6, TMC7 and TMC8 are 0.70, 0.68, 0.50, 0.72 and 0.64 respectively. With the increase of GO content, the friction coefficient decreases to some extent, but when the GO content is 0.5 wt.% and 0.7 wt.%, the friction coefficient increases. The reaction degree between GO and Ti is stronger with the increase of sintering temperature, and the retention of the GO structure is less, so the friction property decreases.

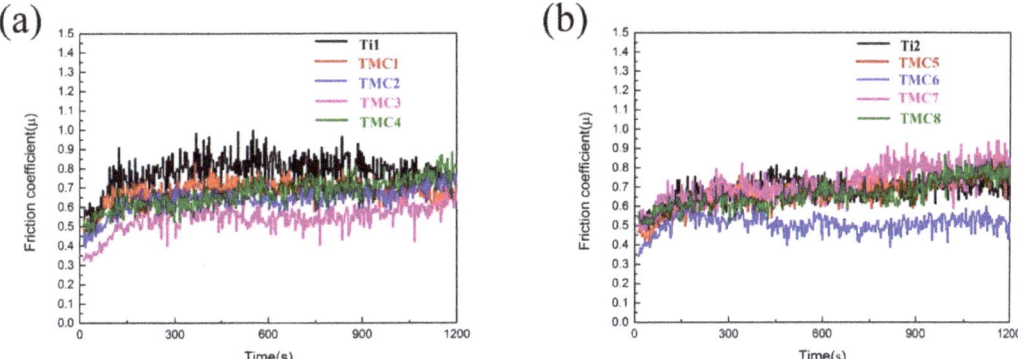

Figure 8. Friction coefficients of Ti and TMCs sintered at (**a**) 1273 K and (**b**) 1373 K.

Table 5. Friction coefficient of Ti and TMCs.

Materials	Friction Coefficient	Materials	Friction Coefficient
Ti1	0.82 ± 0.003	Ti2	0.70 ± 0.002
TMC1	0.70 ± 0.005	TMC5	0.68 ± 0.004
TMC2	0.64 ± 0.004	TMC6	0.50 ± 0.003
TMC3	0.55 ± 0.006	TMC7	0.72 ± 0.007
TMC4	0.67 ± 0.003	TMC8	0.64 ± 0.005

Figure 9 shows the three-dimensional wear trace morphology and wear volume of TMCs under sliding dry friction. Compared with pure Ti, the wear degree of the TMCs is slower, and the wear resistance is consistent with the friction coefficient.

Figure 10 shows the friction surface morphology of TMC6. As is shown in Figure 10a, Obvious friction marks can be observed. Under the action of stress, granular debris is crushed and flattened, and then apparent abrasive wear marks are formed in Figure 10b. Figure 10d is an EDS analysis of the wear marks of the TMC6 in Figure 10c. It shows that there is C of GO is squeezed to the surface due to the friction force. Then, a film with lubrication is formed on the contact surface between the friction pair and the matrix, thus reducing the wear of the friction pair to the matrix. And, there are Fe elements only in friction pairs. Results show that material transfer of friction pair and surface element oxidation occurs in the friction and wear process. The wear mechanism is adhesive wear and oxidation wear.

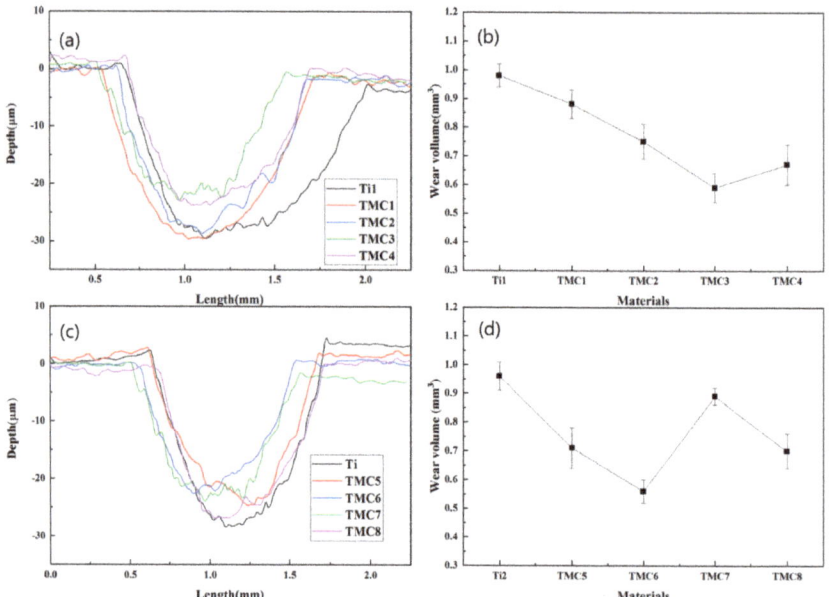

Figure 9. Depth of wear marks and wear volume of Ti and TMCs. (**a**) Depth of wear marks of Ti and TMCs sintered at 1273 K. (**b**) Wear volume of Ti and TMCs sintered at 1273 K. (**c**) Depth of wear marks of Ti and TMCs sintered at 1373 K. (**d**) Wear volume of TMCs sintered at 1373 K.

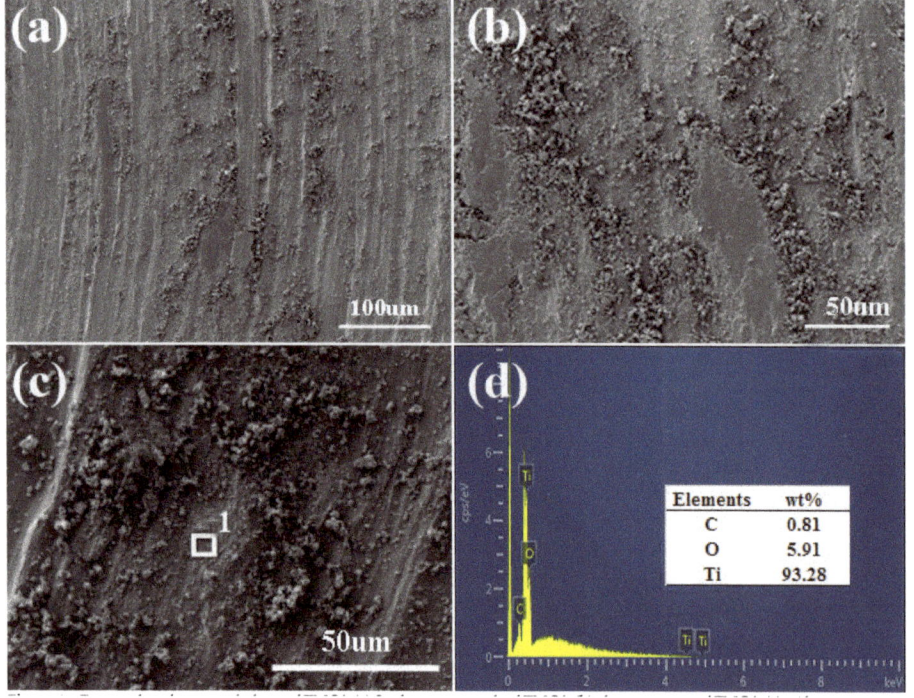

Figure 10. Frictional surface morphology of TMC6. (**a**) Surface wear mark of TMC6; (**b**) abrasive wear of TMC6; (**c**) a Abrasive particle of TMC6; (**d**) EDS of TMC6 in (**c**).

4. Conclusions

0.1–0.7 wt.% of GO /Ti composites were prepared by hot-pressed sintering at 1273 K and 1373 K. The results demonstrated that this method is simple and highly efficient to fabricate GO/Ti composites with remarkably high mechanical properties, which will meet the industrial development of lightweight, high-strength materials. The results of the current investigation are as follows:

1. The microstructure of Ti and TMCs is equiaxed α. The average grain size of TMCs decreases with GO increasing. GO and TiC exists in TMCs simultaneously.
2. The hardness of TMCs is higher than that of pure Ti. The hardness of TMCs with 0.7 wt.% GO sintered at 1373 K is the biggest, 320 HV, which is 43% higher than that of pure Ti. The yield strength of 0.5 wt.% GO sintered at 1373 K is 1324 MPa, 77% more than pure Ti. The strengthening mechanism is that reinforcement bears the load from matrix and fine-grain strengthening.
3. Compared with pure Ti under the same condition, the friction coefficient of TMCs decreases. The friction coefficient of TMCs containing 0.5 wt.% GO sintered at 1273 K and 0.3 wt.% GO sintered at 1373 K is 0.55 and 0.5, which are 0.27 and 0.2 lower than that of pure Ti. The wear mechanism is abrasive wear, adhesive wear and oxidation wear.

Author Contributions: Z.W., investigation, performing the experiment, writing—original draft preparation; S.H., investigation, part of writing—original draft preparation; J.L., conceptualization, review & editing, supervision. D.Y., conceptualization, review & editing. All authors have read and agreed to the published version of the manuscript.

Funding: This research received no external funding.

Institutional Review Board Statement: Not applicable.

Informed Consent Statement: Not applicable.

Data Availability Statement: The data supporting the finding of this study are available within the article.

Conflicts of Interest: The authors declare no conflict of interest.

References

1. Hua, K.; Wan, Q.; Zhang, Y.; Kou, H.; Li, J. Crystallography and microstructure of the deformation bands formed in a metastable β titanium alloy during isothermal compression. *Mater. Charact.* **2021**, *176*, 111119. [CrossRef]
2. Lu, H.F.; Wang, Z.; Cai, J.; Xu, X.; Lu, J. Effects of laser shock peening on the hot corrosion behaviour of the selective laser melted Ti6Al4V titanium alloy. *Corros. Sci.* **2021**, *188*, 109558. [CrossRef]
3. Yang, D.; Liu, Y.L.; Xie, F.; Wu, M. Identification and Quantitative Analysis of the Isolated and Adhesive Beta Phases in Titanium Alloy Ti-6Al-4V. *Mater. Trans.* **2020**, *61*, 1220–1229. [CrossRef]
4. Fourest, T.; Bouda, P.; Fletcher, L.C.; Notta-Cuvier, D.; Markiewicz, E.; Pierron, F.; Langrand, B. Image-Based Inertial Impact Test for Characterisation of Strain Rate Dependency of Ti6Al4V Titanium Alloy. *Exp. Mech.* **2020**, *60*, 235–248. [CrossRef]
5. Bansal, P.; Singh, G.; Sidhu, H.S. Plasma-Sprayed HA/Sr Reinforced Coating for Improved Corrosion Resistance and Surface Properties of Ti13Nb13Zr Titanium Alloy for Biomedical Implants. *J. Mater. Res.* **2021**, *36*, 431–442. [CrossRef]
6. Wang, X.; Li, S.; Han, Y.; Huang, G.; Lu, W. Roles of reinforcements in twin nucleation and nano-α precipitation in the hybrid TiB/TiC-reinforced titanium matrix composites during high-temperature fatigue. *Scr. Mater.* **2021**, *196*, 113758. [CrossRef]
7. Mcintyre, D.J.; Hirschman, R.K.; Puchades, I.; Landi, B.J. Enhanced copper–carbon nanotube hybrid conductors with titanium adhesion layer. *J. Mater. Sci.* **2020**, *55*, 6610–6622. [CrossRef]
8. Li, H.; Jia, D.; Yang, Z.; Zhou, Y. Achieving near equiaxed α-Ti grains and significantly improved plasticity via heat treatment of TiB reinforced titanium matrix composite manufactured by selective laser melting. *J. Alloys Compd.* **2020**, *836*, 155344. [CrossRef]
9. Li, A.; Shi, L.; Zhang, W.; Zhou, S.Q.; Sun, Y.J.; Ma, S.; Liu, M.B.; Sun, Y. A simple way to fabricate Ti6Al4V matrix composites reinforced by graphene with exceptional mechanical properties. *Mater. Lett.* **2019**, *257*, 126750. [CrossRef]
10. Hu, C.L.; Sun, R.L. Microstructure of B_4C/TiC/TiB_2 reinforced surface titanium matrix composite produced by laser cladding. *IOP Conf. Ser. Mater. Sci. Eng.* **2020**, *770*, 012003. [CrossRef]
11. Ying, W.; Peng, X.S. Graphene oxide nanoslit-confined $AgBF_4$/ionic liquid for efficiently separating olefin from paraffin. *Nanotechnology* **2020**, *31*, 085703. [CrossRef]

12. Yao, W.; Tang, L.L.; Nong, J.P.; Wang, J.; Yang, J.; Jiang, Y.D.; Shi, H.F.; Wei, X.Z. Electrically tunable graphene metamaterial with strong broadband absorption. *Nanotechnology* **2021**, *32*, 075703. [CrossRef] [PubMed]
13. Wang, F.Z.; Cai, X.X. Improvement of mechanical properties and thermal conductivity of carbon fiber laminated composites through depositing graphene nanoplatelets on fibers. *J. Mater. Sci.* **2019**, *54*, 3847–3862. [CrossRef]
14. Kumar, S.; Singh, K.K.; Ramkumar, J. Comparative study of the influence of graphene nanoplatelets filler on the mechanical and tribological behavior of glass fabric-reinforced epoxy composites. *Polym. Compos.* **2020**, *41*, 5403–5417. [CrossRef]
15. Xavior, M.A.; Kumar, H.P. Processing and Characterization Techniques of Graphene Reinforced Metal Matrix Composites (GRMMC); A Review. *Mater. Today Proc.* **2017**, *4*, 3334–3341. [CrossRef]
16. Rashad, M.; Pan, F.S.; Asif, M.; Tang, A.T. Powder metallurgy of Mg–1%Al–1%Sn alloy reinforced with low content of graphene nanoplatelets (GNPs). *J. Ind. Eng. Chem.* **2014**, *20*, 4250–4255. [CrossRef]
17. Zhou, W.; Mikulova, P.; Fan, Y.C.; Kikuchi, K.; Nomura, N.; Kawasaki], A. Interfacial reaction induced efficient load transfer in few-layer graphene reinforced Al matrix composites for high-performance conductor. *Compos. Part B* **2019**, *179*, 107463. [CrossRef]
18. Han, X.Q.; Yang, L.Z.; Zhao, N.Q.; He, C.N. Copper-Coated Graphene Nanoplatelets-Reinforced Al–Si Alloy Matrix Composites Fabricated by Stir Casting Method. *Acta Metall. Sin.* **2020**, *34*, 111–124. [CrossRef]
19. Ali, S.; Ahmad, F.; Yusoff, P.S.M.M.; Muhamad, N.; Qnate, E.; Raza, M.R.; Malik, K. A review of graphene reinforced Cu matrix composites for thermal management of smart electronics. *Compos. Part A* **2021**, *144*, 106357. [CrossRef]
20. Shin, S.E.; Choi, H.J.; Shin, J.H.; Bae, D.H. Strengthening behavior of few-layered graphene/aluminum composites. *Carbon* **2015**, *82*, 143–151. [CrossRef]
21. Chen, F.Y.; Ying, J.M.; Wang, Y.F.; Du, S.Y.; Liu, Z.P.; Huang, Q. Effects of graphene content on the microstructure and properties of copper matrix composites. *Carbon* **2016**, *96*, 836–842. [CrossRef]
22. Liu, J.H.; Khan, U.; Coleman, J.; Fernandez, B.; Rodriguez, P.; Naher, S.; Brabazon, D. Graphene oxide and graphene nanosheet reinforced aluminium matrix composites: Powder synthesis and prepared composite characteristics. *Mater. Des.* **2016**, *94*, 87–94. [CrossRef]
23. Shuai, C.J.; Wang, B.; Yang, Y.W.; Peng, S.P.; Gao, C.D. 3D honeycomb nanostructure-encapsulated magnesium alloys with superior corrosion resistance and mechanical properties. *Compos. Part B* **2019**, *162*, 611–620. [CrossRef]
24. Mu, X.N.; Zhang, H.M.; Cai, H.N.; Fan, Q.B.; Zhang, Z.H.; Wu, Y.; Fu, Z.J.; Yu, D.H. Microstructure evolution and superior tensile properties of low content graphene nanoplatelets reinforced pure Ti matrix composites. *Mater. Sci. Eng. A* **2017**, *687*, 164–174. [CrossRef]
25. Cao, Z.; Li, J.L.; Zhang, H.P.; Li, W.B.; Wang, X.D. Mechanical and tribological properties of graphene nanoplatelets-reinforced titanium composites fabricated by powder. *J. Iron Steel Res. Int.* **2020**, *21*, 1357–1362. [CrossRef]
26. Dong, L.L.; Xiao, B.; Jin, L.H.; Lu, J.W.; Liu, Y.; Fu, Y.Q.; Zhao, Y.Q.; Wu, G.H.; Zhang, Y.S. Mechanisms of simultaneously enhanced strength and ductility of titanium matrix composites reinforced with nanosheets of graphene oxides. *Ceram. Int.* **2019**, *45*, 19370–19379. [CrossRef]
27. Cao, H.C.; Liang, Y.L. The microstructures and mechanical properties of graphene-reinforced titanium matrix composites. *J. Alloys Compd.* **2020**, *812*, 152057. [CrossRef]
28. Liu, J.; Wu, M.X.; Yang, Y.; Yang, G.; Yan, H.X.; Jiang, K.L. Preparation and mechanical performance of graphene platelet reinforced titanium nanocomposites for high temperature applications. *J. Alloys Compd.* **2018**, *765*, 1111–1118. [CrossRef]
29. Dong, L.L.; Xiao, B.; Liu, Y.; Li, Y.L.; Fu, Y.Q.; Zhao, Y.Q.; Zhang, Y.S. Sintering effect on microstructural evolution and mechanical properties of spark plasma sintered Ti matrix composites reinforced by reduced graphene oxides. *Ceram. Int.* **2018**, *44*, 17835–17844. [CrossRef]
30. Chen, H.; Mi, G.B.; Li, P.J.; Huang, X.; Cao, C.X. Microstructure and Tensile Properties of Graphene-Oxide-Reinforced High-Temperature Titanium-Alloy-Matrix Composites. *Materials* **2020**, *13*, 3358. [CrossRef]
31. Haghighi, M.; Shaeri, M.H.; Sedghi, A.; Djavanroodi, F. Effect of Graphene Nanosheets Content on Microstructure and Mechanical Properties of Titanium Matrix Composite Produced by Cold Pressing and Sintering. *Nanomaterials* **2018**, *8*, 1024. [CrossRef] [PubMed]
32. Liu, J.Q.; Hu, N.; Liu, X.Y.; Liu, Y.L.; Lv, X.W.; Wei, L.X.; Zheng, S.T. Microstructure and Mechanical Properties of Graphene Oxide-Reinforced Titanium Matrix Composites Synthesized by Hot-Pressed Sintering. *Nanoscale Res. Lett.* **2019**, *14*, 114. [CrossRef] [PubMed]
33. Mu, X.N.; Cai, H.N.; Zhang, H.M.; Fan, Q.B.; Wang, F.C.; Cheng, X.W.; Zhang, Z.H.; Li, J.B.; Jiao, X.L.; Ge, Y.X.; et al. Size effect of flake Ti powders on the mechanical properties in graphene nanoflakes/Ti fabricated by flake powder metallurgy. *Compos. Part A* **2019**, *123*, 86–96. [CrossRef]

Article

Formation of Bioresorbable Fe-Cu-Hydroxyapatite Composite by 3D Printing

Valentina Vadimovna Chebodaeva [1,*], Nikita Andreevich Luginin [1,2], Anastasiya Evgenievna Rezvanova [1], Natalya Valentinovna Svarovskaya [1], Konstantin Vladimirovich Suliz [1], Ludmila Yurevna Ivanova [1], Margarita Andreevna Khimich [1], Nikita Evgenievich Toropkov [1], Ivan Aleksandrovich Glukhov [1], Andrey Aleksandrovich Miller [1], Sergey Olegovich Kazantsev [1] and Maksim Germanovich Krinitcyn [1]

[1] Institute of Strength Physics and Materials Science of Siberian Branch Russian Academy of Sciences (ISPMS SB RAS), 634055 Tomsk, Russia
[2] Research School of High-Energy Physics, National Research Tomsk Polytechnic University, 634050 Tomsk, Russia
* Correspondence: valentinoch@ispms.ru

Abstract: Studies of the microstructure, phase composition and mechanical characteristics, namely the microhardness of metal–ceramic composites made of Fe 90 wt.%–Cu 10 wt.% powder and hydroxyapatite (Fe-Cu-HA), are presented in the manuscript. The composite material was obtained using additive manufacturing based on the 3D-printing method, with different content levels of powder (40, 45 and 50%) and polymer parts (60, 55 and 50%). It is shown that varying the proportion of Fe-Cu-HA powder does not significantly affect the elemental and phase compositions of the material. The X-ray phase analysis showed the presence of three phases in the material: alpha iron, copper and hydroxyapatite. It is shown in the experiment that an increase in the polymer component of the composite leads to an increase in the defectiveness of the structure, as well as an increase in microstresses. An increase in the mechanical properties of the composite (Vickers microhardness), along with a decrease in the percentage of Fe-Cu-HA powder from 50 to 40%, was established. At the same time, the composite containing 45% Fe-Cu-HA powder demonstrated the maximum increase in the microhardness of the composite by ~26% compared to the composite containing 50% Fe-Cu-HA powder, which is due to the more uniform distribution of components.

Keywords: composite; additive manufacturing; 3D printing; bioresorbable Fe-Cu-hydroxyapatite composite

1. Introduction

In the past few decades, biodegradable metallic materials have served as one of the most promising strategies in regenerative medicine [1]. Biodegradable metals have excellent mechanical properties, providing sufficient temporary support to resist the applied load, while the potential risk of long-term complications is effectively eliminated due to the progressive degradation of metals in the body [2]. In addition, there is no need to per-form a second operation to remove the implant, since biodegradable bone implants in the form of rods, plates, screws and anchors should provide initial mechanical support and gradually dissolve in the physiological environment without causing infection, since they do not contain toxic components. Therefore, biodegradable metal implants are best fit for the stabilization of damaged bone tissue and directed bone healing. However, from the point of view of biological safety, there are strict requirements for the choice of material, such as a certain resorption rate, a balanced decrease in mechanical properties during bone tissue regeneration and the metabolism of resorption by-products in the body [2].

Iron is the most abundant metal in the human body and is involved in a wide range of metabolic processes, such as oxygen transport, energy metabolism, enzyme function and DNA synthesis [3]. In particular, it is known for its vital role in bone homeostasis, and iron deficiency causes bone disease and impairs bone mineralization [4]. Despite its great potential for use in biodegradable orthopedic implants, its use is limited by its low corrosion rate in physiological environments [5,6].

The doping of iron alloys with copper makes it possible to increase the rate of resorption due to the formation of a Fe-Cu galvanic pair and to provide the antibacterial properties of the material implanted in the body [7]. In addition, bicomponent Fe-Cu nanoparticles can suppress antibiotic resistance in bacteria [7]. However, copper in large quantities can cause allergies to the implanted product [8].

In order to allow bone ingrowth and promote a stable implant–bone interface, the surface of the artificial material must be bioactive to promote osteoconductivity through bone cell growth and biological apatite generation [9,10]. Hydroxyapatite (HA) with stoichiometry $Ca_{10}(PO_4)_6(OH)_2$ is widely used in the field of medical materials science as a bioactive bone substitute due to its excellent biocompatibility and chemical similarity with the mineral phase of human bone [11,12]. Hydroxyapatite has already been successfully added to iron [13,14] and magnesium [15,16] matrices to improve the bioactivity of these materials and use them in medicine.

According to the literature data, the Vickers hardness of iron is 0.6 GPa [17], and that of copper is ~0.3 GPa, which is insufficient when the Fe–Cu composite material is used as a bone implant for the reconstruction of damaged bone tissues. Hydroxyapatite not only has excellent biocompatibility and a similar chemical and phase composition to bone tissue but also has a significantly higher hardness of ~5–6 GPa [18,19]. The addition of HA particles to the Fe-Cu composition will make it possible to obtain a composite material with higher strength characteristics.

However, in the case of iron implants, the problem of a low resorption rate remains relevant even in the case of alloying such materials with copper and hydroxyapatite. This problem can be solved by the formation of porous structures, using the 3D printing meth-od from a material based on iron powder alloyed with copper and hydroxyapatite to im-prove the biological properties of the final product. In addition, the manufacture of im-plants by this method allows us to avoid further mechanical processing of these products, which will positively affect the cost of the final medical device. However, for the formation of a metal product by this method, the addition of plasticizing binders is required.

The most commonly used controlled-release polymer in implant materials is ethylene vinyl acetate (EVA), which is a biocompatible, insoluble and non-toxic thermoplastic copolymer of ethylene and vinyl acetate (VA). The content of VA in the EVA copolymer can vary from 0 to 40%. The EVA brand in this work was chosen after the following considerations. The properties of copolymers differ depending on the concentration of VA. Higher content results in increased polarity, adhesion, impact resistance, flexibility and compatibility of EVA with other polymers. At the same time, a higher VA content causes a decrease in the crystallinity, stiffness, softening and melting point of the copolymer [20]. Based on these two competing factors, the prospective polymer is EVA EA28025 with a content of 28 wt.% VA [21].

In the manufacture of composite materials, tall oil rosin is also widely used as a plasticizing component. Rosin acids, present in tall oil rosin, provide it with unique properties, such as solubility in many organic solvents, good compatibility with many polymeric materials, plasticity and relative adhesion. Tall oil rosin is characterized as a linear oligomer whose hydrogen bonds increase its mechanical strength, while maintaining elasticity [22].

The aim of this work was to produce a porous biodegradable composite material based on iron, copper and hydroxyapatite powders that has antibacterial and bioactive properties, using the 3D printing method, with the addition of plasticizing binders in the form of polymers.

2. Materials and Methods

As a feedstock for the production of composites, we used a nanopowder with a nominal composition: Fe—90 wt.% and Cu—10 wt.% (Fe90-Cu10), obtained by the method of electric explosion of wire (EEW), with a diameter of d = 0.35 and d = 0.10 mm grades ST1 and M1, respectively, with the same length of 120 mm. The detailed EEW process and the schematic diagram of the installation are described in detail in References [23,24]. The production of nanopowder was carried out in an argon atmosphere, at a pressure of 3 atm., with a capacitor discharge of 2.4 µF and a voltage of 19 kV.

The obtained theoretical density of the powder Fe-Cu was 8 g/cm^3. The powder of stoichiometric hydroxyapatite $Ca_{10}(PO_4)_6(OH)_2$ (ISSCM SB RAS, Novosibirsk, Russia) [25] was added to the metal part of the composite in the amount of 5 wt.%, at a density of 3.16 g/cm^3. The total density of the powder cermet mixture in these ratios was 7.76 g/cm^3.

For the subsequent 3D printing, a polymer component was produced with the following nominal composition: 75 wt.%—tall oil rosin; 15 wt.%—1.6 hexanediol; and 10 wt.%—ethylene vinyl acetate.

The cermet and polymer parts were mixed in the following ratios: 50 to 50 (50Fe-Cu-HA), 45 to 55 (45Fe-Cu-HA) and 40 to 60 wt.% (40Fe-Cu-HA), respectively. For the uniform elements' distribution, the resulting mixture was mixed in a Schatz laboratory mixer model C 2.0 "Turbula" (Vibrotechnik, St. Petersburg, Russia) for 30 min. To prevent further oxidation of the metal part of the powder, the polymer part was plasticized by heating under acetone at a temperature of 50 °C in an ultrasonic bath, Ferroplast VU-09- "Ya-FP"-03 (Ferroplast, Yaroslavl, Russia), which allowed us to limit the direct contact of the atmosphere with Fe and Cu particles.

The resulting material was passed through a screw extruder 4 times, with a gradual decrease in the nozzle diameter from 2.0 to 0.8 mm, which also increases the uniformity of the mixture's composition. The desired shape for 3D printing was a cylinder with a diameter of 20 mm and a height of 2.5 mm. Printing was carried out on a 3D printer, Prusa i3 (Prusa Research, Prague, Czech Republic), with a modified wire feed system for printing powders with a polymer component. The following printing parameters were used in the manufacture of the "green part": layer height, 0.25 mm; print speed, 60 mm/min; nozzle diameter, 0.5 mm; substrate temperature, 90 °C; and nozzle temperature, 140 °C.

To remove the polymer component of the workpiece, the product was subjected to solution debinding in acetone for 24 h. At this stage, tall oil rosin and 1.6 hexanediol were removed, while ethylene vinyl acetate is insoluble in acetone and allows us to keep the shape of the "green product" until the sintering of the cermet part.

Sintering was carried out in a Nabertherm vacuum furnace. Exposure at the temperature T1 (450 °C) ensures the removal of ethylene vinyl acetate from the final product, and at the temperature T2 (1000 °C), the particles of the ceramic–metal part are consolidated, and the product is finally formed. A general flowchart of the composite material samples is shown in Figure 1.

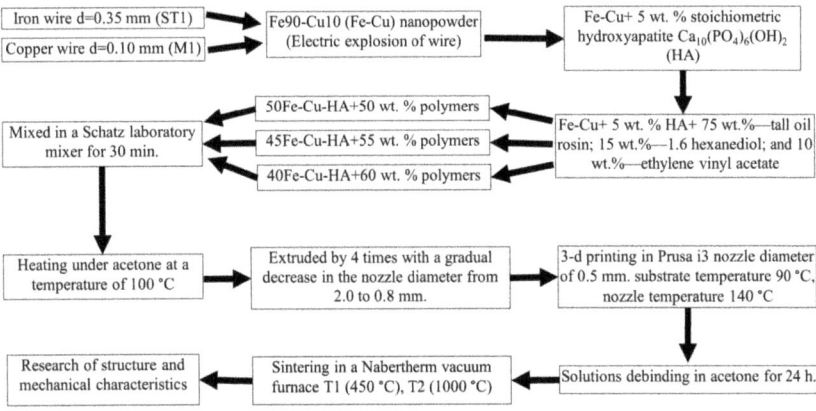

Figure 1. General scheme of sample production.

Sample preparation was carried out by grinding on SiO_2 abrasive with a gradual decrease in grain size from M180 to M1000 (GOST 6456-82). The polishing of the sample was carried out with a diamond suspension and abrasive 1/0 (GOST 25593-83). The sample was washed with ethanol to remove the remaining particles of a larger fraction at each stage of sample preparation. The microstructure was revealed by etching the sample surface for 5 s with a reagent of the following composition: 90 vol.%, H_2O; and 10 vol.%, HNO_3.

The study of the structure and the elemental analysis of materials were carried out using the following methods: optical microscopy (Altami MET 1MT (Altami, St. Petersburg, Russia)), scanning electron microscopy (Apreo S LoVac (Thermo Fisher Scientific, Waltham, USA) and LEO EVO 50 (Zeiss, Jena, Germany)) with attachments for energy-dispersive analysis and transmission electron microscopy (JEOL JEM-2100 (Tokyo Boeki Ltd., Tokyo, Japan)). The phase composition of the samples was identified by X-ray diffraction analysis (DRON 8N (Bourevestnik, St. Petersburg, Russia)). The processing of X-ray patterns was performed in the Match! 3 software, using the database COD from 18.07.2022. The phase-unit cell parameter was calculated in the FullProf additional software package. To determine the instrumental FWHM, the data from SiO_2 specimen were used, the diffraction peaks of which were approximated using the Laplace function. The crystalline size and microstrains were calculated using the Williamson–Hall method. The distribution of Fe90-Cu10 powder particles was studied by sedimentation analysis on a CPS DC24000 UHR centrifuge (Analytik Ltd., Cambridge, UK).

The hardness tests of the materials were carried out by indentation, using an Affri DM8 microhardness tester with a Vickers diamond tetrahedral tip at various indentation loads in the range from 0.25 to 9.8 N for 10 s. For each composition, 3 samples were used. For each load cycle, 10 measurements were carried out in different parts of the sample, after which the results were averaged.

The Vickers hardness values were determined through the diagonal of the print obtained after the indenter was pressed in, according to Equation (1):

$$HV = 1.854 \cdot \frac{P}{d^2} \qquad (1)$$

where P is the applied load, N; and d is the print diagonal, μm.

The measurement of porosity in sintered composites was carried out using the method of hydrostatic weighing in kerosene.

3. Results and Discussion

Figure 2 represents the TEM image and selected area electron diffraction with the Miller indices of the corresponding phases of the initial metal powder Fe90-Cu10. The selected area electron diffraction pattern indicates the presence of two phases in the powder particles. The first component of the picture is the point diffraction, for which the lattice parameters are close to the α-Fe phase. The second component is the ring-shaped diffraction of the Cu phase. The general diffraction pattern indicates that the particles of nanopowders consist of a matrix of the α-Fe phase, where Cu clusters are distributed, and the sizes do not exceed several tens of nanometers, since the ring shape of the diffraction pattern indicates multiple misorientations of the Cu lattice.

The EDS spectrum analysis showed (Figure 3) that the composition of the powder is close to the nominal (Fe, 86.2 wt.%; Cu, 13.8 wt.%). Copper clusters are mainly located in the bulk of α-Fe phase particles, forming with it single metal particles.

Figure 4 shows the X-ray diffraction pattern of the Fe90-Cu10 nanopowder obtained by the EEW.

The presence of two α-Fe and Cu phases was established in the metal powder, and the exact quantitative content of the second phase cannot be established due to the low intensity of reflections in relation to the first phase. Table 1 represents the calculated lattice parameters of the α-Fe phase. The parameters of the Cu phase were not calculated due to the low intensity.

(a) (b)

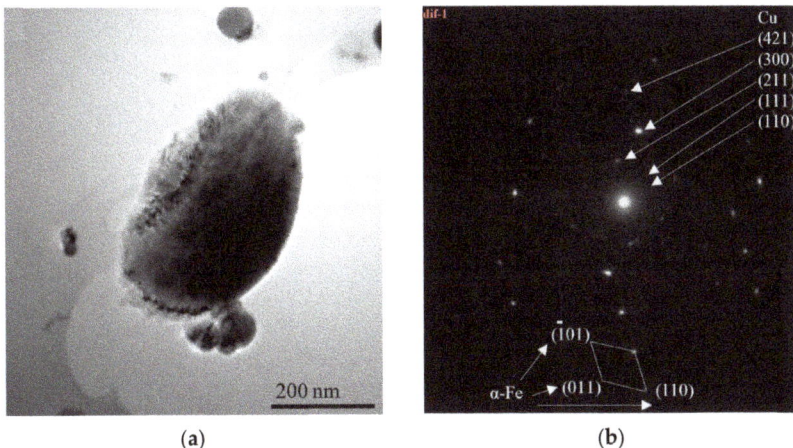

Figure 2. TEM image of Fe90-Cu10 powder (**a**) with corresponding selected area electron diffraction (**b**).

(a) (b)

(c)

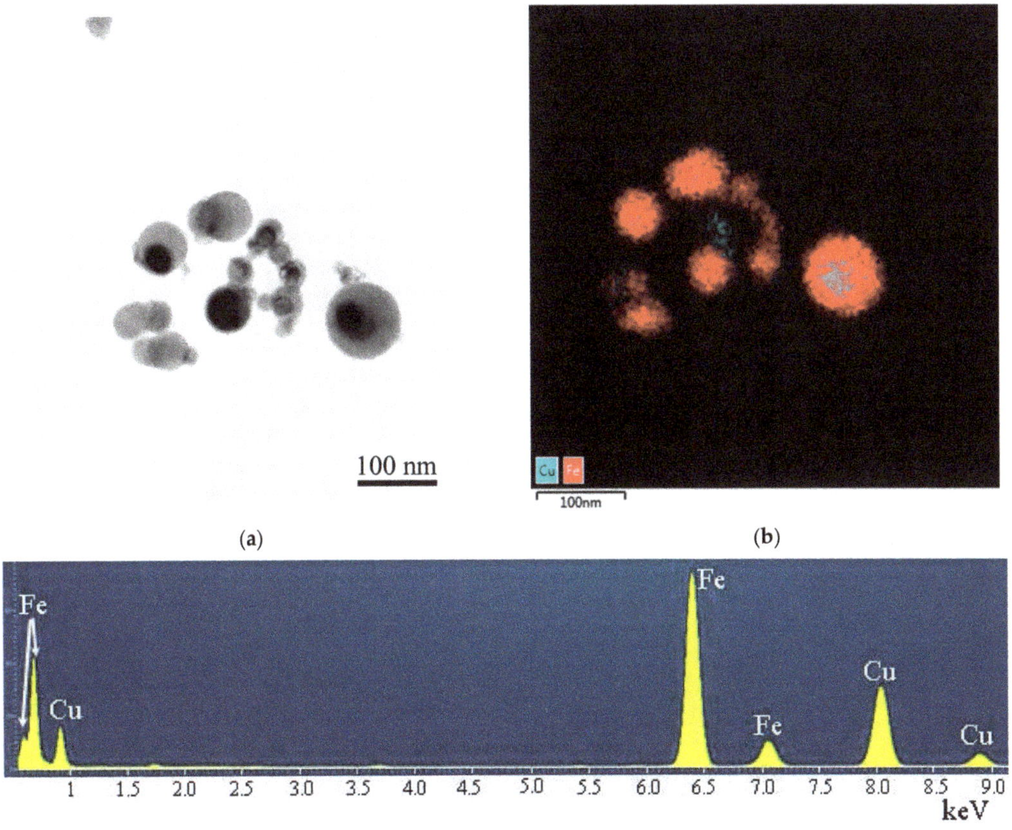

Figure 3. TEM image of Fe90-Cu10 powder (**a**) with element distribution map (**b**) and overall EDS spectrum (**c**).

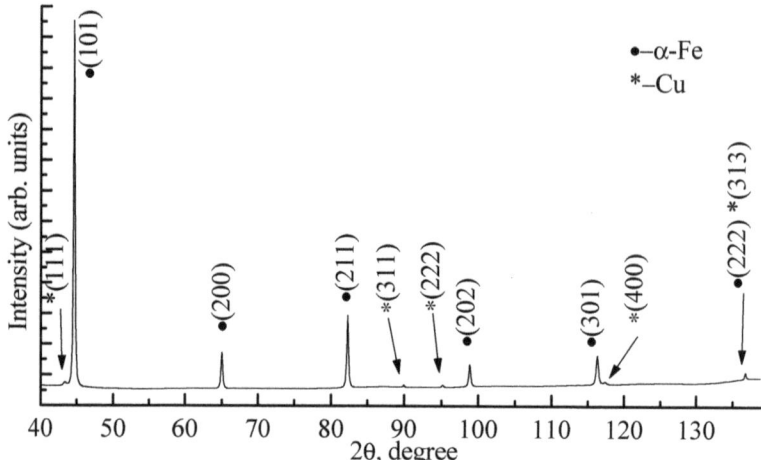

Figure 4. X-ray diffraction pattern of the Fe90-Cu10 powder.

Table 1. Calculated lattice parameter of phase in Fe90-Cu10 powder.

Phase	Sample	Lattice Parameter a, Å	Unit Cell Volume, Å³	Crystallite Size, nm	ε^*, %
α-Fe	Reference data (COD 96-110-0109)	2.868	23.590	-	-
	Fe90-Cu10	2.869 ± 0.001	23.62 ± 0.02	33	14.3

ε^*—microstrains.

The calculated lattice parameter for the α-Fe phase coincides with the reference data within the measurement error. The powder has a high level of microstresses due to the method of production, which produces a metal with a high level of defects.

Figure 5 shows the particle distribution of the 90Fe-10Cu metal powder obtained by the EEW method. The distribution has a monomodal character, and the average particle size was 50 ± 20 nm. The powder contained particles larger than 100 nm, but their volume fraction did not exceed 5%. The data from X-ray analysis are comparable with the EEW method. The difference in crystallite size values is due to the presence of dislocations and atoms in the powder particles.

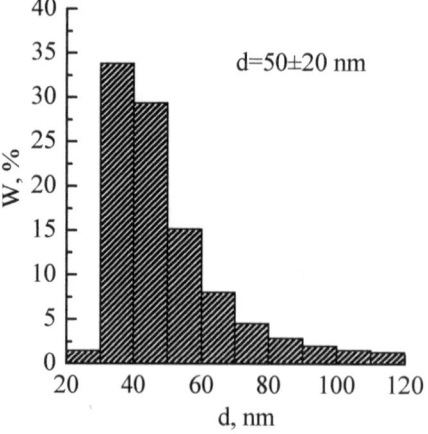

Figure 5. Particle size distribution of 90Fe-10Cu powder.

Figure 6 represents SEM images and maps of the elements' distribution on the surface of samples produced with the addition of different amounts of Fe-Cu and HA powders (50, 45 and 40%). The surface of all samples is characterized by the presence of pores and HA particles. From the image and maps of the calcium and phosphorus distribution, it can be seen that, in the sample with 50% Fe-Cu-HA (Figure 6a), there are small pores with an average size of 1.5 µm and a small number of HA particles with a larger size of 6.8 µm. In addition, copper is uniformly distributed on the surface of the composite in the form of small inclusions not exceeding 1 µm (Figure 6b).

Figure 6 showed SEM images of the surface and maps of the elements (Fe, Cu, Ca and P) distribution of the produced samples: 50Fe-Cu-HA (a), 45Fe-Cu-HA (b) and 40Fe-Cu-HA (c).

Reducing the proportion of Fe-Cu and HA powders to 45% (Figure 6b) leads to a more uniform distribution of HA powder particles and an increase in their amount. At the same time, the average pore size does not change with a decrease in the powder concentration to 45%, and the average size of HA particles decreases to 5.5 µm. The EDS maps show a uniform distribution of iron and copper, while calcium and phosphorus are concentrated in places where large HA particles are found. The surface of the sample with 40% Fe-Cu and HA (Figure 6c) is characterized by large HA particles with an average size of 10 ± 5 µm and pores with an average size of 1.8 µm. The maximum amount of the polymer component in this sample led to less homogeneous mixing of the Fe-Cu-HA powder in the feedstock.

Figure 7 represents the dependence of the area occupied by HA particles and pores on the surface of composites vs. content of Fe-Cu-HA powder in the feedstock. With an increase in the Fe-Cu-HA powder content from 40 to 50% and a decrease in the polymer component of the feedstock, the total area occupied by HA particles and pores decreases from 16 to 10%. At the same time, as follows from the SEM results, the pattern of particles distribution in the composite varies depending on the amount of Fe-Cu-HA powder added. The hydrostatic weighing method showed that the open porosity of the composites does not depend on the amount of the polymer part ($\approx$14%).

Optical images of the microstructure of composites after etching are shown in Figure 8.

The structure of the samples after sintering is represented by the grains of α-Fe phase (light areas) with an average size of $d = 10 \pm 5$ µm, with small inclusions of the Cu phase (orange areas), with an average size of $d = 1 \pm 0.5$ µm, and some grains are separated from each other by wide grain boundaries (dark elongated areas). In addition, their fraction increases with the increasing of the polymer component before sintering.

Figure 9 shows X-ray diffraction patterns of composites based on Fe-Cu-HA after sintering. As a result of the X-ray diffraction analysis, intense peaks of the α-Fe phase were found. The presence of copper and hydroxyapatite in any state was not identified by this method. Table 2 represents the calculated lattice parameters of the main phase of iron, as well as the values of crystallite sizes and microstrains of the corresponding phases.

Powder consolidation leads to a decrease in microstrain values and an increase in crystallite size for the samples 50Fe-Cu-HA and 45Fe-Cu-HA. The 40Fe-Cu-Ha sample is characterized by the opposite change in values, which is due to the more defective structure of the α-Fe phase (Table 2). The unit cell volume and the lattice parameter of the main phase in all composites correspond to the reference data within the measurement error. However, with a decrease in the Fe-Cu-HA powder concentration and an increase in the polymer component concentration in the composite after sintering, an increase in the microstrains and a decrease in the crystallite size occur (Table 2). As mentioned above, the composites are characterized by the same value of open porosity; however, the values of the crystallite size and microstrains indicate an increase in the defectiveness of the structure, implying an increase in the amount or average size of closed pores with a change in the concentration of the polymer component of composites before consolidation.

Table 2. Calculated lattice parameters of α-Fe phase according to X-ray diffraction data for compo-sites.

Sample	Lattice Parameter a, Å	Unit Cell Volume, Å³	Crystallite Size, nm	ε, %
Reference (COD 96-110-0109)	2.868	23.590	-	-
50Fe-Cu-HA	2.867 ± 0.001	23.58 ± 0.02	53	11.7
45Fe-Cu-HA	2.868 ± 0.001	23.59 ± 0.02	45	13.6
40Fe-Cu-HA	2.869 ± 0.001	23.60 ± 0.02	37	15.7

Figure 10 and Table 3 represent the obtained dependences of average Vickers microhardness (HV) of composites on applied loads (P). Microhardness measurements were carried out on three materials of different composition. The microhardness of samples was measured by the Vickers indentation method for three samples in each composition. For each sample, 10 indentations were performed for each load. A multiply coefficient of 9.807 was used to convert Vickers hardness values from HV to MPa. The statistical analysis shows that the distribution of the random value of microhardness corresponds to the normal (Gaussian) law. The plot shows that with an increase in load from 0.25 to 10 N, the microhardness of all the samples changes from 1655 to 860 MPa.

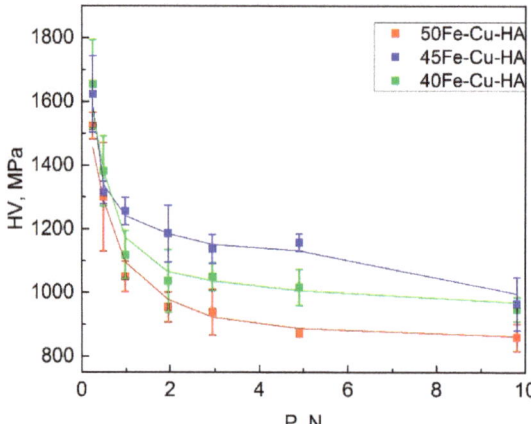

Figure 10. Vickers microhardness vs. load applied to composite specimens.

Table 3. Vickers microhardness values for composites.

P*, H	HV**, MPa		
	50Fe-Cu-HA	45Fe-Cu-HA	40Fe-Cu-HA
0.25	1525 ± 40	1625 ± 120	1655 ± 140
0.49	1300 ± 170	1315 ± 35	1380 ± 110
0.98	1050 ± 50	1255 ± 45	1120 ± 75
1.96	955 ± 50	1185 ± 90	1040 ± 100
2.94	940 ± 70	1140 ± 45	1050 ± 45
4.9	870 ± 7	1160 ± 30	1020 ± 55
9.8	860 ± 40	965 ± 85	950 ± 40

P*—indenter load; HV**—Vickers microhardness.

The 50Fe-Cu-HA sample is characterized by the lowest microhardness; it decreases from 1525 to 860 MPa. This is due to the small amount of HA particles nonuniformly distributed in the sample bulk.

The 40Fe-Cu-HA sample is characterized by a higher value of microhardness. As the load is increased, the microhardness of the sample decreases from 1655 to 950 MPa. It should be noted that a great number of large HA particles separated from each other were observed in this sample, indicating a heterogeneous structure. At the same time, a high concentration of HA particles in the sample contributes to an increase in microhardness compared to the 50Fe-Cu-HA sample.

The highest microhardness values were observed during the 45Fe-Cu-HA samples' tests. The microhardness of these samples decreased from 1625 to 965 MPa as the load was increased from 1 to 10 N. This result corresponds to the SEM results. A more uniform distribution of iron and HA particles in the samples contributed to an increase in the microhardness compared to the 50Fe-Cu-HA and 40Fe-Cu-HA samples.

In this case, the maximum scatter occurs in the lower range of loads and decreases as the applied load increases. At the same time, under such a load, the largest scatter of values is also visible, which indicates the inhomogeneity (HA particle distribution, porosity and feedstock component mixing) of the surface.

As a result of the research, the influence of the concentration of Fe-Cu-HA powder in the feedstock on the average Vickers hardness of the obtained composites was established.

In particular, the maximum values of microhardness under all loads were observed for the samples with the 45Fe-Cu-HA composition. Similar hardness values were obtained for other implant materials in the literature. In the reference [26]. the vickers microhardness of β Ti-40Nb alloy for implants at the indentation load 4.9 N varies between 266 and 181 HV (2608 and 1775 MPa).

The microhardness of these bioresorbable composites is higher than that of magnesium alloys, but in acceptable ranges for medical devices [27].

4. Conclusions

In the work, the bulk composite based on powders of the Fe-Cu system and hydroxyapatite was obtained using additive manufacturing based on the extrusion of materials.

Varying the amount of added Fe-Cu-HA powder from 50 to 40% led to the change in the morphology, phase and elemental compositions, as well as the microhardness of the resulting composites. The 45Fe-Cu-HA samples were characterized by the highest microhardness values, 1623-965 MPa at a load of 0.25-10 N, which is, on average, 26% higher than for the 50Fe-Cu-HA samples with the lowest microhardness values. This result is in a good agreement with the results obtained by the SEM method. In the 45Fe-Cu-HA composite, the most uniform distribution of iron and HA particles was observed, which contributed to an increase in its mechanical. Thus, a composite with a ratio of 45 to 55 wt.% of the cermet and polymer parts showed a higher microhardness compared to other composites and a more uniform distribution of antibacterial copper particles and bioactive hydroxyapatite particles, which can improve the biological properties of the material. In the future, a comparative study of the in vitro biological properties of three groups of composites is planned.

The study showed that the creation of a biocomposite material by 3D printing based on Fe-Cu bimetallic powder and hydroxyapatite is a promising approach for the manufacture of biodegradable implants for medical use.

Author Contributions: V.V.C., conceptualization, methodology, investigation, writing—original draft preparation, visualization, supervision, project administration, funding acquisition; N.A.L., methodology, investigation, writing—original draft, visualization; A.E.R., formal analysis, methodology, investigation, data curation, visualization; N.V.S., K.V.S. and L.Y.I., methodology, validation, formal analysis, investigation; M.A.K., formal analysis, validation, methodology, investigation, funding acquisition; N.E.T., I.A.G., A.A.M., S.O.K. and M.G.K., formal analysis, methodology, investigation, data curation. All authors have read and agreed to the published version of the manuscript.

Funding: This work was financially supported by the Russian Science Foundation, No. 22-73-00207, https://rscf.ru/project/22-73-00207/. Accessed on 7 July 2022.

Institutional Review Board Statement: Not applicable.

Informed Consent Statement: Not applicable.

Data Availability Statement: Not applicable.

Conflicts of Interest: The authors declare that they have no conflicts of interest to report regarding the present study.

References

1. Kang, M.-H.; Cheon, K.-H.; Jo, K.-I.; Ahn, J.-H.; Kim, H.-E.; Jung, H.-D.; Jang, T.-S. An asymmetric surface coating strategy for improved corrosion resistance and vascular compatibility of magnesium alloy stents. *Mater. Des.* **2020**, *196*, 109182. [CrossRef]
2. Cheon, K.-H.; Park, C.; Kang, M.-H.; Kang, I.-G.; Lee, M.-K.; Lee, H.; Kim, H.-E.; Jung, H.-D.; Jang, T.-S. Construction of tantalum/poly(ether imide) coatings on magnesium implants with both corrosion protection and osseointegration properties. *Bioact. Mater.* **2021**, *6*, 1189–1200. [CrossRef] [PubMed]
3. Lin, W.; Zhang, H.; Zhang, W.; Qi, H.; Zhang, G.; Qian, J.; Li, X.; Qin, L.; Li, H.; Wang, X.; et al. In vivo degradation and endothelialization of an iron bioresorbable scaffold. *Bioact. Mater.* **2021**, *6*, 1028–1039. [CrossRef] [PubMed]
4. Liao, Y.; Cao, H.; Xia, B.; Xiao, Q.; Liu, P.; Hu, G.; Zhang, C. Changes in Trace Element Contents and Morphology in Bones of Duck Exposed to Molybdenum or/and Cadmium. *Biol. Trace Elem. Res.* **2017**, *175*, 449–457. [CrossRef] [PubMed]
5. Lin, W.; Qin, L.; Qi, H.; Zhang, D.; Zhang, G.; Gao, R.; Qiu, H.; Xia, Y.; Cao, P.; Wang, X.; et al. Long-term in vivo corrosion behavior, biocompatibility and bioresorption mechanism of a bioresorbable nitrided iron scaffold. *Acta Biomater.* **2017**, *54*, 454–468. [CrossRef]
6. Lee, M.-K.; Lee, H.; Park, C.; Kang, I.-G.; Kim, J.; Kim, H.-E.; Jung, H.-D.; Jang, T.-S. Accelerated biodegradation of iron-based implants via tantalum-implanted surface nanostructures. *Bioact. Mater.* **2022**, *9*, 239–250. [CrossRef]
7. Gutmanas, E.Y.; Gotman, I.; Sharipova, A.; Psakhie, S.G.; Swain, S.K.; Unger, R. Drug loaded biodegradable load-bearing nanocomposites for damaged bone repair. *AIP Conf. Proc.* **2017**, *1882*, 020025. [CrossRef]
8. Kręcisz, B.; Kieć-Świerczyńska, M.; Chomiczewska-Skóra, D. Allergy to orthopedic metal implants—A prospective study. *Int. J. Occup. Med. Environ. Health* **2012**, *25*, 463–469. [CrossRef]
9. Radulescu, D.-E.; Neacsu, I.A.; Grumezescu, A.-M.; Andronescu, E. Novel Trends into the Development of Natural Hydroxyapatite-Based Polymeric Composites for Bone Tissue Engineering. *Polymers* **2022**, *14*, 899. [CrossRef]
10. Hou, X.; Zhang, L.; Zhou, Z.; Luo, X.; Wang, T.; Zhao, X.; Lu, B.; Chen, F.; Zheng, L. Calcium Phosphate-Based Biomaterials for Bone Repair. *J. Funct. Biomater.* **2022**, *13*, 187. [CrossRef]
11. LeGeros, R.Z. Calcium Phosphate-Based Osteoinductive Materials. *Chem. Rev.* **2008**, *108*, 4742–4753. [CrossRef] [PubMed]
12. Yamasaki, H.; Sakai, H. Osteogenic response to porous hydroxyapatite ceramics under the skin of dogs. *Biomaterials* **1992**, *13*, 308–312. [CrossRef] [PubMed]
13. Lozhkomoev, A.S.; Lerner, M.I.; Pervikov, A.V.; Kazantsev, S.O.; Fomenko, A.N. Development of Fe/Cu and Fe/Ag Bimetallic Nanoparticles for Promising Biodegradable Materials with Antimicrobial Effect. *Nanotechnol. Russ.* **2018**, *13*, 18–25. [CrossRef]
14. Ulum, M.; Arafat, A.; Noviana, D.; Yusop, A.; Nasution, A.; Kadir, M.A.; Hermawan, H. In vitro and in vivo degradation evaluation of novel iron-bioceramic composites for bone implant applications. *Mater. Sci. Eng. C* **2014**, *36*, 336–344. [CrossRef]
15. Razavi, M.; Huang, Y. Effect of hydroxyapatite (HA) nanoparticles shape on biodegradation of Mg/HA nanocomposites processed by high shear solidification/equal channel angular extrusion route. *Mater. Lett.* **2020**, *267*, 127541. [CrossRef]
16. Gu, X.; Zhou, W.; Zheng, Y.; Dong, L.; Xi, Y.; Chai, D. Microstructure, mechanical property, bio-corrosion and cytotoxicity evaluations of Mg/HA composites. *Mater. Sci. Eng. C* **2010**, *30*, 827–832. [CrossRef]
17. Tobita, T.; Nakagawa, S.; Takeuchi, T.; Suzuki, M.; Ishikawa, N.; Chimi, Y.; Saitoh, Y.; Soneda, N.; Nishida, K.; Ishino, S.; et al. Effects of irradiation induced Cu clustering on Vickers hardness and electrical resistivity of Fe–Cu model alloys. *J. Nucl. Mater.* **2014**, *452*, 241–247. [CrossRef]
18. Ramesh, S.; Tan, C.; Sopyan, I.; Hamdi, M.; Teng, W. Consolidation of nanocrystalline hydroxyapatite powder. *Sci. Technol. Adv. Mater.* **2007**, *8*, 124–130. [CrossRef]
19. Zyman, Z.Z.; Tkachenko, M.V.; Polevodin, D.V. Preparation and characterization of biphasic calcium phosphate ceramics of desired composition. *J. Mater. Sci. Mater. Med.* **2008**, *19*, 2819–2825. [CrossRef]
20. Genina, N.; Holländer, J.; Jukarainen, H.; Mäkilä, E.; Salonen, J.; Sandler, N. Ethylene vinyl acetate (EVA) as a new drug carrier for 3D printed medical drug delivery devices. *Eur. J. Pharm. Sci.* **2016**, *90*, 53–63. [CrossRef]
21. Shelenkov, P.G.; Pantyukhov, P.V.; Popov, A. Highly filled biocomposites based on ethylene-vinyl acetate copolymer and wood flour. *IOP Conf. Ser. Mater. Sci. Eng.* **2018**, *369*, 012043. [CrossRef]
22. Aslamazova, T.R.; Lomovskoy, V.A.; Shorshina, A.S.; Zolotarevskii, V.I.; Kotenev, V.A.; Lomovskaya, N.Y. Temperature–Frequency Domains of Inelasticity in the Rosin–Copper and Rosin–Cellulose Composites. *Russ. J. Phys. Chem. A* **2022**, *96*, 222–229. [CrossRef]
23. Pervikov, A.; Glazkova, E.; Lerner, M. Energy characteristics of the electrical explosion of two intertwined wires made of dissimilar metals. *Phys. Plasmas* **2018**, *25*, 070701. [CrossRef]

24. Lerner, M.I.; Pervikov, A.V.; Glazkova, E.A.; Svarovskaya, N.V.; Lozhkomoev, A.S.; Psakhie, S.G. Structures of binary metallic nanoparticles produced by electrical explosion of two wires from immiscible elements. *Powder Technol.* **2016**, *288*, 371–378. [CrossRef]
25. Chaikina, M.V.; Bulina, N.V.; Vinokurova, O.B.; Prosanov, I.Y.; Dudina, D.V. Interaction of calcium phosphates with calcium oxide or calcium hydroxide during the "soft" mechanochemical synthesis of hydroxyapatite. *Ceram. Int.* **2019**, *45*, 16927–16933. [CrossRef]
26. Santos, R.F.M.; Ricci, V.P.; Afonso, C.R.M. Influence of Swaging on Microstructure, Elastic Modulus and Vickers Microhardness of β Ti-40Nb Alloy for Implants. *J. Mater. Eng. Perform.* **2021**, *30*, 3363–3369. [CrossRef]
27. Elambharathi, B.; Kumar, S.D.; Dhanoop, V.; Dinakar, S.; Rajumar, S.; Sharma, S.; Kumar, V.; Li, C.; Eldin, E.M.T.; Wojciechowski, S. Novel insights on different treatment of magnesium alloys: A critical review. *Heliyon* **2022**, *8*, e11712. [CrossRef]

Disclaimer/Publisher's Note: The statements, opinions and data contained in all publications are solely those of the individual author(s) and contributor(s) and not of MDPI and/or the editor(s). MDPI and/or the editor(s) disclaim responsibility for any injury to people or property resulting from any ideas, methods, instructions or products referred to in the content.

Article

Ca–Zn Phosphate Conversion Coatings Deposited on Ti6Al4V for Medical Applications

Diana-Petronela Burduhos-Nergis [1], Nicanor Cimpoesu [1], Elena-Luiza Epure [2], Bogdan Istrate [3], Dumitru-Doru Burduhos-Nergis [1] and Costica Bejinariu [1,*]

1. Faculty of Materials Science and Engineering, Gheorghe Asachi Technical University, 700050 Iasi, Romania; nicanor.cimpoesu@academic.tuiasi.ro (N.C.)
2. Department of Natural and Synthetic Polymers, Faculty of Chemical Engineering and Environmental Protection, "Gheorghe Asachi" Technical University of Iaşi, Str. Prof.dr.doc. D. Mangeron nr. 73, 700050 Iasi, Romania; lepure@tuiasi.ro
3. Faculty of Mechanical Engineering, "Gheorghe Asachi" Technical University of Iasi-Romania, Blvd. Dimitrie Mangeron, No. 61–63, 700050 Iasi, Romania; bogdan.istrate@academic.tuiasi.ro
* Correspondence: costica.bejinariu@tuiasi.ro

Abstract: This paper aims to study the possibility of improving the chemical and surface characteristics of the Ti6Al4V alloy by depositing phosphate layers on its surface. Accordingly, an innovative phosphating solution was developed and used in a chemical conversion process to obtain Ca–Zn phosphate layers on the base material surface. Moreover, the chemical composition of the phosphate solution was chosen considering the biocompatibility of the chemical elements and their possibility of contributing to the formation of phosphate compounds. The obtained layer was characterized by optical microscopy (OM), scanning electron microscopy (SEM), energy dispersive spectroscopy (EDS), X-ray diffraction (XRD), Fourier-transform infrared spectroscopy (FTIR), and potentiodynamic polarization tests. The wetting of the Ca–Zn sample surface was also investigated using water and two liquids similar to body fluids, namely, Ringer and Dulbecco solutions. According to the surface energy study, the polar component is almost two times larger compared with the dispersive one. The SEM and EDS tests revealed a uniformly coated surface with intercalated crystals leading to a rough surface. Furthermore, the XRD results showed not only the presence of hopeite and scholzite but also of phosphophyllite. By the vibrations of the PO_4^{-3} groups, the FTIR test confirmed the presence of these phases. The potentiodynamic tests revealed that the samples coated with the Ca–Zn phosphate layer present better corrosion resistance and a lower corrosion rate compared with the uncoated ones.

Keywords: titanium alloy; phosphate layer; conversion coating; biomaterial; zinc phosphate; scholzite

1. Introduction

The number of people who need implants is increasing over time, either in response to trauma caused by accidents or degenerative bone diseases. Bone tissue engineering is concerned primarily with finding new biomaterials with superior properties or improving existing ones on the market [1]. Titanium is one of the most popular materials used in the medical field; it is found in hip prostheses, knee prostheses, dental implants, etc. This is due to the fact that, in addition to being biocompatible, it has superior mechanical properties as well as acceptable corrosion resistance [2,3].

However, there is a risk of implant rejection or failure due to titanium's bioinert surface [4]. Therefore, over time, different methods have been tried to promote the osteogenic activity and stability of titanium (or titanium alloy) implants. These methods include mechanical processing [5–7] (grinding, polishing, sandblasting, laser cavitation, etc.) and acidic treatments [8,9] that are conducted to obtain a surface with high roughness. The surface roughness is important to promote tissue regeneration and the anchoring of further deposited layers. Several studies were also investigated the protection of Ti implants against corrosion by coating their surfaces with different types of thin layers. For example, many researchers have been studying the deposition of hydroxyapatite on the Ti surface since the 1980s until now [10,11]. Hydroxyapatite layers can be deposited by different methods, including micro-arc oxidation [12], the sol–gel method [13], the plasma spraying method [14], electrodeposition [15], the ion-beam deposition method, etc. [16–18]. Other researchers have deposited nitride layers, such as zirconium nitride, titanium nitride, etc., using different methods such as the Physical Vapor Deposition method (PVD) [19–21]. Using the same techniques as previously mentioned, the osseointegration of titanium can also be promoted by coating its surface with different types of oxides, such as CuO, ZnO, ZrO_2, etc. [22]. Considering the advantages related to their ease of use, electrodeposition and PVD are two of the most used methods to cover a Ti surface.

Due to the importance of human health and the high use of titanium implants, it is still of great interest to find new solutions or to improve existing ones that contribute to Ti implant acceptance and biointegration. Accordingly, in recent years, numerous studies have focused on the improvement of titanium alloy characteristics by depositing different types of phosphate layers. Due to its multiple advantages, this chemical conversion process has garnered great interest from different industries (automotive, construction sector, etc.), but its suitability for bioengineering was only discovered recently [23]. Some of the strong points of this process are its low economic cost, high adherence to the substrate, and, of course, the high corrosion resistance of the resultant coating [24,25]. Considering the versatility of the phosphate solutions (they can incorporate different types of metals), biocompatible elements that can promote the osseointegration of the substrate can be deposited onto its surface while also protecting the base materials against corrosion [26,27]. The characteristics of the phosphate layer and its anchorage to the substrate surface are strongly influenced by multiple factors. Liu et al. [28] studied the influence of the pH of the phosphate solution on the structure of the Ca–Zn phosphate layer deposited on pure titanium. According to their publication, both hopeite and scholzite phases can be deposited on the substrate surface. Hopeite is created in higher amounts at pHs between 2.50–3.25, while higher quantities of the scholzite phase can be observed at pHs between 3.50–4.25. In this case, corrosion tests revealed that the samples with a higher content of hopeite had superior corrosion resistance to those where scholzite was predominant. The same group of researchers also published another study regarding the influence of temperature on the crystals' structure [29]. In their publication, they observed that by increasing the immersion time and temperature, the coating structure (compactness) and thickness could be improved. Zhao et al. [30] studied the effect of voltage and temperature on the formation of Ca–Zn phosphate layers. Compared with the previously cited studies, in this research, the coating was obtained through an electrolysis-induced phosphate chemical conversion method. Accordingly, the authors observed that the corrosion resistance and the structural characteristics of the deposited layer can be improved if the voltage is decreased and the temperature is increased. Liu et al. [31] also studied the biocompatibility of scholzite coatings deposited on the surfaces of pure Ti and Ti6Al4V, demonstrating that this type of process and coating can be used in order to promote the osteointegration of titanium and its alloys.

According to the reviewed studies, the deposition of phosphate layers by chemical conversion is a suitable method for promoting the biointegration of Ti implants; however, there are very few studies that consider Ti6Al4V as a substrate, and many of the parameters that influence layer formation were not considered. Furthermore, the literature is very poor in terms of the types of phosphating solutions that can be used to obtain protective layers for Ti implants. In this study, an innovative phosphating solution was developed, which was further used in a chemical conversion process to obtain Ca–Zn phosphate layers on the Ti6Al4V surface. Moreover, the chemical composition of the phosphate solution was chosen considering the biocompatibility of the chemical elements and their possibility of contributing to the formation of phosphate compounds. Accordingly, a new phase was confirmed in the structure of the deposited layer, while relevant tests were employed to analyze the layer morphology and its corrosion resistance in two types of corrosive media (Ringer and Dulbecco solutions) relevant for bioengineering applications.

2. Materials and Methods

2.1. Materials

Due to their biocompatibility and good mechanical properties, titanium alloys are widely used for implant manufacturing. In this study, Ti6Al4V purchased from AEMMetal (Hunan, China) was used as a substrate. The material was supplied in a round bar form, and its chemical composition, provided by the supplier, is presented in Table 1.

Table 1. Chemical composition of Ti6AL4V used as substrate.

Element	Al	V	Fe	O	C	Ti
wt.%	6.14	4.22	0.12	0.11	0.028	bal.

2.2. Sample Preparation

The Ti6Al4V bars were cut into specimens with a size of Ø10 mm and a thickness of 3 mm. In order to promote the biointegration of the titanium alloy, a phosphate layer was deposited onto the surface of the obtained samples by a chemical conversion process (phosphating). First, the samples were ground with 400, 600, 800, 1000, and 1200 grit SiC abrasive paper. After a homogenous surface was obtained, the samples were degreased by immersing them successively in an ultrasonic bath with acetone, ethyl alcohol, and distilled water for 10 min. Following that, the surface was pickled and activated by immersing the samples in a 2% HF solution for 15 s, followed by 30 s in a 3 g/L titanium colloidal solution ($Na_4TiO(PO)_2$ 0–7H_2O). Then, a Ca–Zn phosphate layer was deposited on the samples surfaces by immersing them in a phosphating solution. The phosphating step took place for 60 min at 90 °C. The chemical composition of the phosphating solution used in this research is different from the ones previously presented in the literature. The solution used here contains the following accelerators and inhibitors: HNO_3, NaOH, $NaNO_2$, $Na_5P_3O_{10}$, and NaF in different quantities—the influence of each component was described in Ref. [32]. In addition, in this solution, $Ca(NO_3)_2$, Zn chips, Fe powder, and H_3PO_4 were added to obtain the metal cations and compounds that will conduct to the phosphate layer formations. Afterwards, the samples were rinsed in cold water and dried at room temperature. A schematic representation of the phosphating process is presented in Figure 1.

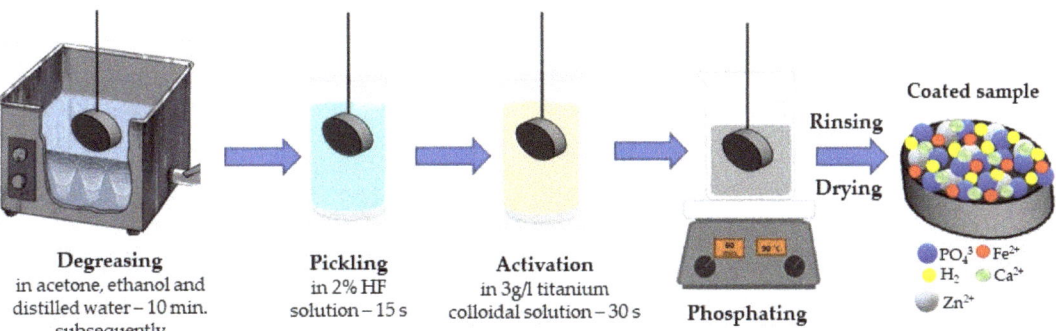

Figure 1. The process flow chart of sample coating.

2.3. Methods

The morphology and the chemical composition of the deposited layer were studied using an optical microscope (Zeiss Imager Axio a1M) and a scanning electron microscope (Vega Tescan LMH II) equipped with an energy dispersive detector (EDAX Bruker). The chemical composition of the layer was determined by making 5 measurements at different points of the surface of the samples; in this study, the measurement closest to the average is presented.

The type of crystals which were formed on the surface of the titanium alloy Ti6Al4V was determined using an X-ray diffractometer (PANalytical X'Pert PRO MPDR) using Cu-Kα1 radiation. This study used a 2θ angle ranging between 5°–70°, the number of steps was 6474, and the step size was 0.0131° every 60 s with a scanning speed of 0.054 °/s. In order to highlight the compounds specific to the obtained phosphate layer, a Fourier-transform infrared spectroscope (Bruker Hyperion 1000 FTIR spectrometer) was used with a wavenumber range between 4000 cm^{-1} and 600 cm^{-1} at a spectral resolution of 4 cm^{-1} and several 64 scans for each analyzed surface.

Because this material was designed to be used as an implant, an indispensable property is the hydrophilic surface. To determine this kind of characteristic, the contact angle was evaluated. For this determination, a Drop Shape Analyzer (DSA100, Kruss) with a software-controlled dosing system was used, and the contact angle was measured through the sessile-drop method. The wettability of the metal surface was tested by deionized water, ethylene glycol, and Ringer and Dulbecco solutions at room temperature. The volume of the drop deposited on the metal piece was 4 µL. After the drop was deposited, the contact angle was recorded for 60 s. The metal surface was cleaned with high-purity ethanol and dried before the contact angle tests. For each liquid sample, measurements were carried out at least five times. The average of all the individual measurements was used to compute the contact angle value [33]. Additionally, the Owens–Wendt–Rabel–Kaelble (OWRK) method [34] based on contact angle was used to calculate the sample's surface energy.

The corrosion properties of the Ca–Zn phosphate layer were analyzed by potentiodynamic polarization in two different corrosion media. In this study, Ringer solution (pH = 6.18) and Dulbecco's phosphate-buffered saline solution without Ca and Mg (pH = 7.4) were used to evaluate the corrosion resistance of the obtained samples. The potentiodynamic polarization tests were conducted using an OrigaFlex potentiostat equipped with a three-electrode cell. The cell includes a reference electrode (saturated calomel electrode), a platinum counter electrode, and the working electrode (the analyzed sample). The analyzed sample has an exposed area of 0.785 cm^2. The obtained results were processed with OrigaMaster 5 software (Version 2.5.0.3.). The polarization curves were obtained at a potential range of (−300) to (+300) mV vs. o... pen circuit potential (OCP), with a scan rate of 0.5 mV/s. The electrochemical tests were repeated 3 times to ensure their repeatability.

3. Results and Discussion

3.1. Characterization of the Ca–Zn Phosphate Layer

Optical and scanning electron microscopy were used to study the structure and the uniformity of the coating. Figure 2 presents the images obtained by OM at a magnification of 5×, 10×, and 20×, while Figure 3 shows the SEM micrographs for the Ca–Zn phosphate layer deposited on the Ti6Al4V surface.

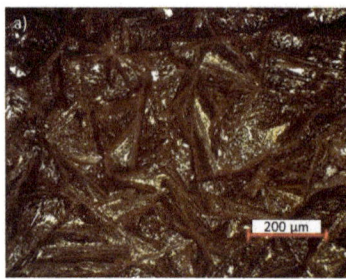 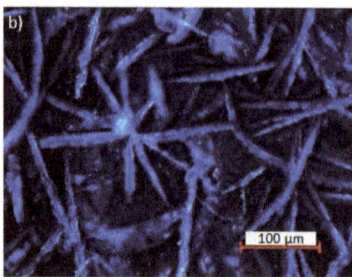 

Figure 2. Optical microstructure of the Ca–Zn phosphate layer deposited on the Ti6Al4V surface viewed at (**a**) 5×, (**b**) 10×, and (**c**) 20× magnification.

 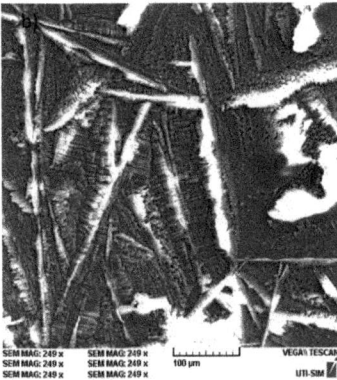 

Figure 3. The morphology of the Ca–Zn phosphate layer deposited on the Ti6Al4V surface viewed at (**a**) 500×, (**b**) 250× and (**c**) 100× magnification.

A morphology specific to many crystals of different sizes can be observed uniformly deposited all over the surface of the titanium alloy (Figure 3c). Due to the areas between the crystals, called inter-crystalline areas, the surface of the layer is rough, which is specific to phosphate layers deposited by the chemical conversion process. The average roughness value (Ra) of the obtained coatings, determined by profilometry, was around 1.34 μm. This aspect represents an advantage of these layers since they can facilitate cell adhesion, similarly to the porous structure of bones. The formed crystals are compact, having an acicular shape in some areas (Figure 3b) and a flower shape in others (Figure 2b,c). The flower-like structure is specific to the hopeite compound created after phosphating, and its presence was confirmed by XRD analysis. Compared with other phosphate layers based on Zn and Ca, the morphology of those obtained in this study is different from that obtained by Liu et al. [31]. In their study, the authors observed that the crystals they obtained have the typical morphology of scholzite, being composed of smaller and thicker lamellar crystals. Another study [29] shows that the structure of crystals obtained on a pure Ti surface looks like flakes, with their dimensions depending on the temperature (25 °C, 55 °C, 75 °C) of the phosphate solution. Therefore, the structure of the Ca–Zn phosphate layer differs depending on the process parameters (immersion time, phosphating temperature, coupled systems, coupling voltage, sample preparation process, etc.) and very much on the chemical composition of the phosphating solution. In this case, the crystal's dimensions could be higher than those presented previously due to the high phosphating temperature (90 °C).

In order to observe the uniformity of the chemical composition of the deposited layer, the sample surface was analyzed by EDS and examined from five different points of view from the chemical composition point of view. Figure 4 shows the chemical element distribution in the area of the Ca–Zn phosphate layer, where the chemical composition is close to the average of the five different points. In addition, the chemical composition and the energy spectra of this zone are presented in Figure 5. The SEM and EDS analyses were also used to evaluate the thickness of the obtained phosphate layer. Figure 6 shows the morphology of the coated samples in cross-section; the average thickness layer is around 63 μm.

As can be seen from Figure 4, the elemental distribution confirms the presence of P, O, Zn, Ca, and Fe all over the analyzed surface. The presence of Ti, V, and Al is explained by the penetration of X-rays up to the interface between the phosphate layer and the titanium alloy substrate in those areas where the phosphate layer is thinner. This aspect was also observed by other researchers looking at phosphate conversion coatings deposited on titanium or other alloys [30,32].

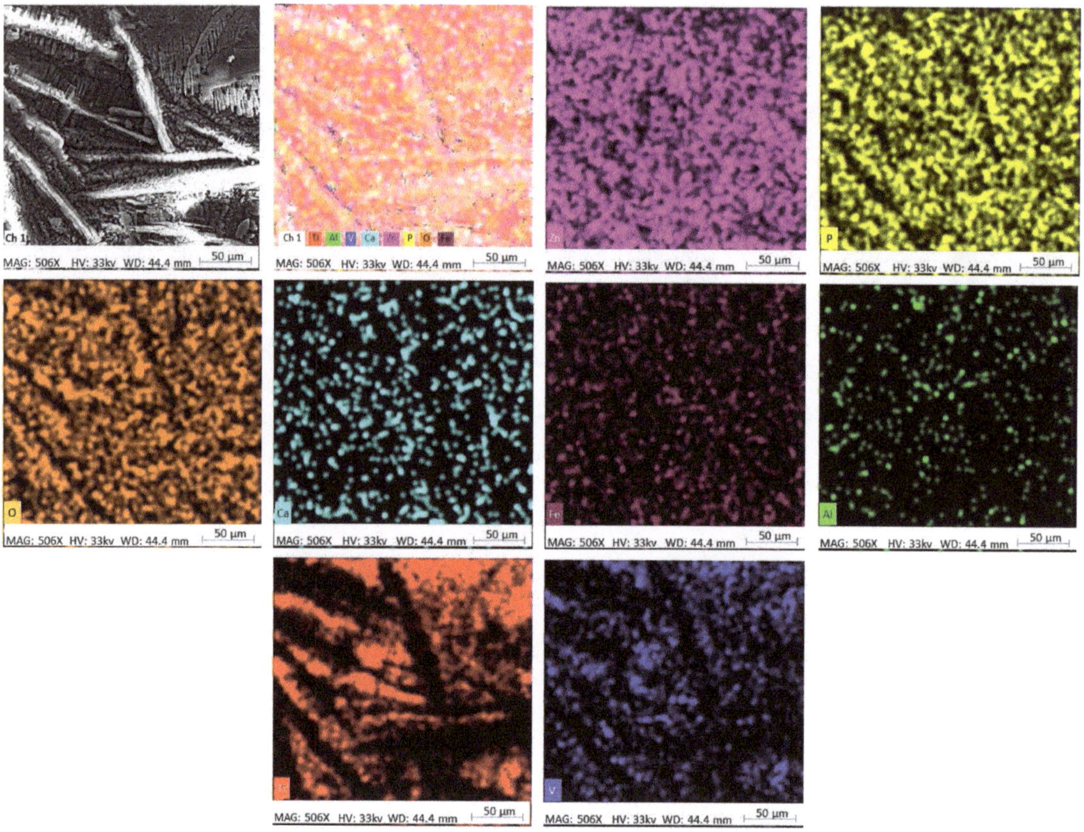

Figure 4. EDS area scan of the Ca–Zn phosphate layer.

Element	Norm. wt.,%	Norm. at., %	Error, %
Zinc	54.24	27.78	1.23
Oxygen	24.30	50.88	4.28
Phosphorus	16.76	18.12	0.66
Iron	2.16	1.30	0.11
Titanium	1.91	1.33	0.10
Aluminum	0.25	0.31	0.09
Calcium	0.19	0.16	0.06
Vanadium	0.19	0.12	0.06
	100	100	

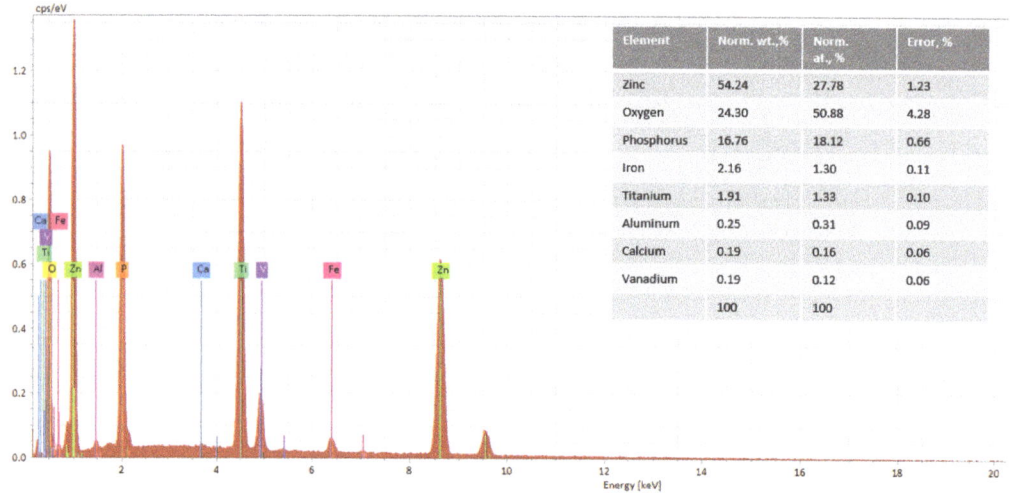

Figure 5. Energy spectra and the chemical composition of the Ca–Zn phosphate layer.

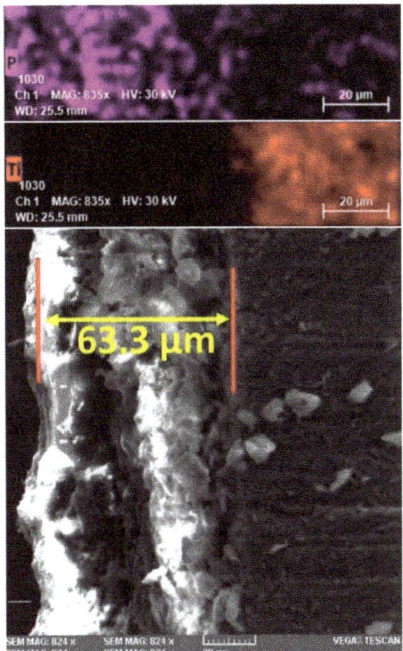

Figure 6. SEM–EDS image of the cross-section of the coated sample.

In order to determine the phase formed on the surface of the Ti6Al4V by depositing a Ca–Zn phosphate layer through the chemical conversion process, an XRD analysis was conducted. The XRD diffractogram of the coated sample is presented in Figure 7.

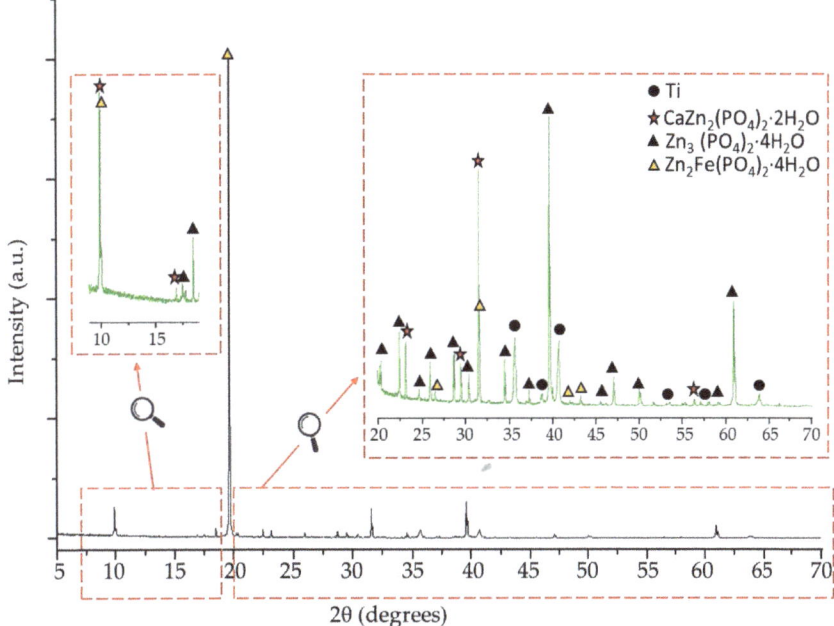

Figure 7. XRD diffractogram of the Ca–Zn phosphate layer.

The XRD pattern presents significant peaks of specific phases which can be indexed as scholzite (CaZn$_2$(PO$_4$)$_2$·2H$_2$O, PDF 04-010-2736), hopeite (Zn$_3$(PO$_4$)$_2$·4H$_2$O, PDF 04-015-0707), and phosphophyllite (Zn$_2$Fe(PO$_4$)$_2$·4H$_2$O, PDF 01-074-6617 and PDF 04-011-6924).

As can be observed, the strongest diffraction intensity is specific to the phosphophyllite phase at a 2θ degree of 19.7° corresponding to the (002) crystal plane. Furthermore, the Zn$_2$Fe(PO$_4$)$_2$·4H$_2$O phase is detected at 2θ degree of 9.9°, 26.4°, 31.5°, 41.2°, 43.4° corresponding to the (100), (112), ($\bar{3}$12), (400), and ($\bar{4}$14) crystal planes. Regarding the zinc phosphate tetrahydrate compound, hopeite shows many significant peaks, including one with high intensity at a 2θ degree of 39.6° corresponding to the (080) crystal plane. Due to the introduction of Ca in the phosphate solution, another phase formed is scholzite, which has two peaks with remarkable intensity at a 2θ degree of 10.1° and 31.7° corresponding to the (200) and (1422) crystal planes, respectively. According to the XRD analysis, the highest number of peaks is specific to hopeite. Thus, the addition of Ca and Fe in the phosphate solution formed specific compounds which will improve the characteristics of the Ti6Al4V substrate.

An FTIR spectrum of the Ca–Zn phosphate layer is shown in Figure 8. As can be seen, the vibration band between 600–700 cm^{-1} has a significant peak at 607 cm^{-1}; this band can be assigned to the stretching vibration of hydroxyl groups or PO$_4$$^{-3}$ groups [35]. The O-H group can appear due to the motion of the absorbed water [36]. In addition, significant peaks are presented between 3100–3600 cm^{-1}, which can also be attributed to O-H groups that can indicate the existence of crystal water [30]. This aspect is available also for the peak between 1500–1700 cm^{-1}. The peak at 1041 cm^{-1} can be attributed to the PO$_4$$^{-3}$ group, which is part of all the identified phases from the XRD results. Furthermore, the peak at 964 cm^{-1} and the vibration band between 900–1200 cm^{-1} correspond to the P-O stretching in HPO$_4$^{2} [37].

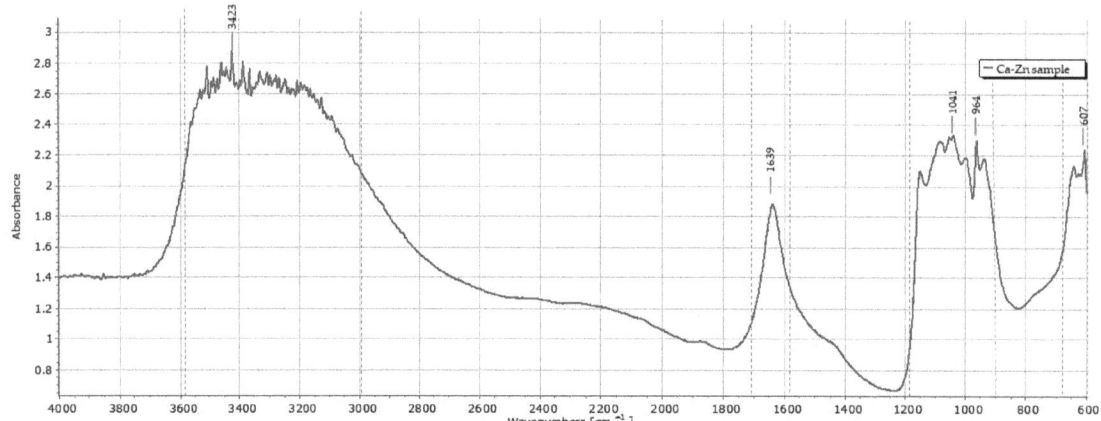

Figure 8. FTIR spectra of the Ca–Zn phosphate layer.

3.2. Corrosion Characteristics

Figure 9 shows the potentiodynamic polarization curves of Ti6Al4V coated with a Ca–Zn phosphate layer (Ca–Zn) and Ti6Al4V uncoated samples, tested in Dulbecco solution (Figure 9a) or Ringer solution (Figure 9b). The corrosion parameters of E$_{corr}$, I$_{corr}$, β$_a$, β$_c$, polarization resistance (R$_p$) and corrosion rate (C$_R$) were calculated with formulas presented in a previous study [25] and are summarized in Table 2. In both cases, Ti6Al4V specimens were tested as control samples in order to reveal the influence of the coating on the corrosion properties of the titanium alloy.

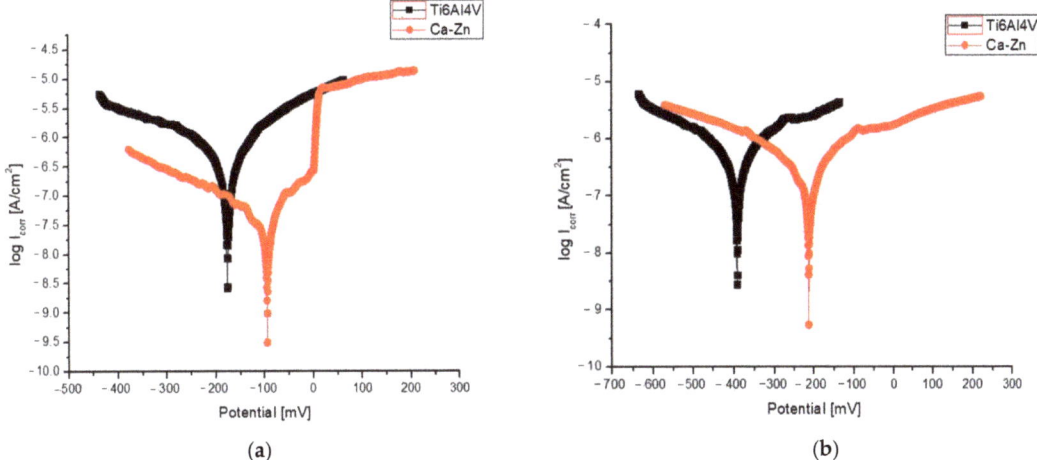

Figure 9. Potentiodynamic polarization curves of uncoated Ti6Al4V or Ti6Al4V coated with a Ca–Zn phosphate layer tested in Dulbecco solution (**a**) or Ringer solution (**b**).

Table 2. Electrochemical corrosion parameters determined for the Ca–Zn phosphate layer and uncoated Ti6Al4V tested in the two corrosive media.

Sample	Corrosive Media	E_{corr} mV	I_{corr} µA/cm^2	β_a mV/dec.	β_c mV/dec.	R_p kΩ·cm^2	C_R µm/Year
Ti6Al4V	Dulbecco	−177	0.63	172	−270	87.8	13.15
Ca–Zn		−95	0.02	53	−131	419.5	0.46
Ti6Al4V	Ringer	−448	0.86	100	−106	17.9	18.03
Ca–Zn		−213	0.04	38	−59	101.1	0.95

As can be seen from Figure 9, the anodic curve specific to the Ca–Zn sample presents a slope change near 0 mV, indicating the start of a passivation plateau within the increase of potential. Moreover, for both cases, the curves of the phosphate sample shifted to a positive direction of corrosion potential while current density decreased significantly compared with the values of the uncoated Ti6Al4V samples. These results show that the phosphate samples present a higher corrosion resistance due to the small values of I_{corr} (0.02 and 0.04 µA/cm^2) and more positive values of E_{corr} (−95 and −213 mV) compared with the high values of I_{corr} (0.63 and 0.86 µA/cm^2) and more negative values of E_{corr} (−177 and −448 mV) for uncoated samples. It is also well known that a high value of polarization resistance indicates a good corrosion resistance; therefore, the values of R_p for the Ca–Zn sample are about five times higher compared with Ti6Al4V samples. At the same time, the corrosion rate decreased significantly from 13.15 and 18.03 µm/year for uncoated samples to 0.46 and 0.95 µm/year for the coated ones. Furthermore, regardless of the sample type (coated or uncoated), the results show that the Ringer solution is more corrosive compared with the Dulbecco solution.

Another study [31] regarding the deposition of scholzite and hopeite on the surface of titanium by phosphating revealed an improvement of corrosion resistance in 0.9% NaCl solution. As in this research, the corrosion potential for the coated samples exhibited more positive values (from −440 mV to −310 mV) and lower current density values (from 0.53 to 0.37). Compared with the results of our study, even though the corrosive media are different, we can observe an improvement of corrosion resistance due to the presence of the phosphophyllite phase.

3.3. Surface Wettability

The contact angle values can be influenced by surface texture, chemistry, and cleanliness. Other variables that affect the contact angle measurement include the experimental temperature, the volume, and the kind of liquid employed. The contact angle was tested for water, ethylene glycol, and Ringer and Dulbecco solutions at room temperature. The measurements were carried out for 60 s following/after the drop's deposition. For all liquids, it was found that the contact angle dramatically decreases in the first 10 s, then stabilizes in the next 20–60 s, reaching a plateau value. The decreasing contact angle values could be explained by a very low spreading of the drop until this reaches thermodynamic equilibrium. In this case, the droplet spreading occurs on a fast-time scale with minimal fluid penetration into the surface layer, the surface being porous as revealed by SEM. After the spreading phase is through, the volume loss from the droplet to the substrate occurs on a slow-time scale [38]. The average of all measurements for each liquid sample is shown in Table 3, along with its standard deviation.

Table 3. Contact angle values of the liquids deposited on the Ca–Zn sample.

No.	Solution	Contact Angle (°)
1	Water	58.3 ± 2.1
2	Ringer solution	56.0 ± 5.2
3	Dulbecco solution	68.5 ± 5.2
4	Ethylene glycol	34.5 ± 3.2

Images of the water, Ringer, and Dulbecco solution droplets deposited on the metal surface are shown in Figure 10.

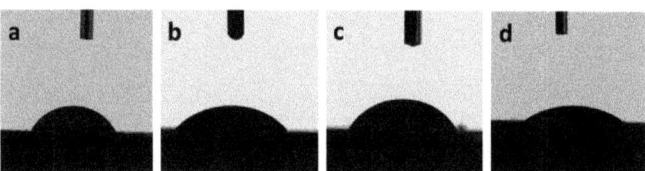

Figure 10. The sessile drop of water (**a**), Ringer solution (**b**), Dulbecco solution (**c**) and ethylene glycol (**d**).

All testing solutions, having a contact angle less than 90°, wet the Ca–Zn coated sample (Table 3). Comparing the liquids related to the body fluids, the capacity to wet the metal surface is almost the same for the water and Ringer solution, with a slightly better value for the latter one. Amongst the three body fluids, the Dulbecco solution has the lowest surface-wetting capacity, presenting the highest contact angle value. Because of its lower contact angle, the Ringer solution is more attracted to and retained on the sample's surface. The coating appears to have grown permeable over time, enabling liquid to pass through it and moisture to reach the underlying material and start corrosion processes. The test corrosion revealed that the Ringer solution is more corrosive than the Dulbecco solution, which has a bigger contact angle than the Ringer solution. An inverse correlation between contact angle values and corrosion rates was reported for super-hydrophobic surfaces [39], but with a different explanation for this phenomenon. The metal surface is hydrophilic, as evidenced by the contact angle between the water and the sample, which is 58.3 ± 2.1°.

The spread of the liquid drops over the metal surface is due to cohesive (the magnitude of the interactions among a material's molecules) and adhesive forces (intensity of the interactions with molecules from other materials) that develop on the liquid's molecules. In this regard, surface energy is a reliable indicator of the degree of wettability [34]. Based on contact-angle values, the sample's surface energy was calculated using the Owens–Wendt–Rabel–Kaelble (OWRK) method [34], based on Fowkes' theory. According to this theory, the surface energy is the sum of dispersion, polar, hydrogen, induction, and acid-base interactions. In the OWRK methods, the interactions were divided into two types: dispersive (van der Waals) and polar (Equation (1)).

$$\gamma_s = \gamma_s^d + \gamma_s^p \tag{1}$$

where γ_s^d and γ_s^p are the dispersive and polar components of the surface energy of the solid, respectively.

At least two test liquids with well-known polar and dispersive interactions are necessary for the OWRK model (Equation (2)) [40]. Thus, two liquids with different polarities were used to determine the surface energy of the metallic sample: water (polar solvent) and ethylene glycol (partially polar solvent).

$$\sqrt{\gamma_s^d \gamma_l^d} + \sqrt{\gamma_s^p \gamma_l^p} = 0.5[\gamma_l(1 + cos\theta)] \tag{2}$$

where γ_s^d, γ_s^p are dispersive components and γ_l^d, γ_l^p are polar components of the solid and liquid surface energies, respectively; γ_l is the surface tension of the liquid; and θ is the contact angle.

It was calculated that the total surface energy (γ_s) of the coated sample is 42.49 ($\pm$1.64) (mN/m). The polar and dispersive components of the total surface tension of the solid are as follows: γ_s^p = 27.57 ($\pm$0.79) (mN/m) and γ_s^d = 14.92 ($\pm$0.85) (mN/m). The higher polar component of the surface energy is due to the Ca–Zn phosphate layer, which was also highlighted by the XRD and FTIR studies. The long-range London dispersion forces are responsible for the dispersive forces. The significant part of the polar component can explain the affinity of the metal sample for polar liquids.

4. Conclusions

In this study, a new phosphating solution for obtaining a Ca–Zn phosphate layer on the Ti6Al4V surface was developed. In addition, relevant analyses were conducted to validate the formation of the coating layer and its potential improvements in the corrosion resistance of the base material.

A Ca–Zn phosphate layer was deposited on the surface of Ti6Al4V in order to study the possibility of improving the osteointegration of implants by increasing the corrosion resistance and the surface roughness. The SEM micrographs revealed that the deposited layer has a crystalline structure, specific to the zinc phosphate layer, where crystals are intercalated. The EDX, XRD, and FTIR analyses showed that hopeite, phosphophyllite, and scholzite crystals were formed all over the titanium alloy surface.

Both the coated and the base materials have been tested in two corrosion media (Ringer and Dulbecco solutions). The electrochemical tests revealed that the Ti6Al4V corrosion resistance was highly improved by depositing a Ca–Zn phosphate layer onto its surface, and the polarization resistance significantly increased by almost five times for the coated samples in both corrosion media. Moreover, the experiments revealed that the Ringer solution is a more aggressive corrosion media than the Dulbecco solution.

All solutions considered show satisfactory surface wettability. The wettability gradually increased from the Ringer solution to water and finally to the Dulbecco solution. From surface energy calculations, it was found that the coated Ti6Al4V has a predominantly polar surface, with the polar component being almost two times larger compared with the dispersive one. This polarity of the surface created by phosphating can be suitable for cell adhesion; therefore, future studies should be conducted on this topic.

Author Contributions: Conceptualization, writing, and investigation: D.-P.B.-N.; writing the original draft, project administration, and scientific supervision: D.-P.B.-N. and C.B.; methodology, investigation, data curation, and validation: E.-L.E., N.C., B.I. and D.-D.B.-N.; data curation, validation, and writing—reviewing and editing: D.-P.B.-N. and D.-D.B.-N.; resources and formal analysis: C.B. All authors have read and agreed to the published version of the manuscript.

Funding: This research received no external funding.

Institutional Review Board Statement: Not applicable.

Informed Consent Statement: Not applicable.

Data Availability Statement: Not applicable.

Acknowledgments: This paper was realized with the support of COMPETE 2.0 project nr.27PFE/2021, financed by the Romanian Government, Minister of Research, Innovation and Digitalization. This paper was also supported by "Gheorghe Asachi" Technical University from Iași (TUIASI), through the Project "Performance and Excellence in postdoctoral research 2022".

Conflicts of Interest: The authors declare no conflict of interest.

References

1. Bandyopadhyay, A.; Mitra, I.; Goodman, S.B.; Kumar, M.; Bose, S. Improving Biocompatibility for next Generation of Metallic Implants. *Prog. Mater. Sci.* **2023**, *133*, 101053. [CrossRef] [PubMed]
2. de Viteri, V.S.; Fuentes, E.; de Viteri, V.S.; Fuentes, E. Titanium and Titanium Alloys as Biomaterials. *Tribol.—Fundam. Adv.* **2013**, *1*, 154–181. [CrossRef]
3. Sidambe, A.T. Biocompatibility of Advanced Manufactured Titanium Implants—A Review. *Materials* **2014**, *7*, 8168. [CrossRef] [PubMed]
4. Silva, R.C.S.; Agrelli, A.; Andrade, A.N.; Mendes-Marques, C.L.; Arruda, I.R.S.; Santos, L.R.L.; Vasconcelos, N.F.; Machado, G. Titanium Dental Implants: An Overview of Applied Nanobiotechnology to Improve Biocompatibility and Prevent Infections. *Materials* **2022**, *15*, 3150. [CrossRef] [PubMed]
5. Kuji, C.; Soyama, H. Mechanical Surface Treatment of Titanium Alloy Ti6Al4V Manufactured by Direct Metal Laser Sintering Using Laser Cavitation. *Metals* **2023**, *13*, 181. [CrossRef]
6. Tsuji, A.; Jia, P.; Takizawa, M.; Murata, J. Improvement in the Polishing Characteristics of Titanium-Based Materials Using Electrochemical Mechanical Polishing. *Surf. Interfaces* **2022**, *35*, 102490. [CrossRef]
7. Xue, T.; Attarilar, S.; Liu, S.; Liu, J.; Song, X.; Li, L.; Zhao, B.; Tang, Y. Surface Modification Techniques of Titanium and Its Alloys to Functionally Optimize Their Biomedical Properties: Thematic Review. *Front. Bioeng. Biotechnol.* **2020**, *8*, 1261. [CrossRef]
8. Jemat, A.; Ghazali, M.J.; Razali, M.; Otsuka, Y. Surface Modifications and Their Effects on Titanium Dental Implants. *Biomed. Res. Int.* **2015**, *2015*, 791725. [CrossRef]
9. Doe, Y.; Ida, H.; Seiryu, M.; Deguchi, T.; Takeshita, N.; Sasaki, S.; Sasaki, S.; Irie, D.; Tsuru, K.; Ishikawa, K.; et al. Titanium Surface Treatment by Calcium Modification with Acid-Etching Promotes Osteogenic Activity and Stability of Dental Implants. *Materialia* **2020**, *12*, 100801. [CrossRef]
10. Albrektsson, T.; Lekholm, U. Osseointegration: Current State of the Art. *Dent. Clin. N. Am.* **1989**, *33*, 537–554. [CrossRef]
11. Satomi, K.; Akagawa, Y.; Nikai, H.; Tsuru, H. Tissue Response to Implanted Ceramic-Coated Titanium Alloys in Rats. *J. Oral. Rehabil.* **1988**, *15*, 339–345. [CrossRef] [PubMed]
12. Wang, Y.; Yu, H.; Chen, C.; Zhao, Z. Review of the Biocompatibility of Micro-Arc Oxidation Coated Titanium Alloys. *Mater. Des.* **2015**, *85*, 640–652. [CrossRef]
13. Jaafar, A.; Schimpf, C.; Mandel, M.; Hecker, C.; Rafaja, D.; Krüger, L.; Arki, P.; Joseph, Y. Sol–Gel Derived Hydroxyapatite Coating on Titanium Implants: Optimization of Sol–Gel Process and Engineering the Interface. *J. Mater. Res.* **2022**, *37*, 2558–2570. [CrossRef]
14. Kuo, T.Y.; Chin, W.H.; Chien, C.S.; Hsieh, Y.H. Mechanical and Biological Properties of Graded Porous Tantalum Coatings Deposited on Titanium Alloy Implants by Vacuum Plasma Spraying. *Surf. Coat Technol.* **2019**, *372*, 399–409. [CrossRef]
15. León, M.; Alvarez, D.; Valarezo, A.; Bejarano, L.; Viteri, D.; Giraldo-Betancur, A.L.; Muñoz-Saldaña, J.; Alvarez-Barreto, J. Electrodeposition of Chitosan on Ti-6Al-4V Surfaces: A Study of Process Parameters. *Mater. Res.* **2022**, *25*, e20210552. [CrossRef]
16. Sharma, A. Hydroxyapatite Coating Techniques for Titanium Dental Implants—An Overview. *Qeios* **2023**. [CrossRef]
17. Huang, C.-H.; Yoshimura, M. Biocompatible Hydroxyapatite Ceramic Coating on Titanium Alloys by Electrochemical Methods via Growing Integration Layers [GIL] Strategy: A Review. *Ceram Int.* **2023**, *in press*. [CrossRef]
18. Schwartz, A.; Kossenko, A.; Zinigrad, M.; Gofer, Y.; Borodianskiy, K.; Sobolev, A. Hydroxyapatite Coating on Ti-6Al-7Nb Alloy by Plasma Electrolytic Oxidation in Salt-Based Electrolyte. *Materials* **2022**, *15*, 7374. [CrossRef]
19. Kurup, A.; Dhatrak, P.; Khasnis, N. Surface Modification Techniques of Titanium and Titanium Alloys for Biomedical Dental Applications: A Review. *Mater. Today Proc.* **2021**, *39*, 84–90. [CrossRef]
20. Shenhar, A.; Gotman, I.; Radin, S.; Ducheyne, P.; Gutmanas, E.Y. Titanium Nitride Coatings on Surgical Titanium Alloys Produced by a Powder Immersion Reaction Assisted Coating Method: Residual Stresses and Fretting Behavior. *Surf. Coat. Technol.* **2000**, *126*, 210–218. [CrossRef]
21. Gabor, R.; Cvrček, L.; Doubková, M.; Nehasil, V.; Hlinka, J.; Unucka, P.; Buřil, M.; Podepřelová, A.; Seidlerová, J.; Bačáková, L. Hybrid Coatings for Orthopaedic Implants Formed by Physical Vapour Deposition and Microarc Oxidation. *Mater. Des.* **2022**, *219*, 110811. [CrossRef]

22. Pesode, P.; Barve, S. Surface Modification of Titanium and Titanium Alloy by Plasma Electrolytic Oxidation Process for Biomedical Applications: A Review. *Mater. Today Proc.* **2021**, *46*, 594–602. [CrossRef]
23. Darband, G.B.; Aliofkhazraei, M. Electrochemical Phosphate Conversion Coatings: A Review. *Surf. Rev. Lett.* **2017**, *24*, 1730003. [CrossRef]
24. Burduhos-Nergis, D.P.; Vizureanu, P.; Sandu, A.V.; Bejinariu, C. Evaluation of the Corrosion Resistance of Phosphate Coatings Deposited on the Surface of the Carbon Steel Used for Carabiners Manufacturing. *Appl. Sci.* **2020**, *10*, 2753. [CrossRef]
25. Burduhos-Nergis, D.-P.; Vizureanu, P.; Sandu, A.V.; Bejinariu, C. Phosphate Surface Treatment for Improving the Corrosion Resistance of the C45 Carbon Steel Used in Carabiners Manufacturing. *Materials* **2020**, *13*, 3410. [CrossRef]
26. Zhao, D.W.; Liu, C.; Zuo, K.Q.; Su, P.; Li, L.B.; Xiao, G.Y.; Cheng, L. Strontium-Zinc Phosphate Chemical Conversion Coating Improves the Osseointegration of Titanium Implants by Regulating Macrophage Polarization. *Chem. Eng. J.* **2021**, *408*, 127362. [CrossRef]
27. Zhao, D.W.; Zuo, K.Q.; Wang, K.; Sun, Z.Y.; Lu, Y.P.; Cheng, L.; Xiao, G.Y.; Liu, C. Interleukin-4 Assisted Calcium-Strontium-Zinc-Phosphate Coating Induces Controllable Macrophage Polarization and Promotes Osseointegration on Titanium Implant. *Mater. Sci. Eng. C Mater. Biol. Appl.* **2021**, *118*, 111512. [CrossRef]
28. Liu, B.; Xiao, G.; Lu, Y. Effect of PH on the Phase Composition and Corrosion Characteristics of Calcium Zinc Phosphate Conversion Coatings on Titanium. *J. Electrochem. Soc.* **2016**, *163*, C477–C485. [CrossRef]
29. Liu, B.; Xiao, G.Y.; Chen, C.Z.; Lu, Y.P.; Geng, X.W. Hopeite and Scholzite Coatings Formation on Titanium via Wet-Chemical Conversion with Controlled Temperature. *Surf. Coat. Technol.* **2020**, *384*, 125330. [CrossRef]
30. Zhao, X.C.; Dong, S.F.; Ge, B.; Huang, B.X.; Ma, J.; Chen, H.; Hao, X.H.; Wang, C.Z. Effects of Temperature and Voltage on Formation of Electrolysis Induced Chemical Conversion Coating on Titanium Surface. *Surf. Coat. Technol.* **2018**, *354*, 330–341. [CrossRef]
31. Liu, B.; Shi, X.M.; Xiao, G.Y.; Lu, Y.P. In-Situ Preparation of Scholzite Conversion Coatings on Titanium and Ti-6Al-4V for Biomedical Applications. *Colloids Surf. B Biointerfaces* **2017**, *153*, 291–299. [CrossRef] [PubMed]
32. Burduhos-Nergis, D.P.; Bejinariu, C.; Sandu, A.V. *Phosphate Coatings Suitable for Personal Protective Equipment*; Materials Research Forum LLC: Millersville, PA, USA, 2021; Volume 89, ISBN 9781644901113.
33. Axinte, M.; Vizureanu, P.; Cimpoesu, N.; Nejneru, C.; Burduhos-Nergis, D.P.; Epure, E.L. Analysis of Physicochemical Properties of W1.8507 Steel Parts with Sharp Edges, Thermochemically Treated by Plasma Nitriding with and without Polarized Screens. *Coatings* **2023**, *13*, 177. [CrossRef]
34. Owens, D.K.; Wendt, R.C. Estimation of the Surface Free Energy of Polymers. *J. Appl. Polym. Sci.* **1969**, *13*, 1741–1747. [CrossRef]
35. Xiao, G.Y.; Zhao, X.C.; Zhang, X.; Xu, W.H.; Lu, Y.P. Electric Field Induced Rapid Formation of Novel Structural Hopeite Coating on Titanium. *Mater. Lett.* **2015**, *144*, 30–32. [CrossRef]
36. Li, J.; Li, J.; He, N.; Fu, Q.; Feng, X.; Li, Q.; Wang, Q.; Liu, X.; Xiao, S.; Jin, W.; et al. In Situ Growth of Ca-Zn-P Coatings on the Zn-Pretreated WE43 Mg Alloy to Mitigate Corrosion and Enhance Cytocompatibility. *Colloids Surf. B Biointerfaces* **2022**, *218*, 112798. [CrossRef]
37. Zhao, X.C.; Xiao, G.Y.; Zhang, X.; Wang, H.Y.; Lu, Y.P. Ultrasonic Induced Rapid Formation and Crystal Refinement of Chemical Conversed Hopeite Coating on Titanium. *J. Phys. Chem. C* **2014**, *118*, 1910–1918. [CrossRef]
38. Alleborn, N.; Raszillier, H. Spreading and Sorption of a Droplet on a Porous Substrate. *Chem. Eng. Sci.* **2004**, *59*, 2071–2088. [CrossRef]
39. Trisnanto, S.R.; Setiawan, I.; Sunnardianto, G.K.; Triawan, F. Stearic Acid-Modified CuO Coating Metal Surface with Superhydrophobicity and Anti-Corrosion Properties. *J. Eng. Res.* **2019**, *2019*, 63–75.
40. Annamalai, M.; Gopinadhan, K.; Han, S.A.; Saha, S.; Park, H.J.; Cho, E.B.; Kumar, B.; Patra, A.; Kim, S.W.; Venkatesan, T. Surface Energy and Wettability of van Der Waals Structures. *Nanoscale* **2016**, *8*, 5764–5770. [CrossRef]

Disclaimer/Publisher's Note: The statements, opinions and data contained in all publications are solely those of the individual author(s) and contributor(s) and not of MDPI and/or the editor(s). MDPI and/or the editor(s) disclaim responsibility for any injury to people or property resulting from any ideas, methods, instructions or products referred to in the content.

Article

Transparent, High-Strength, and Antimicrobial Polyvinyl Alcohol/Boric Acid/Poly Hexamethylene Guanidine Hydrochloride Films

Shaotian Zhang, Dafu Wei *, Xiang Xu and Yong Guan *

School of Materials Science and Engineering, East China University of Science and Technology, Shanghai 200237, China; 17717919604@163.com (S.Z.); xiangxu@ecust.edu.cn (X.X.)
* Correspondence: dfwei@ecust.edu.cn (D.W.); yguan@ecust.edu.cn (Y.G.)

Abstract: It is still crucial to improve the mechanical characteristics of polyvinyl alcohol (PVA) films without resorting to chemical cross-linking. In this study, boric acid (BA) was used to enhance the mechanical characteristics of PVA films while maintaining their excellent transparency and biodegradability. The hydrogen bond interaction between PVA and BA resulted in a 70% increase in tensile strength (from 48.5 to 82.1 MPa) and a 46% increase in elongation at break (from 150 to 220%). To introduce antibacterial properties, polyhexamethylene guanidine hydrochloride (PHMG) was incorporated into PVA/BA composite films resulting in PVA/BA/PHMG composite films. The PVA/BA/PHMG films exhibited 99.99% bacterial inhibition against *Escherichia coli* and *Staphylococcus aureus* with negligible leaching of PHMG. The PVA/BA/PHMG films maintained a tensile strength of 75.3 MPa and an elongation at a break of 208%. These improved mechanical and antimicrobial properties make PVA/BA and PVA/BA/PHMG films promising for applications in food and medicinal packaging.

Keywords: PVA films; boric acid; polyhexamethylene guanidine hydrochloride; mechanical properties; antimicrobial properties

Citation: Zhang, S.; Wei, D.; Xu, X.; Guan, Y. Transparent, High-Strength, and Antimicrobial Polyvinyl Alcohol/Boric Acid/Poly Hexamethylene Guanidine Hydrochloride Films. *Coatings* 2023, 13, 1115. https://doi.org/10.3390/coatings13061115

Academic Editor: Julie M. Goddard

Received: 17 May 2023
Revised: 10 June 2023
Accepted: 14 June 2023
Published: 17 June 2023

Copyright: © 2023 by the authors. Licensee MDPI, Basel, Switzerland. This article is an open access article distributed under the terms and conditions of the Creative Commons Attribution (CC BY) license (https:// creativecommons.org/licenses/by/ 4.0/).

1. Introduction

In recent years, there has been an increasing interest among researchers of various sciences in polymers that exhibit excellent biocompatibility, biodegradability, and potential for modification [1,2]. Polyvinyl alcohol (PVA) has been widely used in films [3–5], hydrogels [6–9], fibers [10–12], nanobody-based beads [13], and other fields due to its good mechanical properties, oxygen barrier [14–16], optical transparency, biocompatibility, and biodegradability. Water-soluble PVA film is an emerging green and environmentally friendly material that is non-polluting, non-toxic, and suitable for food and pharmaceutical packaging [17–20].

A crucial need is for the PVA film to have excellent mechanical characteristics. The mechanical characteristics of PVA film were typically enhanced by physical mixing and cross-linking modification. One of the key techniques for producing high-performance polymer compounds is physical mixing [21,22]. The additives, such as graphene oxide (GO) [22], silver nanoparticles [23], and silica [24], might interact with the hydroxyl group on PVA. Zhu et al. [22] combined GO and single-walled carbon nanotubes with PVA films to boost their tensile strength from 34.6 to 62.8 MPa. Fan et al. [23] found that adding silver-loaded nano-cellulose enhanced the tensile strength of PVA films, and the nanocomposite films displayed reduced moisture absorption and efficient antibacterial properties. The cross-linking of PVA films includes radiation cross-linking and chemical cross-linking [25,26]. Irradiation cross-linking can form new linkage bonds via the free radicals generated by radiation. In order to create three-dimensional porous foam structures, Sabourian et al. [25] exposed PVA/polyvinylpyrrolidone mixes to γ rays. This enhanced the tensile strength of PVA foams from 0.43 to 0.54 MPa and the elongation at break from 30 to 205 percent. Macromolecular chains are connected to create a network structure using chemical cross-linking. According to Chen et al. [26] the PVA that had been cross-linked with sodium borate under these circumstances increased in tensile strength (from 23.3 to 66.5 MPa) but decreased in elongation at break (from 60% to 30%), and its degradable characteristics deteriorated. The antimicrobial properties of PVA films are also in great demand. Blending [27–29] and graft modification of PVA [30] are the common methods to obtain antimicrobial PVA films.

A quick and effective way to achieve antibacterial characteristics is by physical mixing. Antimicrobial PVA was created using a variety of antimicrobial substances, including chitosan [27,28], TiO_2 [29], and silver [30]. Growing interest is being paid to TiO_2 and other photo-catalytical antibacterial substances with potent bactericidal properties. Ma et al. [29] created a TiO_2/N-halamine nanoparticles/PVA composite film that, in 30 min, rendered 99.97% of *S. aureus* and 100% of *E. coli* inactive. The antibacterial composite PVA films also showed outstanding storage stability, regeneration potential, and UV light stability. Physical blending has the drawback of causing the blended antimicrobial ingredient to seep from the modified film, reducing the antimicrobial characteristics' longevity. Antimicrobial agent leaching may be reduced by graft modification. Mei et al.'s in situ green synthesis approach was used to create N-halamine compound-grafted PVA electrospun nanofibrous membranes with rechargeable antibacterial activity. In order to reduce the amount of *E. coli* by 6 log CFU in only one minute, the chlorinated dimethylol-5,5-dimethylhydantoin (DMDMH)-grafted PVA nanofiber film shows a combination of qualities, including long-term durability and great mechanical strength. However, processing and the complexity of incompletely reacted monomers continue to be issues. As a reinforcing agent in this study, boric acid (BA), a substance with water solubility, non-toxicity, and bacteriostatic qualities, was selected. Because of its aqueous solubility, wide range, and superior antibacterial capabilities, polyhexamethylene guanidine (PHMG) was selected as the antimicrobial agent. To create PVA/BA films with outstanding mechanical characteristics, PVA solution, and BA solution were combined, then dried. While maintaining a similar elongation at break to pure PVA films, the PVA film's tensile strength was increased thanks to the strong hydrogen bonds between BA and PVA. Additionally, PVA, BA, and PHMG solutions were combined to create PVA, BA, and PHMG films that, thanks to hydrogen bonds, had excellent mechanical properties and were transparent, non-leaching, and antibacterial. It is anticipated that PVA/BA and PVA/BA/PHMG films will be used in the domains of food and pharmaceutical packaging [31–33].

To the best of our knowledge, there has not been investigated the potential of the PVA and PVA–BA blends with or without the addition of PHMG as the active agent and their antibacterial and physicochemical properties as a natural preservative in coatings. This work studied the effect of both BA and PHMG (two antibacterial components) on the physicochemical and biological properties of PVA-based materials. The obtained results allowed for an evaluation of a likely increase or decrease in the antibacterial effect of PVA-based materials modified with both biologically active compounds.

2. Experiment

2.1. Materials and Methods

PVA with an average molecular weight of 73,000–78,000 (g/mol) and a hydrolysis degree of 98–99% was purchased from Shanghai Titan Technology Co. (Shanghai, China). The boric acid (AR $\geq$ 99.5 percent) was provided by Shanghai Aladdin Biotechnology Co. (Shanghai, China). The PHMG with a molecular weight of 740 Da was synthesized using the method described by Wei et al., 2009 [34] (as measured by ESI-TOF-MS).

2.2. Preparation of PVA/BA Films and PVA/BA/PHMG Films

A 10-weight percent PVA aqueous solution was made by mixing 50 g of PVA with 450 g of deionized (DI) water. The mixture was then agitated at 95 °C for four hours. The boric acid (BA) was mixed with the DI water for an hour at room temperature using magnetic stirring. The mixed solution, which was then cast onto glass Petri dishes, defoamed, and dried to make PVA/BA films, was made by mixing PVA and BA solutions in a variety of mass ratios. Films having a PVA/BA concentration of 1.0 wt% BA were recorded as PVA-1.0 wt%, and so on. Next, a group of the dried films was assembled and put to the test. At room temperature, PHMG was dissolved in DI water using magnetic stirring for 4 h. The three solutions, PVA, BA, and PHMG, were combined in various mass ratios, agitated for one hour, then cast on glass Petri plates, and maintained at room temperature. PVA/BA/PHMG-0.1 wt% was the designation for the PVA/BA/PHMG film that included 0.5 wt% BA and 0.1 wt% PHMG, and so on. The dried films were then gathered and examined [22].

2.3. Characterization

2.3.1. Fourier-Transform Infrared Spectroscopy (FTIR)

In order to evaluate the functional groups on the surface, PVA films were scanned using a Nicolet 5700 spectrometer in attenuated total reflectance (ATR) mode across the 4000–400 cm^{-1} wavenumber range following the method described by Abedinia et al. [35].

2.3.2. Water Solubility (WS)

To calculate the WS, the specimens were initially cut into 5 cm × 5 cm strips. The films were then dried at 60 °C for 24 h to determine their initial dry matter (M_i). The samples were then baked at 20 °C for 28 days in 250 mL of DI water. The samples were dried again for 24 h at 60 °C to reach their final dry weight (M_f). Equation (1) was used to calculate the WS%. The samples were examined three times, and the findings were represented as WS% [35].

$$\text{WS (\%)} = \frac{M_i - M_f}{M_i} \tag{1}$$

2.3.3. Light Transmission

The quantity of light that went through the films in the range of 200–800 nm was measured using an ultraviolet-visible (UV-vis) spectrophotometer (Lambda 950, Perkin Elmer, Waltham, MA, USA). The film samples were cut into 4 cm × 1 cm strips and placed into cuvettes. An empty cuvette was used as a reference. The transparency of the film was calculated with the following Equation (2):

$$\text{Transparency} = -\log T/x \tag{2}$$

where T is the transmission (%) at 600 nm, and x is the thickness of the film (cm).

2.3.4. X-ray Diffraction (XRD)

Wide-angle XRD was used to analyze the crystalline structures of PVA films using a rotating anode X-ray powder diffractometer (18 KW/D/max2550 VB/PC, Rigaku Corporation, Tokyo, Japan) with Cu Kα radiation (λ = 1.542 Å) at a voltage of 30 kV and current of 15 mA. Film samples were cut into 3 cm × 3 cm and placed on the glass slides, then secured with tapes before being placed in the diffractometer chamber for measurement. The angle diffraction ranges and scanning times were set between 2θ = 1 − 60° and 2° per minute, respectively.

2.3.5. Thermogravimetric Analysis (TGA)

The thermal stability of the PVA sheets was evaluated using TGA (STA409PC, NETZSCH, Selb, Germany). All experiments used samples weighing around 10 mg and were conducted at temperatures ranging from 20 to 600 °C at a rate of 10 °C/min under a N_2 (purity ≥ 99.999%.) flow rate of 50 mL/min.

2.3.6. Differential Scanning Calorimeter (DSC)

The thermal properties of the PVA/BA films were evaluated using DSC (Perkin Elmer, Waltham, MA, USA). The sample pan for the DSC apparatus received a sample of about 5 mg. An empty aluminum pan was used as a reference. While being scanned from 0 to 250 °C, samples were heated at a rate of 5 °C/min.

2.3.7. Morphology Observations by Scanning Electron Microscopy (SEM)

The morphology of the films was examined using a SEM (Hitachi S-3400N, Tokyo, Japan) at an accelerating voltage of 15 kV. The samples were coated with gold prior to inspection. Scans were performed by considering the magnification of 4000× [35].

2.3.8. Mechanical Properties

PVA films' tensile strength and elongation at break were measured using a universal electrical testing apparatus at a speed of 50 mm/min and a temperature of 23 ± 2 °C. (CMT-2203, MTS, Eden Prairie, MN, USA). Each sample had a gauge length of 25 mm. On each sample, at least five tests were run, and the average result was calculated [35].

2.3.9. Antimicrobial Tests

The bactericidal properties of PVA films were assessed using the shaking flask method [32] and the ring diffusion technique [33].

The flask-shaking method is a kind of quantitative test. Before being further diluted to 10^5 CFU/mL, *Staphylococcus aureus* (*S. aureus*) and *Escherichia coli* (*E. coli*) were cultured in nutritive broth at 37 °C for 24 h. Then, 0.10 g of the sample and 5 mL of the bacterial culture (10^5 CFU/mL) were combined, and they were shaken for an hour at 250 rpm at 37 °C. Following some shaking, this culture was diluted repeatedly, and 0.1 mL of it was then seeded on LB agar in a Petri dish. After the dishes had been incubated at 37 °C for 24 h, the colonies were counted. The inhibition rate of cell growth was calculated from Equation (3):

$$\text{Growth inhibition rate} = (A - B)/A \times 100\% \quad (3)$$

where A and B stand for the number of bacterial colonies discovered in the control and composite film samples, respectively. The average values of the inhibition rates were calculated after three measurements of each sample were taken.

The ring diffusion test was used to explain the leaching property. The bacterial culture medium was nutrient agar. Agar plates were inoculated with 0.1 mL solutions of *E. coli* or *S. aureus* containing 10^8 CFU/mL. Circular pieces of film with a diameter of 0.5 cm were placed on the agar plates. Following a 24 h incubator incubation period at 37 °C, the inhibition zones' widths were measured. On each test, duplicate tests were conducted.

2.3.10. Leaching Tests

A 1 g PVA film was soaked in 50 g DI water for seven days at different temperatures. Then, using an ultraviolet (UV) spectrophotometer (Lambda 950, Perkin Elmer, Waltham, MA, USA) operating in the 190–400 nm wavelength range, the PHMG leaching rates were determined from the soaking solution. The calibration Equation (4) [36,37] of the absorbance at 192 nm of PHMG is as follows:

$$A_{192} = 0.06107 + 0.06805 C_{PHMG} \quad (4)$$

where A_{192} represents the absorbance at 192 nm, C_{PHMG} represents the concentration of PHMG in μg/mL. The leaching rate was calculated by the following Equation (5):

$$\text{Leaching rate} = C_{PHMG} V_W / W_A \times 100\% \quad (5)$$

where W_A is the weight (g) of incorporated PHMG, V_W is the volume (mL) of the soaked solution, and C_{PHMG} represents the concentration of PHMG in μg/mL of soaked solution.

2.4. Gaussian Simulation

The intra- and intermolecular hydrogen bonding in the PVA/BA system was evaluated using Gaussian simulation. Gaussian 09W is the Gaussian variation. The structural optimization approach is b3lyp/6–31g**, and the energy of the H-bond is then determined using def2Tzvpp/m062x based on the electron density of the bond-critical point. The equation is as follows:

$$E_{HB} = -223.08 \times \rho(BCP) + 0.7423 \quad (6)$$

where $\rho(BCP)$ is electron density at bond-critical points, E_{HB} is the energy of the H-bond [38]. The bond-critical point is a positive saddle point on the potential energy surface.

2.5. Statistical Analysis

All the tests on each sample were performed in triplicate. Data were analyzed using one-way analysis of variance (ANOVA) in SPSS 22.0 (Chicago, IL, USA). Significant differences in the mean values were examined using Duncan's multiple range test at $p < 0.05$. The results are expressed as mean and standard deviation.

3. Results and Discussion

3.1. Transparency and UV(Ultraviolet–Visible) Absorption of Films

High optical transparency is a crucial characteristic of PVA films. The transmittance-wavelength curves from the UV-vis spectrophotometer used to evaluate the transparency of PVA/BA films are shown in Figure 1. In the 250–800 nm range, pure PVA films show excellent transparency, especially in the visible region where the transmittance approaches 91%. The PVA and BA films both have equivalent optical transparency. The great compatibility of PVA and BA, which is most likely brought on by their strong H-bond interaction, is what accounts for the high transparency.

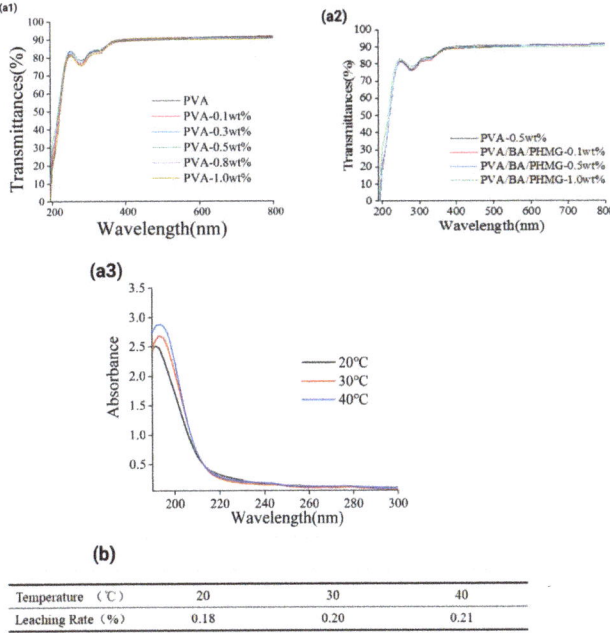

Figure 1. (**a1**–**a3**) UV-vis curves (200–800 nm) of PVA, PVA/BA (**a1**), and PVA/BA/PHMG (**a2**) films and the UV absorption spectra of soaking solution of PVA/BA/PHMG-1 wt% samples after immersion in deionized water for seven days (**a3**). (**b**) The dissolution rates of PVA/BA/PHMG-1 wt% samples at different temperatures immersion.

Leaching tests revealed that the UV absorption spectra of the PVA/BA/PHMG-1 wt% film soaking solution after seven days of immersion are shown in Figure 1b. The breakdown rate of PHMG as determined from the UV absorption spectra is shown in the attached table to Figure 1(a3). The strength at 192 nm rose with the rise in immersion temperature, and the absorption peak at that wavelength is attributed to PHMG. At 20 °C, 30 °C, and 40 °C of immersion temperature, the dissolving rates of PVA/BA/PHMG-1 wt% films were 0.18 percent, 0.20 percent, and 0.21 percent, respectively. The PVA/BA/PHMG-1 wt% film's low PHMG dissolution rate suggested that PHMG had been integrated with the PVA matrix. Guo et al. [32] also obtained similar results after treating PVA with PHMB to increase the antibacterial performance of cotton dressings with ultraviolet radiation for seven days and were able to maintain the antibacterial rate of 99.99%.

3.2. Mechanical and Water Solubility Properties of Films

The pure PVA films had tensile strength and elongation at a break of 48.5 MPa and 150.3%, respectively (Figure 2). As more BA was added up to 0.5 weight percent, the tensile strength of PVA/BA films increased, reaching 82.1 MPa for PVA-0.5 wt percent (Figure 2a). However, as the amount of BA was increased above 0.5 wt%, the tensile strength of PVA/BA films decreased, reaching 66.7 MPa for PVA-1.0 wt%. The maximum tensile strength for PVA-0.5 wt% may have indicated that the H-bond networks and physical cross-linking sites at 0.5 wt% BA may be at an ideal state. The elongation at break is shown in Figure 2b. As the amount of BA within 0.5 weight percent increased, the PVA/BA film's elongation at break increased. However, the elongation at break of the PVA/BA films decreased as the amount of BA increased above 0.5 wt percent. The change in elongation at break should also be connected to the H-bonds.

The PVA-0.5 wt% film had a tensile strength of 81.2 MPa and an elongation at a break of 220 percent. With the addition of PHMG, PVA/BA/PHMG-0.1 wt% film had a tensile strength of 75.3 MPa and an elongation at a break of 208 percent. The reduction in its mechanical properties may be due to the inclusion of PHMG, which altered the well-ordered structure between PVA and BA.

The water solubility of the films is given in the table below, Figure 2. The PVA-1.0 wt%BA film was barely soluble, whereas the pure PVA and BCPVA-0.1 films were entirely soluble. At greater BA levels, the water solubility rapidly decreased. The possibility of these hydroxyl groups forming bonds with water molecules was reduced due to the cross-linking between BA and the hydroxyl groups of PVA. For a crosslinker inserted into a polymer, similar outcomes were reported [32].

3.3. H-Bond Interaction between PVA and BA

The IR spectra of PVA and PVA/BA films were used to analyze the H-bond interaction between PVA and BA. During the formation of an H-bond, the density of the electron cloud is averaged, which reduces the frequency of stretching vibrations. Pure PVA exhibits a strong and wide absorption band at 3000–3600 cm^{-1} focused at 3264 cm^{-1} due to the stretching vibration of hydroxyl groups (O-H), including the massive H-bond (free hydroxyls is usually only observed at 3620 cm^{-1}) Figure 3. Two nearby substantial absorption peaks at 2937 cm^{-1} and 2907 cm^{-1} were seen because of the methylene's stretching vibration. The absorption peak for BA at 3180 cm^{-1} is assumed to be caused by stretching vibrations of the O-H group. For PVA/BA films, the BA absorption peak at 3180 cm^{-1} disappeared. The peak of the O-H group is moved from 3264 cm^{-1} for pristine PVA film to 3241 cm^{-1} for PVA-0.1 wt%, 3240 cm^{-1} for PVA-0.3 wt%, 3239 cm^{-1} for PVA-0.5 wt%, 3240 cm^{-1} for PVA-0.8 wt%, and 3243 cm^{-1} for PVA-1.0 wt%. The altered O-H group absorption peak, however, does not demonstrate an H-bond between PVA and BA.

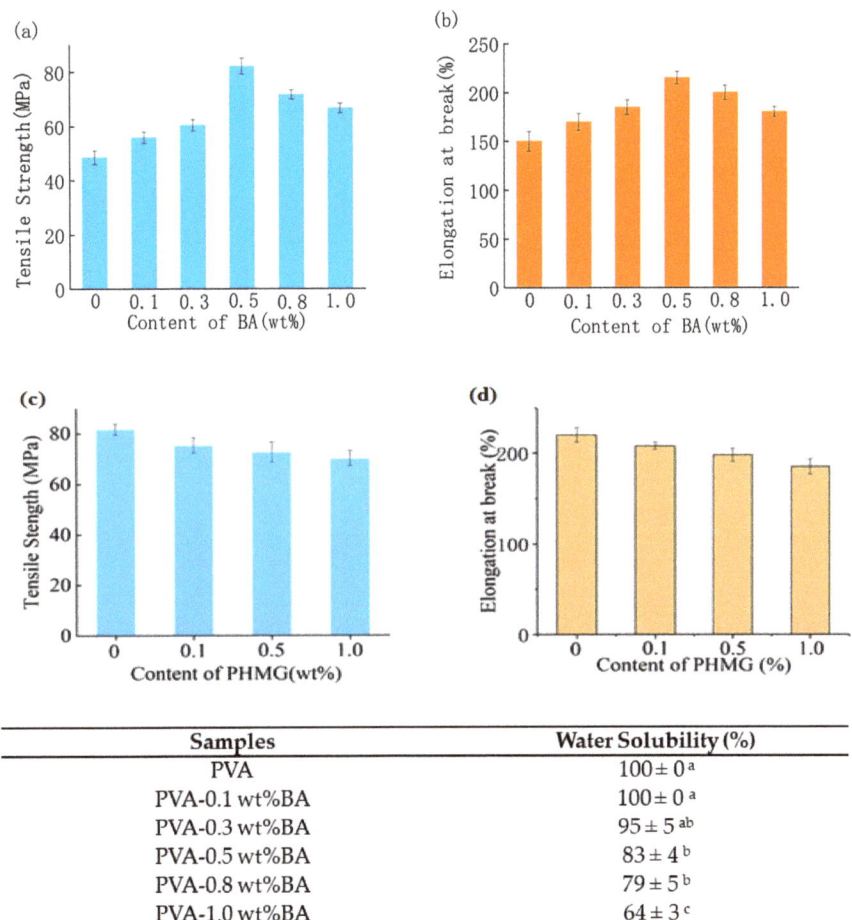

Samples	Water Solubility (%)
PVA	100 ± 0 [a]
PVA-0.1 wt%BA	100 ± 0 [a]
PVA-0.3 wt%BA	95 ± 5 [ab]
PVA-0.5 wt%BA	83 ± 4 [b]
PVA-0.8 wt%BA	79 ± 5 [b]
PVA-1.0 wt%BA	64 ± 3 [c]

Figure 2. Mechanical properties of PVA and PVA/BA films. (**a**) Tensile strength, (**b**) elongation at break; PVA/BA and PVA/BA/PHMG films, (**c**) tensile strength, (**d**) elongation at break and stability of PVA films during 28 days in DI water 15 °C (Significant difference analysis abc letter notation).

Gaussian simulation may provide the theoretical evidence of the H-bond. The stronger the H-bond, the closer the relative distance between the atoms. In order to reduce the difficulty in the simulation, 1.3-propanediol was chosen as the model compound of PVA in the Gaussian simulation. In electrically neutral systems, the energy of intermolecular H-bonds ranges from 2.5 to 14.0 kcal/mol [31]. In the 1.3-propanediol/H_2O system, the calculated energy of H-bonds between water molecules (Figure 4a) was 20.48 kJ/mol, and the energy of the H-bond between 1.3-propanediol molecules (Figure 4d) was 32.31 kJ/mol (Table 1). However, the energy of the H-bond between 1.3-propanediol and BA reached 40.16 kJ/mol, and the total intermolecular force was 72.38 kJ/mol (Table 1). The strength of the H-bond between 1.3-propanediol and BA is stronger than that of the H-bond between 1.3-propanediol and 1.3-propanediol. In addition, the H-bond energy between two 1.3-propanediol molecules and one BA molecule was also simulated, indicating that the biggest energy of the H-bond between 1.3-propanediol and BA reached 43.55 kJ/mol. It was thus predicted that the intermolecular force between PVA and BA should be greater than that between PVA and PVA, which might account for the improvement of PVA/BA films. The elongation at break of PVA/BA films is also increased by this physical improvement, in contrast to the chemical cross-linking, which reduces elongation at break. The hypothesized physical cross-linking network of H-bonds between PVA and BA was based on the outcomes of IR and Gaussian simulations (Figure 5). Due to the interaction of the H-bonds, BA is thought of as a reinforcing agent to enhance both the tensile strength and elongation at break.

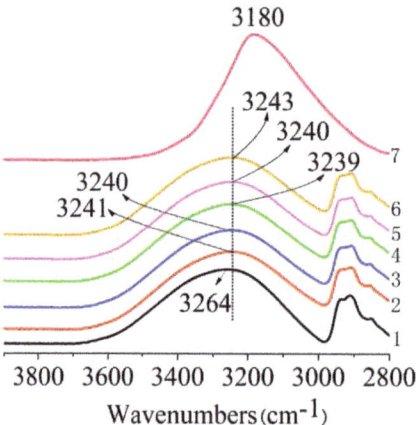

Figure 3. FTIR spectra of BA, PVA, and PVA/BA films of 2800–3900 cm^{-1}. (1) PVA, (2) PVA-0.1 wt%, (3) PVA-0.3 wt%, (4) PVA-0.5 wt%, (5) PVA-0.8 wt%, (6) PVA-1.0 wt%, and (7) BA.

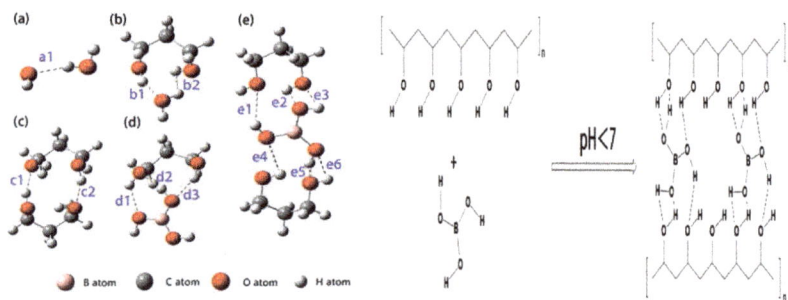

Figure 4. Schemes of H-bonds simulated by Gaussian software. (**a**) H_2O-H_2O, (**b**) PVA-H_2O, (**c**) PVA-PVA, (**d**) PVA-BA, (**e**) PVA-BA-PVA, and illustration of hydrogen bonds interaction among PVA and BA.

3.4. Crystallinity of PVA/BA Films

To determine how the mechanical characteristics of PVA/BA films were enhanced, XRD analysis may be performed. The X-ray diffraction patterns of pure BA, PVA, and various PVA/BA composite films are shown in Figure 5. The neat PVA's diffraction peaks, which are centered at 2 = 19.6°, serve as a representation of the (101) plane of PVA crystals. A particular crystal is BA (Figure 5b). The PVA diffraction peaks in PVA/BA films did not move when BA was added, indicating that the crystal had not changed in any way. However, as BA concentration rose to 0.5 wt percent of BA, the regions of diffraction peaks at 2 = 19.6° steadily grew before declining in the opposite direction as BA content rose. Consequently, crystallinity peaked when BA content reached 0.5 wt%.

Table 1. H-bond length and energy of the intermolecular interaction of H_2O-H_2O, PVA-H_2O, PVA-BA, PVA-PVA, and PVA-BA-PVA calculated by the Gaussian simulation.

	H_2O-H_2O			PVA-H_2O			PVA-PVA	
	Length	Energy (kJ/mol)		Length	Energy (kJ/mol)		Length	Energy (kJ/mol)
a1	1.91	20.48	b1	1.84	28.13	c1	1.78	32.31
			b2	1.84	28.13	c2	1.78	32.31

	PVA-BA			PVA-BA-PVA				
	Length	Energy (kJ/mol)		Length	Energy (kJ/mol)		Length	Energy (kJ/mol)
d1	2.16	11.82	e1	2.41	9.43	e4	2.41	9.45
d2	1.71	40.16	e2	1.61	43.55	e5	1.61	43.52
d3	1.92	20.40	e3	1.83	24.87	e6	1.83	24.86

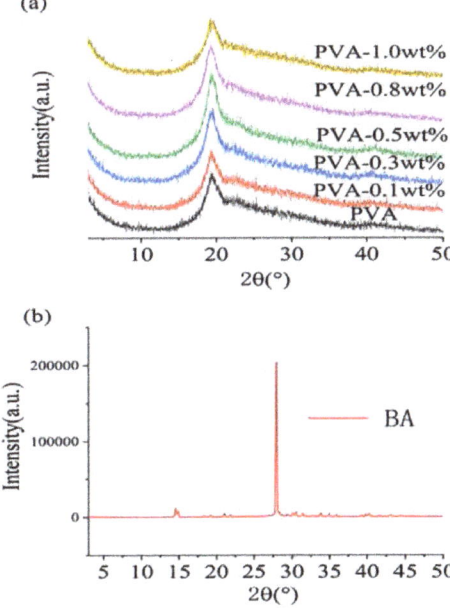

Figure 5. (a) X-ray diffraction of PVA and PVA/BA composite films with various BA contents, (b) X-ray diffraction of BA.

3.5. Thermal Properties of Films

To evaluate the effect of BA and H-bonds on the thermal properties of PVA, DSC, and TGA were utilized. Figure 6 displays the TGA and DTG curves for BA, PVA, and PVA/BA films. The data are summarized below in the figure. Peak temperatures for PVA film thermal breakdown were 297 °C (Tmax1) and 430 °C (Tmax2). The elimination of hydroxyl groups from the side chain and the disintegration of the main chain should happen at 297 °C and 430 °C, respectively. The PVA film's initial breakdown temperature (Ti) is 242 °C. Ti of the PVA/BA film tended to rise with the addition of BA, reaching a maximum of 268 °C at 1.0 wt% BA, which is 26 °C higher than that of PVA. Tmax1 was divided into two peaks (DTG curves), whether the BA concentration was 0.3 wt% or 0.5 wt%, suggesting a shift in the structure of the films. The Tmax1 became one peak once again at a temperature that was 30 °C higher than 297 °C when the BA concentration was over 0.8 wt%, suggesting that the PVA/BA films should be more stable. Tmax2 of PVA/BA films did not substantially differ from Tmax2 of PVA film, indicating that the main chain's structural integrity was unaffected.

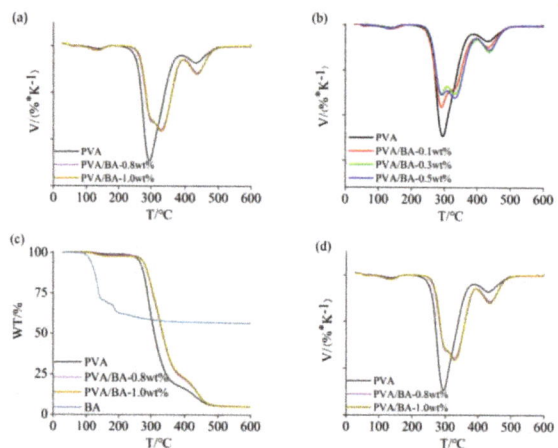

TGA date

Samples	Ti/ °C	T max1/ °C	Tmax2/ °C	Final residual Weight/%
PVA	242	297/-	430	2.4
PVA-0.1 wt%	245	292/-	431	3.1
PVA-0.3 wt%	252	289/329	430	3.4
PVA-0.5 wt%	257	290/330	435	3.6
PVA-0.8 wt%	264	-/328	434	3.7
PVA-1.0 wt%	268	-/330	435	3.6
BA	193	-	-	52.6

DTG data

Samples	PVA (wt%)	BA (wt%)	T_m (°C)	ΔH_m (J/g)	χ_c (%)
PVA	100.0	0	220.3	48.3	34.8
PVA-0.1 wt%BA	99.9	0.1	219.6	50.2	36.2
PVA-0.3 wt%BA	99.7	0.3	218.1	51.3	37.0
PVA-0.5 wt%BA	99.5	0.5	217.4	54.4	39.2
PVA-0.8 wt%BA	99.2	0.8	215.3	50.7	36.6
PVA-1.0 wt%BA	99.0	1.0	213.7	49.8	35.9

Figure 6. TGA (**a**,**c**) and DTG (**b**,**d**) curves of BA, PVA, and PVA/BA films and their summarized data (Tables).

The last table below the figure summarizes the DSC data. Pure PVA has a melting point (Tm) of 220.3 °C and a degree of crystallinity (c) of 34.8 percent based on 138.6 J/g of melting enthalpy (Hm0) of 100% crystalline PVA. As the BA concentration increased, the crystallinity rose, then fell once the BA level rose above 0.5 weight percent. A maximum crystallinity was observed at 0.5 weight percent BA, according to the XRD data. In addition, X-ray diffraction (XRD) patterns or relative crystallinity decrease with increasing concentration of added materials that interact with the starting material [39].

Although the contents of BA were within 1.0 wt%, BA still had significantly positive effects on the thermal stability and crystallinity of PVA/BA composite films.

3.6. Surface and Cross-Section Morphologies of PVA and PVA/BA Films

The surfaces and cross-sections of the PVA and PVA/BA films are shown in SEM pictures in Figure 7. PVA/BA films have very smooth surfaces with a surface morphology that is comparable to that of pure PVA films. PVA and PVA/BA film cross-sections were likewise in the same, consistent condition. It suggested that PVA and BA worked well together.

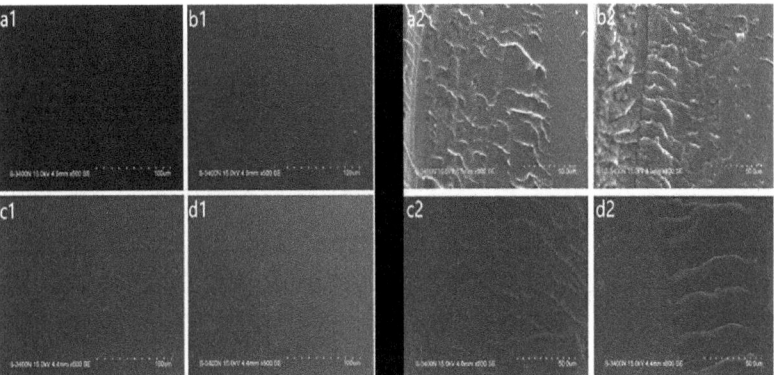

Figure 7. SEM images of films for surfaces (**a1,b1,c1,d1**) and cross-sections (**a2,b2,c2,d2**). (**a1,a2**) pure PVA film, (**b1,b2**) PVA-0.1 wt%, (**c1,c2**) PVA-0.3 wt%, and (**d1,d2**) PVA-0.5 wt%.

3.7. Antimicrobial Properties of PVA/BA/PHMG Films

Since infections caused by pathogenic and resistant bacteria continue to endanger public health, the evaluation and production of antibacterial films, especially against them, are still of interest to scientists [40]. Figure 8 displays the antibacterial graphs of PVA/BA/PHMG and pure PVA film. Significant bacterial growth is seen in Figure 8's panels a1 and b1, demonstrating that the pure PVA film had no antibacterial effect on *E. coli* and *S. aureus*. Figure 8(a2,b2) show the test results for the PVA/BA/PHMG-0.1 wt% film against *E. coli* and *S. aureus*. No bacterial growth was seen on the PVA/BA/PHMG-0.1 wt% film in the Petri dishes.

Table 2 shows the results of bacterial inhibition rates of pure PVA films and PVA/BA/PHMG films after water washing. The PVA film and PVA/BA film had no antibacterial properties, while the PVA/BA/PHMG films achieved 99.99% antibacterial rates even after ten cycles of water washing. This indicates that the PHMG is non-leaching.

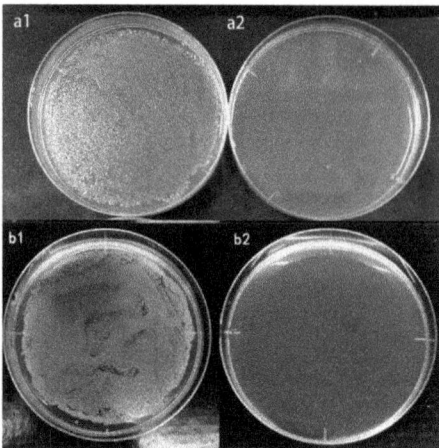

Figure 8. Antimicrobial photos of pure PVA film against *E. coli* (**a1**) and against *S. aureus* (**b1**). Antimicrobial photos of PVA/BA/PHMG-0.5 wt% film against *E. coli* (**a2**) and against *S. aureus* (**b2**) after 10 cycles of DI water washing.

Table 2. Antimicrobial rates of PVA/BA/PHMG films after washing with DI water.

Samples	Antimicrobial Rates against *E. coli* (%)				Antimicrobial Rates against *S. aureus*(%)			
	Before Washing	Cycles of Water Washing			Before Washing	Cycles of Water Washing		
		1	5	10		1	5	10
Neat PVA	0	0	0	0	0	0	0	0
PVA/BA-0.5 wt%	0	0	0	0	0	0	0	0
PVA/BA/PHMG-0.1 wt%	99.99	99.99	99.99	99.99	99.99	99.99	99.99	99.99
PVA/BA/PHMG-0.5 wt%	99.99	99.99	99.99	99.99	99.99	99.99	99.99	99.99
PVA/BA/PHMG-1.0 wt%	99.99	99.99	99.99	99.99	99.99	99.99	99.99	99.99

4. Conclusions

Novel bioactive materials based on poly(vinyl alcohol) (PVA)/BA and PVA/BA/PHMG were successfully synthesized and characterized using X-ray and FTIR techniques. The components were found to be compatible and interacted through hydrogen bonds. The mechanical analysis showed that PHMG acted as a plasticizer, increasing the elasticity of the materials. Similarly, the films containing PVA/PHMG exhibited enhanced bactericidal properties. Introducing poly(hexamethylene guanidine) or PHMG into PVA and PVA/BA mixture improved the thermal stability, attributed to the nitrogen atoms hindering oxygen diffusion. The PHMG also played a role as a dye, as evident from color changes. Surface analysis revealed that PVA/BA films had smoother surfaces due to strong hydrogen bonding, while the addition of PHMG increased hydrophilicity and reduced bacterial adhesion. The PVA/BA and PVA/BA/PHMG films in this investigation were strengthened and transparent via solution casting. Elongation at break and tensile strength of PVA/BA films increased by 46 and 70 percent, respectively. These changes were mainly brought about by the formation of an H-bond between PVA and BA. PVA, BA, and PHMG films have exceptional mechanical properties and broad-spectrum antibacterial properties (antimicrobial rates against *Escherichia coli* and *Staphylococcus aureus* more than 99.99 percent). This simple, eco-friendly preparation offers a useful way to produce PVA films that are strong, non-leaching, and antibacterial.

Author Contributions: Conceptualization: D.W., X.X. and Y.G.; Methodology: D.W.; Software: D.W.; Validation: D.W., X.X. and Y.G.; Formal analysis: S.Z.; Investigation: S.Z.; Resources: S.Z.; Data curation: D.W.; Writing—original draft preparation: S.Z.; Writing—review and editing: D.W.; Visualization: D.W.; supervision: D.W. All authors have read and agreed to the published version of the manuscript.

Funding: This research received no external funding.

Institutional Review Board Statement: Not applicable.

Informed Consent Statement: Not applicable.

Data Availability Statement: Not applicable.

Conflicts of Interest: The authors declare no conflict of interest.

References

1. Abdullah, Z.W.; Dong, Y.; Davies, I.J.; Barbhuiya, S. PVA, PVA Blends, and Their Nanocomposites for Biodegradable Packaging Application. *Polym.-Plast. Technol. Eng.* **2017**, *56*, 1307–1344. [CrossRef]
2. Zhao, G.; Shi, L.; Yang, G.; Zhuang, X.; Cheng, B. 3D fibrous aerogels from 1D polymer nanofibers for energy and environmental applications. *J. Mater. Chem. A* **2023**, *11*, 512–547. [CrossRef]
3. Abral, H.; Atmajaya, A.; Mahardika, M.; Hafizulhaq, F.; Kadriadi; Handayani, D.; Sapuan, S.M.; Ilyas, R.A. Effect of ultrasonication duration of polyvinyl alcohol (PVA) gel on characterizations of PVA film. *J. Mater. Res. Technol.* **2020**, *9*, 2477–2486. [CrossRef]
4. Olewnik-Kruszkowska, E.; Gierszewska, M.; Jakubowska, E.; Tarach, I.; Sedlarik, V.; Pummerova, M. Antibacterial Films Based on PVA and PVA-Chitosan Modified with Poly(Hexamethylene Guanidine). *Polymers* **2019**, *11*, 2093. [CrossRef]
5. Kochkina, N.E.; Lukin, N.D. Structure and properties of biodegradable maize starch/chitosan composite films as affected by PVA additions. *Int. J. Biol. Macromol.* **2020**, *157*, 377–384. [CrossRef]
6. Chen, K.; Liu, J.; Yang, X.; Zhang, D. Preparation, optimization and property of PVA-HA/PAA composite hydrogel. *Mater. Sci. Eng. C Mater. Biol. Appl.* **2017**, *78*, 520–529. [CrossRef]
7. Leone, G.; Consumi, M.; Lamponi, S.; Bonechi, C.; Tamasi, G.; Donati, A.; Rossi, C.; Magnani, A. Thixotropic PVA hydrogel enclosing a hydrophilic PVP core as nucleus pulposus substitute. *Mater. Sci. Eng. C Mater. Biol. Appl.* **2019**, *98*, 696–704. [CrossRef] [PubMed]
8. Wu, X.; Xie, Y.; Xue, C.; Chen, K.; Yang, X.; Xu, J.; Qi, J.; Zhang, D. Preparation of PVA-GO composite hydrogel and effect of ionic coordination on its properties. *Mater. Res. Express* **2019**, *6*, 075306. [CrossRef]
9. Chen, Y.; Li, J.; Lu, J.; Ding, M.; Chen, Y. Synthesis and properties of Poly(vinyl alcohol) hydrogels with high strength and toughness. *Polym. Test.* **2022**, *108*, 107516. [CrossRef]
10. Cao, M.; Liu, Z.; Xie, C. Effect of steel-PVA hybrid fibers on compressive behavior of $CaCO_3$ whiskers reinforced cement mortar. *J. Build. Eng.* **2020**, *31*, 101314. [CrossRef]
11. Assaedi, H.; Alomayri, T.; Siddika, A.; Shaikh, F.; Alamri, H.; Subaer, S.; Low, I.M. Effect of Nanosilica on Mechanical Properties and Microstructure of PVA Fiber-Reinforced Geopolymer Composite (PVA-FRGC). *Materials* **2019**, *12*, 3624. [CrossRef]
12. Piacentini, E.; Bazzarelli, F.; Poerio, T.; Albisa, A.; Irusta, S.; Mendoza, G.; Sebastian, V.; Giorno, L. Encapsulation of water-soluble drugs in Poly (vinyl alcohol) (PVA)- microparticles via membrane emulsification: Influence of process and formulation parameters on structural and functional properties. *Mater. Today Commun.* **2020**, *24*, 100967. [CrossRef]
13. Wang, L.; Ding, Y.; Li, N.; Chai, Y.; Li, Q.; Du, Y.; Hong, Z.; Ou, L. Nanobody-based polyvinyl alcohol beads as antifouling adsorbents for selective removal of tumor necrosis factor-alpha. *Chin. Chem. Lett.* **2022**, *33*, 2512–2516. [CrossRef]
14. Abdullah, Z.W.; Dong, Y.; Han, N.; Liu, S. Water and gas barrier properties of polyvinyl alcohol (PVA)/starch (ST)/ glycerol (GL)/halloysite nanotube (HNT) bionanocomposite films: Experimental characterisation and modelling approach. *Compos. Part B Eng.* **2019**, *174*, 107033. [CrossRef]
15. Kim, J.M.; Lee, M.H.; Ko, J.A.; Kang, D.H.; Bae, H.; Park, H.J. Influence of Food with High Moisture Content on Oxygen Barrier Property of Polyvinyl Alcohol (PVA)/Vermiculite Nanocomposite Coated Multilayer Packaging Film. *J. Food Sci.* **2018**, *83*, 349–357. [CrossRef]
16. Virly; Chiu, C.H.; Tsai, T.Y.; Yeh, Y.C.; Wang, R. Encapsulation of beta-Glucosidase within PVA Fibers by CCD-RSM-Guided Coelectrospinning: A Novel Approach for Specific Mogroside Sweetener Production. *J. Agric. Food Chem.* **2020**, *68*, 11790–11801. [CrossRef]
17. Mallakpour, S.; Darvishzadeh, M. Ultrasonic treatment as recent and environmentally friendly route for the synthesis and characterization of polymer nanocomposite having PVA and biosafe BSA-modified ZnO nanoparticles. *Polym. Adv. Technol.* **2018**, *29*, 2174–2183. [CrossRef]
18. Suganthi, S.; Vignesh, S.; Kalyana Sundar, J.; Raj, V. Fabrication of PVA polymer films with improved antibacterial activity by fine-tuning via organic acids for food packaging applications. *Appl. Water Sci.* **2020**, *10*, 100. [CrossRef]

19. Silva, G.G.D.; Sobral, P.J.A.; Carvalho, R.A.; Bergo, P.V.A.; Mendieta-Taboada, O.; Habitante, A.M.Q.B. Biodegradable Films Based on Blends of Gelatin and Poly (Vinyl Alcohol): Effect of PVA Type or Concentration on Some Physical Properties of Films. *J. Polym. Environ.* **2008**, *16*, 276–285. [CrossRef]
20. Zain, N.A.M.; Suhaimi, M.S.; Idris, A. Development and modification of PVA–alginate as a suitable immobilization matrix. *Process Biochem.* **2011**, *46*, 2122–2129. [CrossRef]
21. Mahmood, R.S.; Salman, S.A.; Bakr, N.A. The electrical and mechanical properties of Cadmium chloride reinforced PVA:PVP blend films. *Pap. Phys.* **2020**, *12*. [CrossRef]
22. Zhu, H.; Cao, H.; Liu, X.; Wang, M.; Meng, X.; Zhou, Q.; Xu, L. Nacre-like composite films with a conductive interconnected network consisting of graphene oxide, polyvinyl alcohol and single-walled carbon nanotubes. *Mater. Des.* **2019**, *175*, 107783. [CrossRef]
23. Fan, L.; Zhang, H.; Gao, M.; Zhang, M.; Liu, P.; Liu, X. Cellulose nanocrystals/silver nanoparticles: In-situ preparation and application in PVA films. *Holzforschung* **2020**, *74*, 523–528. [CrossRef]
24. Yao, K.; Cai, J.; Liu, M.; Yu, Y.; Xiong, H.; Tang, S.; Ding, S. Structure and properties of starch/PVA/nano-SiO2 hybrid films. *Carbohydr. Polym.* **2011**, *86*, 1784–1789. [CrossRef]
25. Sabourian, P.; Frounchi, M.; Dadbin, S. Polyvinyl alcohol and polyvinyl alcohol/ polyvinyl pyrrolidone biomedical foams crosslinked by gamma irradiation. *J. Cell. Plast.* **2016**, *53*, 359–372. [CrossRef]
26. Chen, J.; Li, Y.; Zhang, Y.; Zhu, Y. Preparation and characterization of graphene oxide reinforced PVA film with boric acid as crosslinker. *J. Appl. Polym. Sci.* **2015**, *132*, 42000. [CrossRef]
27. Cao, W.; Yue, L.; Wang, Z. High antibacterial activity of chitosan—Molybdenum disulfide nanocomposite. *Carbohydr. Polym.* **2019**, *215*, 226–234. [CrossRef] [PubMed]
28. Lu, J.; Chen, Y.; Ding, M.; Fan, X.; Hu, J.; Chen, Y.; Li, J.; Li, Z.; Liu, W. A 4arm-PEG macromolecule crosslinked chitosan hydrogels as antibacterial wound dressing. *Carbohydr. Polym.* **2022**, *277*, 118871. [CrossRef]
29. Ma, W.; Li, L.; Liu, Y.; Sun, Y.; Kim, I.S.; Ren, X. Tailored assembly of vinylbenzyl N-halamine with end-activated ZnO to form hybrid nanoparticles for quick antibacterial response and enhanced UV stability. *J. Alloys Compd.* **2019**, *797*, 692–701. [CrossRef]
30. Xiong, K.-R.; Liang, Y.-R.; Ou-yang, Y.; Wu, D.-C.; Fu, R.-W. Nanohybrids of silver nanoparticles grown in-situ on a graphene oxide silver ion salt: Simple synthesis and their enhanced antibacterial activity. *Carbon* **2020**, *158*, 426–433. [CrossRef]
31. Liu, M.; Wang, F.; Liang, M.; Si, Y.; Yu, J.; Ding, B. In situ green synthesis of rechargeable antibacterial N-halamine grafted poly(vinyl alcohol) nanofibrous membranes for food packaging applications. *Compos. Commun.* **2020**, *17*, 147–153. [CrossRef]
32. Guo, C.; Zhang, J.; Feng, X.; Du, Z.; Jiang, Y.; Shi, Y.; Yang, G.; Tan, L. Polyhexamethylene biguanide chemically modified cotton with desirable hemostatic, inflammation-reducing, intrinsic antibacterial property for infected wound healing. *Chin. Chem. Lett.* **2022**, *33*, 2975–2981. [CrossRef]
33. Li, Y.; Zheng, H.; Liang, Y.; Xuan, M.; Liu, G.; Xie, H. Hyaluronic acid-methacrylic anhydride/polyhexamethylene biguanide hybrid hydrogel with antibacterial and proangiogenic functions for diabetic wound repair. *Chin. Chem. Lett.* **2022**, *33*, 5030–5034. [CrossRef]
34. Wei, D.; Ma, Q.; Guan, Y.; Hu, F.; Zheng, A.; Zhang, X.; Teng, Z.; Jiang, H. Structural characterization and antibacterial activity of oligoguanidine (polyhexamethylene guanidine hydrochloride). *Mater. Sci. Eng. C* **2009**, *29*, 1776–1780. [CrossRef]
35. Abedinia, A.; Ariffin, F.; Huda, N.; Nafchi, A.M. Preparation and characterization of a novel biocomposite based on duck feet gelatin as alternative to bovine gelatin. *Int. J. Biol. Macromol.* **2018**, *1*, 855–862. [CrossRef] [PubMed]
36. Patil, A.S.; Waghmare, R.D.; Pawar, S.P.; Salunkhe, S.T.; Kolekar, G.B.; Sohn, D.; Gore, A.H. Photophysical insights of highly transparent, flexible and re-emissive PVA @ WTR-CDs composite thin films: A next generation food packaging material for UV blocking applications. *J. Photochem. Photobiol. A Chem.* **2020**, *400*, 112647. [CrossRef]
37. Wei, D.; Zhou, R.; Guan, Y.; Zheng, A.; Zhang, Y. Investigation on the reaction between polyhexamethylene guanidine hydrochloride oligomer and glycidyl methacrylate. *J. Appl. Polym. Sci.* **2013**, *127*, 666–674. [CrossRef]
38. Emamian, S.; Lu, T.; Kruse, H.; Emamian, H. Exploring Nature and Predicting Strength of Hydrogen Bonds: A Correlation Analysis Between Atoms-in-Molecules Descriptors, Binding Energies, and Energy Components of Symmetry-Adapted Perturbation Theory. *J. Comput. Chem.* **2019**, *40*, 2868–2881. [CrossRef]
39. Tarahi, M.; Hedayati, S.; Shahidi, F. Effects of mung bean (*Vigna radiata*) protein isolate on rheological, textural, and structural properties of native corn starch. *Polymers* **2022**, *14*, 3012. [CrossRef]
40. Yang, R.; Hou, E.; Cheng, W.; Yan, X.; Zhang, T.; Li, S.; Yao, H.; Liu, J.; Guo, Y. Membrane-Targeting Neolignan-Antimicrobial Peptide Mimic Conjugates to Combat Methicillin-Resistant Staphylococcus aureus (MRSA) Infections. *J. Med. Chem.* **2022**, *65*, 16879–16892. [CrossRef]

Disclaimer/Publisher's Note: The statements, opinions and data contained in all publications are solely those of the individual author(s) and contributor(s) and not of MDPI and/or the editor(s). MDPI and/or the editor(s) disclaim responsibility for any injury to people or property resulting from any ideas, methods, instructions or products referred to in the content.

Review

Research Progress of Surface Treatment Technologies on Titanium Alloys: A Mini Review

Bingyu Xie * and Kai Gao *

School of Automobile and Traffic Engineering, Wuhan University of Science and Technology, Wuhan 430065, China
* Correspondence: xiebingyu@wust.edu.cn or bingyu_xie@163.com (B.X.); gaokai@wust.edu.cn (K.G.)

Abstract: Titanium alloys are important strategic structural materials with broad application prospects in the industries of aerospace, space technology, automobiles, biomedicine, and more. Considering the different requirements for the diverse applications of titanium alloys, the modification of physicochemical properties, mechanical properties, and biocompatibility are required, including novel composite materials, novel design, novel manufacturing methods, etc. In this review, the surface treatment technologies utilized on titanium alloys are summarized and discussed. Regarding surface modification of titanium alloys, the methods of laser treatment, electron beam treatment, surface quenching, and plasma spraying are discussed, and in terms of the surface coatings on titanium alloys, thermal spraying, cold spraying, physical vapor deposition, and chemical vapor deposition are also summarized and analyzed in this work. After surface treatments, information on microstructures, mechanical properties, and biocompatibility of titanium alloys are collected in detail. Some important results are summarized according to the aforementioned analysis and discussion, which will provide new thinking for the application of titanium alloys in the future.

Keywords: surface treatment; titanium alloys; microstructure; physicochemical properties; mechanical properties

---

Citation: Xie, B.; Gao, K. Research Progress of Surface Treatment Technologies on Titanium Alloys: A Mini Review. *Coatings* **2023**, *13*, 1486. https://doi.org/10.3390/coatings13091486

Academic Editor: Alexander Tolstoguzov

Received: 19 June 2023
Revised: 31 July 2023
Accepted: 17 August 2023
Published: 23 August 2023

Copyright: © 2023 by the authors. Licensee MDPI, Basel, Switzerland. This article is an open access article distributed under the terms and conditions of the Creative Commons Attribution (CC BY) license (https:// creativecommons.org/licenses/by/ 4.0/).

1. Introduction

Titanium alloys have been widely used in the fields of aerospace, metallurgy, chemicals, petroleum, medicine, and other industries, and are considered to be the most promising metal materials due to their low density, high specific strength, high corrosion resistance, and other excellent characteristics [1–5]. However, during the process of manufacturing and subsequent treatments, the generated tensile stresses may cause the fatigue properties of titanium alloys to deteriorate [6]. Some researchers have shown that this fatigue failure often initiates from the surface [6,7], and the surface integrity and the surface stress could effectively improve the resistance to fatigue failure [8]. Thus, the surface properties are very important to titanium alloys' application, including the surface residual stress, the surface hardness, the surface friction properties, the surface corrosion resistance, etc.

The surface treatments of titanium alloys are very important methods needed to improve their unfavorable properties [9–11], and include physicochemical treatment methods and mechanical treatment methods. Physicochemical treatment methods mainly include surface physical vapor deposition (PVD) [12,13], chemical vapor deposition (CVD) [14,15], surface boriding [16–18], carburizing [19], nitriding [20–22], etc. Mechanical treatment methods mainly include surface laser treatment [23–25], surface shot peening [26–28], and others. The previous development of surface treatments for titanium alloys occurred starting from the initial traditional stage (a single mechanical treatment method) and the middle stage of modern surface technology (a single physicochemical treatment method), and now, surface treatment technologies for titanium alloys are being developed using a variety of technologies for comprehensive application. Therefore, the recent progress in the field of surface treatment methods for titanium alloys are reviewed and discussed in this paper from six aspects.

(1) Surface biochemical treatment of biomedical titanium alloys.
(2) Surface nitriding and nitrocarburizing on titanium alloys.
(3) Hybrid surface modification (HSM) on titanium alloys.
(4) Electrochemical/chemical surface treatment of titanium alloys.
(5) Laser surface modification of titanium alloys.
(6) Surface mechanical treatment on titanium alloys.

Through our analysis of the above six aspects, we hope to obtain the collection of the surface treatment technologies for titanium alloys and to provide new thinking for the application of surface treatments in the development of titanium alloys.

2. Recent Developments of Surface Treatment Technologies for Titanium Alloys

2.1. Surface Biochemical Treatment of Biomedical Titanium Alloys

Regarding surface treatment on biomedical titanium alloys, Wang et al. [29] deposited TiO_2 nanotube coatings on a Ti-6Al-4V substrate by means of anodic oxidation and found that the microhardness and elasticity modulus of the coatings decreased as the oxidation time increased, which is beneficial, as it reduces the possibility of stress shielding and improves the biomechanical compatibility. Tan et al. [30] modified the Ti-Zr binary alloys for dental implant materials via sandblasting and sulfuric acid etching, and lower cell attachment levels were observed on Ti-50Zr. The content of Zr influenced the surface properties of Ti-Zr alloys (as shown in Figure 1). The wettability variation of Ti-Zr alloys was verified by the contact angle of water, which on each sample clearly increased from day 14 and day 28 (in Figure 1a,b). Additionally, Figure 1c indicates the fluorescence microscopy of cells stained with calcein, which confirmed the positive and negative effects of Zr alloys on the cells depending on the Zr proportion. The cell attachment level is shown in Figure 1d. In order to improve the osteogenetic and antibacterial properties of Ti-based implants for orthopedic applications, Liu et al. [31] fabricated TiO_2 nanotube arrays covered with chitosan/sodium alginate multilayer films on titanium substrates, which were able to accelerate the growth of osteoblasts via cytocompatibility evaluation in vitro. Receiving inspiration from the slippery surface of the Nepenthes pitcher plant, the lubricated orthopedic implant surface (LOIS) was introduced by Chae et al. [32] to modify the surface properties of orthopedic implants, and the LOIS showed a long-lasting and extreme liquid repellency against diverse liquids and biosubstances (in Figure 2). Figure 2 shows the antibiofouling property of LOIS against bacteria, cells, protein, and calcium. Bionic structures have always been an important research area in the manufacturing and materials industry, and surface treatments are similar. Many research aims can be achieved with the help of plant-based surface structures such as the surface of the lotus leaf.

Based on high-performance organisms in nature, multi-phase, multi-level, and multi-scale hybrid reinforcements were deposited onto the surfaces of titanium alloys by Bai et al. [33]. Ren et al. [34] utilized a method combining ultrasonic acid etching with anodic oxidation to modify the surface of electron-beam-melting Ti-6Al-4V implants, and the hierarchical micro-/nano-structure was beneficial for cell adhesion and expansion, which significantly promoted cell proliferation and enhanced cell differentiation behaviors as well. Peng et al. [35] fabricated the micro-textures and diamond-like carbon (DLC) coatings on the titanium alloy surface using laser and magnetron sputtering technology, and the best tribological properties were obtained: a friction coefficient of 0.0799 was achieved, and the wear volume was decreased by 97.5%. The surface microstructure and the tribological properties are indicated in Figure 3, and the friction coefficient of the titanium alloy surface with textured coatings is significantly decreased as the wear resistance is effectively improved. Meanwhile, micro-textures with appropriate density and depth could further improve the friction coefficient of textured DLC coated surface. Todea et al. [36] investigated the effects of different surface treatments on the bioactivity of porous Ti-6Al-7Nb implants manufactured via selective laser melting. The materials produced by different treatment methods were validated by the self-assembly of an apatite-type layer when the bioactivity was tested in simulated body fluid (SBF). Zhang et al. [37] introduced some advanced surface-modification technologies such as additive manufacturing, selective laser melting, etc. Cui et al. [38] prepared a nanocrystalline TiN-graded coating on Ti-6Al-4V using the DC (direct current) reactive magnetron sputtering method. The surface nanohardness reached 28.5 GPa and the tribocorrosion resistance increased by 100 times compared to the substrate, which demonstrated good corrosion and wear resistance. Different treatment methods introduce different surface microstructures, which will cause the variation of surface mechanical properties and biocompatibility and influence the application of titanium alloys in the biological field. For instance, coatings with high hardness and high toughness could be obtained via coupling and multi-functional response mechanisms between different phases. Some special structures on the surface, as the micro/nano-structure, could promote cell proliferation and enhance cell differentiation behaviors. The surface fine texture could show favorable anti-friction and wear-resistance effects.

The aim of applying surface treatments to biomedical titanium alloys involves the friction and wear properties of treated materials. In addition, the physicochemical properties and biocompatibility of the treated titanium alloys should also be considered in order to obtain more combinations of properties. Although some better results have been obtained with surface treatments of biomedical titanium alloys, such as the nanocrystalline surface structure, the micro/nano surface structure, the laser surface treatment, etc., they were obtained under some special conditions, such as in the laboratory. The popularization and application of some new surface structures need to be further developed, such as the micro/nano surface structure mentioned above.

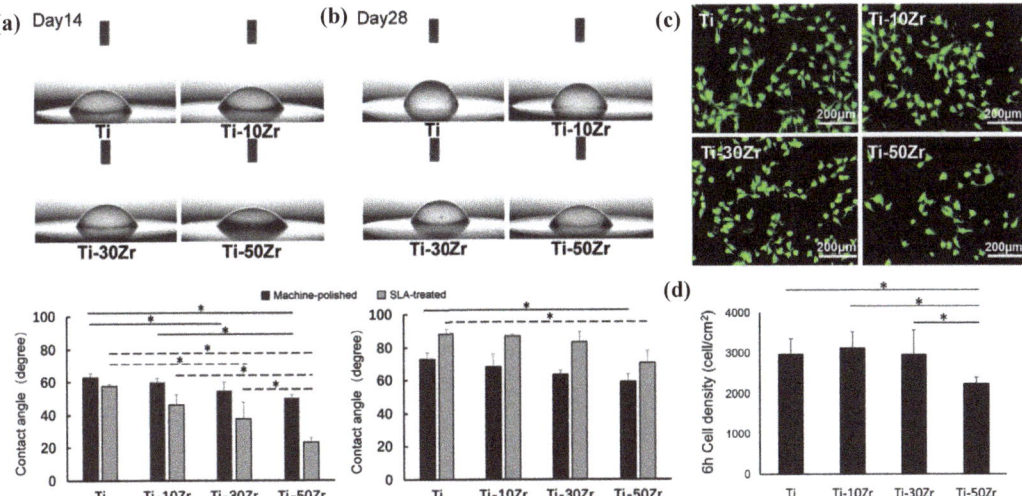

Figure 1. The wettability of Ti-Zr alloys after 14 days (**a**) and 28 days (**b**). Data are expressed in mean ± SD values (n = 5). * $p < 0.05$ indicates a statistically significant difference among the c.p. Ti and Ti–Zr alloy surfaces; (**c**) fluorescence microscopy images of osteoblastic cells stained with calcein after 6 h of culture; (**d**) osteoblastic cellular attachment level, evaluated by fluorescence images. Data are expressed as mean ± SD values (n = 5 × 3 disks.) * $p < 0.05$ indicates a statistically significant difference between the c.p. Ti and Ti–Zr alloy surfaces [30]. (Reused with permission from ref. [30]. Copyright 2022, The Japanese Society for Dental Materials and Devices).

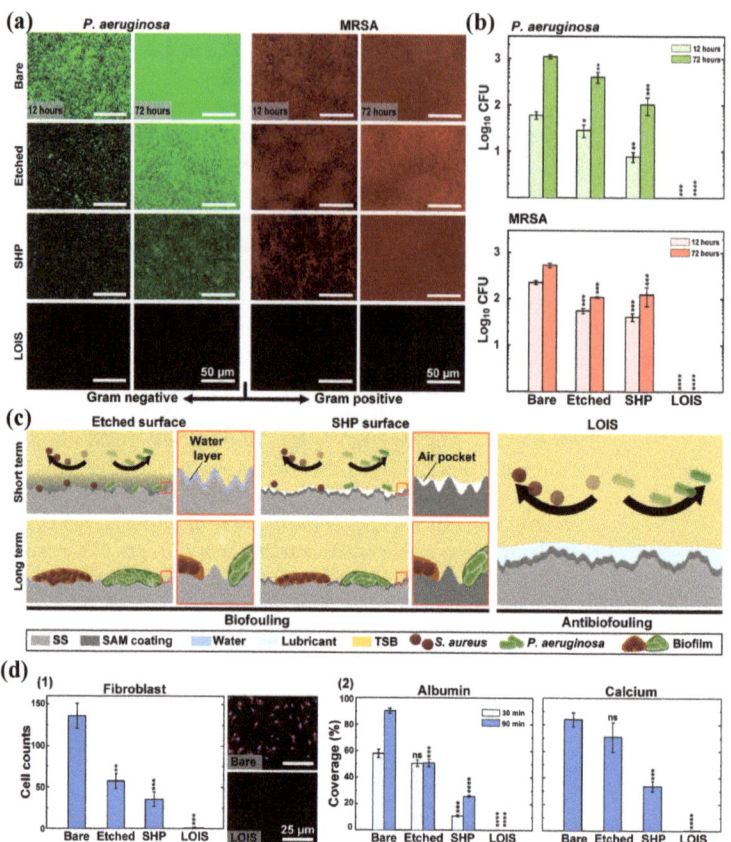

Figure 2. Antibiofouling property of LOIS against bacteria, cells, protein, and calcium. (**a**) Fluorescence microscopy images of each group (bare, etched, superhydrophobic, and LOIS) incubated in *P*. (**b**) The number of adherent colony-forming units of *P. aeruginosa* and MRSA on each group of surfaces. (**c**) Schematics for the antibiofouling mechanisms of etched, superhydrophobic, and LOIS in the short and long term. (**d**) (1) Number of fibroblasts adhered on each substrate and fluorescence microscopy images of the cells adhered on bare and LOIS. (2) Adhesion test for immune-related protein, albumin, and calcium involved in the bone healing process. (* $p < 0.05$, ** $p < 0.01$, *** $p < 0.001$, and **** $p < 0.0001$) [32] (Reused with permission from ref. [32]. Copyright 2020, AAAS publications. Open access article CC BY-NC-ND).

2.2. Surface Nitriding and Nitrocarburizing on Titanium Alloys

The surface treatments of nitriding and nitrocarburizing are popular surface treatment methods for metal materials. Takesue et al. [39] utilized gas blow induction heating (GBIH) to modify the surface microstructure, and the disappearance of the passivation film was caused by the diffusion of oxygen atoms in the passivation film into the substrate. Zhang et al. [40] fabricated core-shell-structured Ti alloys via spark plasma sintering (SPS) with high yield strength (~1.4 GPa) and high thermal stability, which was ascribed to the hard Ti-N shells and the soft Ti cores. Figure 4 indicates the schematic diagram of the core-shell structure, and the formation of this structure can be attributed to occurrence of α and β phase transformations, in contrast to the gradient structure, which has not experienced phase transformations. Li et al. [41] investigated the friction and wear behaviors of Ti/Cu/N coatings on titanium alloys using direct current magnetron sputtering, and the wear resistance of the coatings was obviously improved while the content of Cu reached 10.98 at%. Takesue et al. [42] found that the decrease in Ti-6Al-4V fatigue strength was ascribed to the higher Young's modulus of the compound layer and the grain coarsening of the α-phase. After removing the compound layer formed by GBIH nitriding, the fatigue properties were improved. The microstructure of the fracture surface is shown in Figure 5. Furthermore, the combination of GBIH-nitriding and fine-particle-peening (FPP) pre-treatments improved the wear resistance, because FPP promoted the diffusion of the nitrogen into titanium [42]. Via laser irradiation of pure graphite powder in a nitrogen environment, Seo et al. [43] achieved carbonitriding on a Ti-6Al-4V surface, and Ti (C, N) compounds were formed in the hardening layer on the surface, which was ascribed to the high thermal conductivity of graphite, and it enhanced heat transfer between the laser source and the substrate.

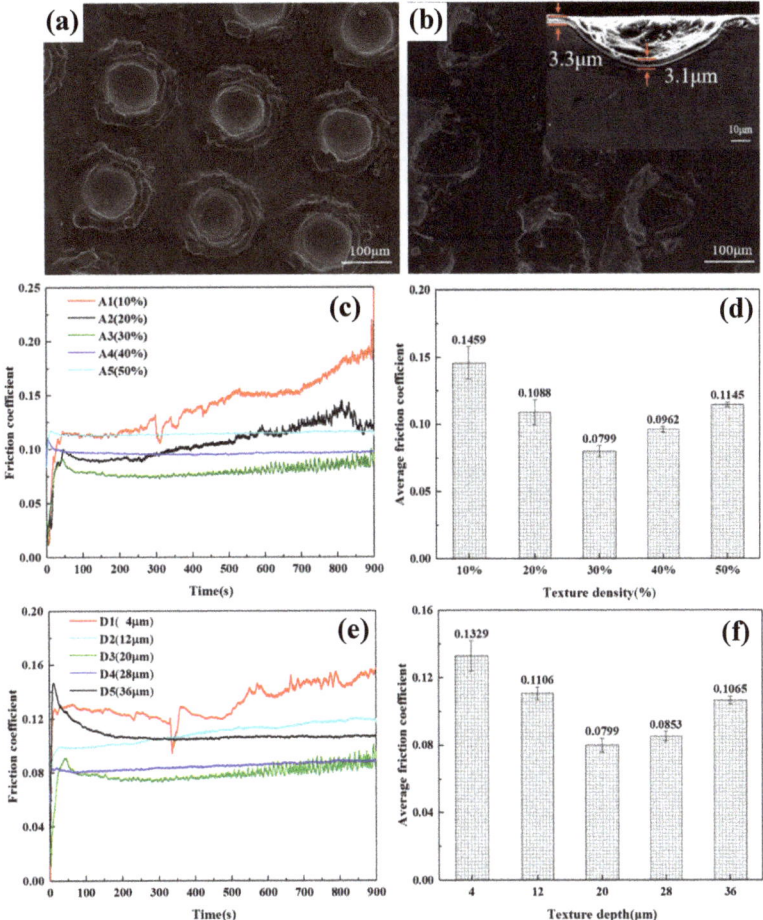

Figure 3. SEM image and EDS analysis of samples: (**a**) textured titanium alloy sample; (**b**) textured diamond-like carbon (DLC)-coated sample. Friction coefficient of samples with different texture densities; (**c**) friction coefficient curve; (**d**) average value of friction coefficient; different texture depths; (**e**) friction coefficient curve; and (**f**) average value of friction coefficient [35] (Reused with permission from ref. [35]. Copyright 2020, IOP Publishing).

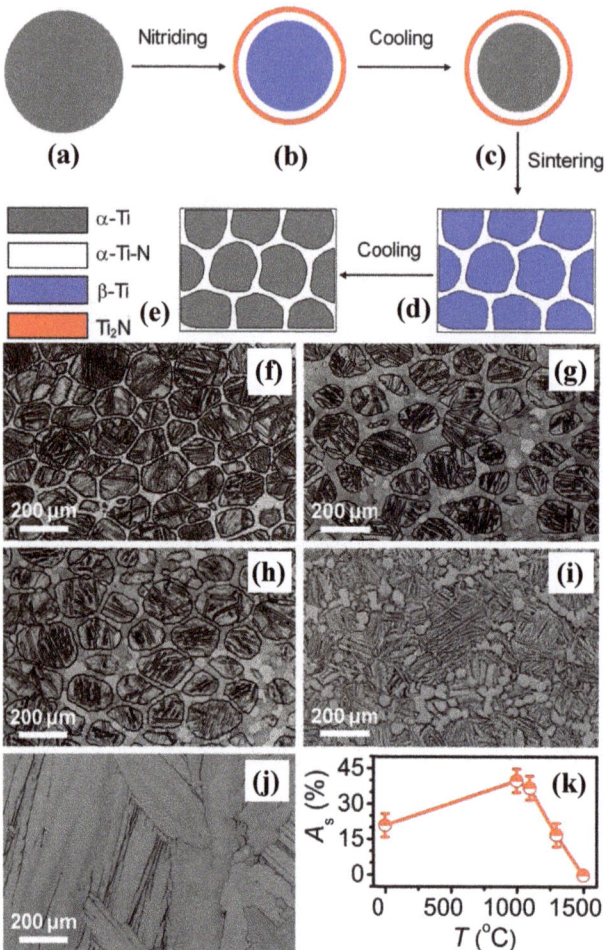

Figure 4. The formation process of the core–shell Ti-N alloys during the nitriding (**a–c**) and SPS sintering (**c–e**) processes. Microstructure variation of the SPS-sintered specimen: (**f**) before annealing; (**g–j**) those annealed under 1000, 1100, 1300, and 1500 °C; (**k**) variations of shell area fraction A_S versus annealing temperature T. [40] (Reused with permission from ref. [40]. Copyright (2017), Springer Nature).

Based on the influence of nitriding and nitrocarburizing on the surface, GBIH nitriding showed the ability to modify the surface microstructure of Ti-6Al-4V within a short time. Some novel treatments, like laser carburizing, laser nitriding, and laser carbonitriding, were introduced, and larger hardness and greater hardening depth were tested after the carbonitriding process. The surface treatments of nitriding and nitrocarburizing can improve the strength and hardness of the surface layer. This is due to the diffusion and carbonitriding reaction of carbon and nitrogen atoms, because reactions occur easily between Ti, C, and N atoms. Although the nitrocarburizing provides a pathway towards the advanced material which combines high strength, good plasticity, and better thermal stability, the influence of the hardening layer's depth on the surface properties is still a difficult point needing to be resolved. Because there are too many complex factors affecting the hardening layer depth, even the current study is difficult to conduct thoroughly and completely.

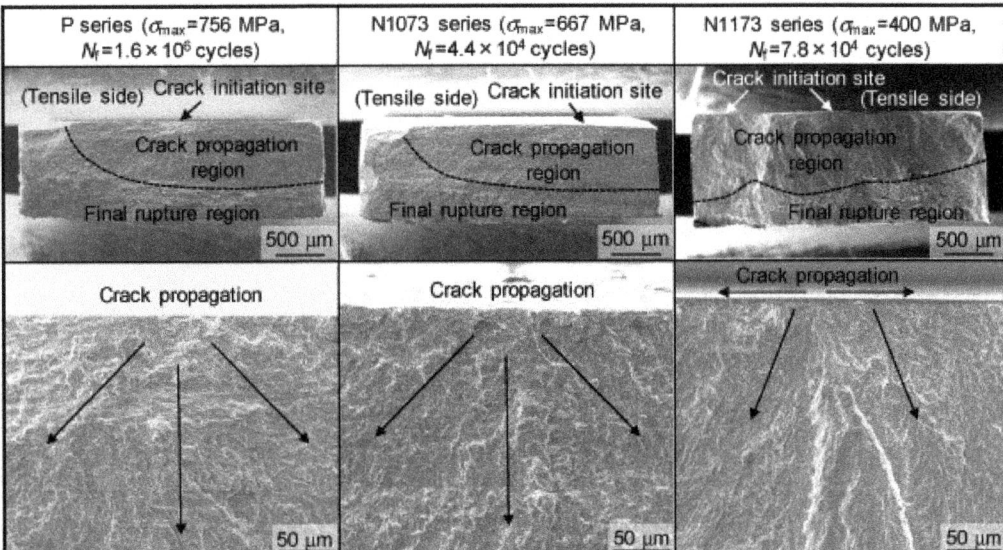

Figure 5. SEM micrographs of the fractured surfaces of failed specimens that were polished (P series) and induction-heated at 1073 K (N1073 series) and 1173 K (N1173 series) in nitrogen [42]. (Reused with permission from ref. [42]. Copyright 2020, Elsevier).

2.3. Hybrid Surface Modification (HSM) on Titanium Alloys

Different surface treatment methods have different advantages, a variety of which can be utilized to maximize their advantages. In the work by Zammit et al. [44], an HSM treatment combining shot peening and the deposition of a tungsten-doped diamond-like carbon coating (WC/C) via PVD on the Ti-6Al-4V surface was developed, with a high surface hardness of 12.79 GPa. In the work by Zhang et al. [45], a novel packed-powder diffusion coating (PPDC) technique was utilized on the surfaces of Ti alloys, and a controllable Al_3Ti intermetallic-based composite coating was formed, which resulted in a significant increase in the oxidization resistance of the Ti alloys. The application of the ultrasonic nanocrystal surface modification (UNSM) technique for strengthening the surfaces of titanium alloys was introduced [46]. The schematic and mechanism of UNSM are shown in Figure 6. Special surface structures like the tracks and the micro-dimples are introduced on the surface, which can improve the surface hardness and the fatigue properties. Via electron beam boriding and composite alloying of titanium alloys and steels, wear-resistant coatings were obtained in Baris' work [47], and the main effect influencing the structure of the coating and its hardness was the effective concentration of alloying elements. Casadei et al. [48] synthesized a Ti/Ti$_x$N$_y$ composite coating on Ti-6Al-4V utilizing a new layered coating system, including the reactive-plasma-sprayed/arc-deposited method. Li et al. found that a surface layer consisting of TiO_2 and TiN was introduced in Ti-6Al-4V via a new surface treatment process based on simultaneous nitriding and thermal oxidation, and the coating fabricated under 800 °C showed the best mechanical properties and corrosion resistance [49].

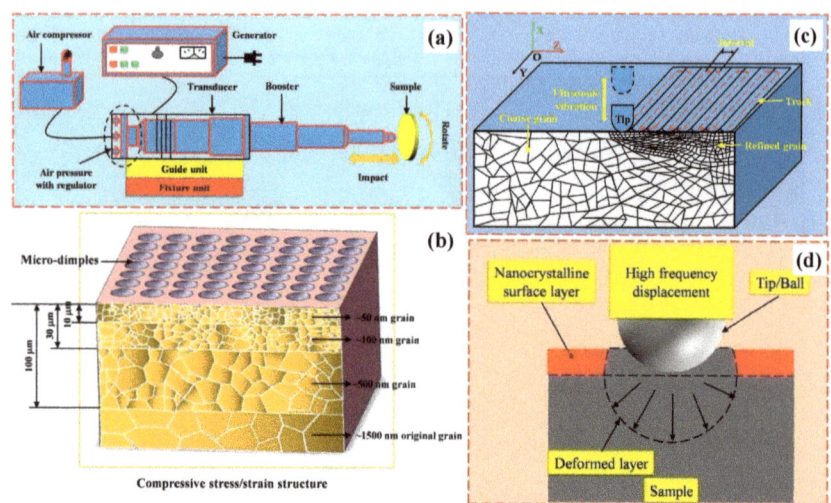

Figure 6. (**a**) Experimental installation diagram of the UNSM technique; (**b**) schematic of the microstructure of the modified layer under UNSM; (**c**) schematic diagram of UNSM; (**d**) schematic diagram of continuous UNSM processing. [46]. (Reused with permission from ref. [46]. Copyright 2021, Elsevier).

Tarbokov et al. [50] manufactured a titanium alloy surface with a typical microroughness height of 2–4 μm via a successive mechanical treatment and irradiation with powerful ion beams (PIB). The machining and PIB were conducted at 1.5 J/cm^2 in a pulse, which was able to improve the adhesion properties of the titanium alloys for application in the medical field. In the work of Takesue et al. [51], nitrided layers with high hardness and compressive residual stress were obtained, which were able to obviously improve the wear resistance and fatigue strength. This was ascribed to the diffusion of nitrogen atoms during nitriding and the fine particle peening. Zammit et al. [52] improved the adhesion and fatigue characteristics of Ti-6Al-4V after shot-peening pre-treatment followed by the deposition of a WC/C coating via PVD. The residual stresses measured on the shot-peened samples reached a maximum of −764 MPa at around 12 μm below the surface and prevailed up to a depth of slightly higher than 75 μm. Song et al. [53] adopted a type of plasma spray–physical vapor deposition technology to synthesize TiN coatings on Ti-6Al-4V, and the longer heating time, lower substrate temperature, and decreasing reactant concentration influenced the hybrid structure of the TiN coating. With increasing spraying distance, the average hardness (H) and elasticity modulus (E) of TiN coatings decreased from 16.3 to 13.4 GPa and from 234.2 to 202.9 GPa, respectively. However, the average H and E of coated Ti-6Al-4V increased to 5.81 and 132.3 GPa, respectively.

Compared with a single treatment method, hybrid surface treatment can allow for better surface performance to be attained and can also provide some special properties which are not available with a single surface treatment. Compared to the usual method of PVD deposition, the chemical reactions between the PVD film and the coating were beneficial for the graded properties and the outstanding adhesion of the multilayer system. Shot peening was able to improve the fatigue life with both LCF (low-cycle fatigue) and HCF (high-cycle fatigue), which was ascribed to the resistance effect of crack initiation due to the high dislocation densities and the resistance effect of crack propagation due to the compressive residual stresses. Combining shot peening with nitriding and nitrocarburizing, the tribological, fatigue-related, and biological adhibition of Ti alloys were further improved because of the advantages of both shot peening and nitrocarburizing.

Of course, hybrid surface treatment requires more complex treatments and preparation methods. Although HSM can bring many advantages, its equipment demands are more complex, and the parameters are more complicated as well. If researchers want to obtain optimal parameters, they need to spend more time exploring reasonable HSM parameters which can achieve the best surface performance.

2.4. Electrochemical/Chemical Surface Treatment of Titanium Alloys

Electrochemical/chemical surface treatment mainly utilizes chemical reagents to react with the metal surface, then generates a surface layer with special functions and special physical and chemical properties. After that, it can improve the physical, chemical, and mechanical properties of the material's surface. The electrochemical/chemical surface treatment method is one of the common methods for titanium alloy surface treatment. Hou et al. [54] used the plasma electrolytic oxidation (PEO) method to form ~12-micron-thick, uniform, adherent, and porous oxide coatings on titanium alloy surfaces under low voltages (120 V), which improved the hardness but introduced stress-induced cracking. Kesik et al. [55] found that alkali-treated oxide layers were formed and showed high bioactivity on anodized titanium alloys (Ti-15Mo, Ti-13Nb-13Zr, and Ti-6Al-7Nb) in Wollastonite suspension. The microstructures of the alkali-treated oxide layers in titanium alloys are shown in Figure 7. Based on the porous oxide layer formed on the titanium, it shows that the optimal condition for the oxide layers' treatment is at temperature of 60 °C for 8 h.

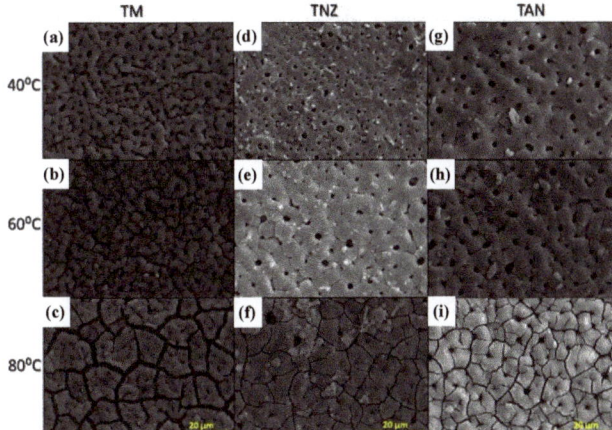

Figure 7. SEM images of Ti-15Mo (TM) (**a–c**), Ti-13Nb-13Zr (TNZ) (**d–f**), and Ti-6Al-7Nb (**g–i**) samples after anodization and immersion in 5 M NaOH at various temperatures for 8 h. [55] (Reused with permission from ref. [55]. Copyright 2017, MDPI).

In the work of Rudawska et al. [56], anodizing followed by vibratory shot peening was able to improve the strength of the titanium alloy sheet adhesive lap joints. This is because the vibrational shot peening increased the curing of the adhesive surface layer, which then increased the strength of the adhesive joint under variable force loads. Four different surface treatments were adopted on a Ti-35Nb-7Zr-5Ta surface by Vlcak et al. [57]. During the corrosion process, the charge transfer influenced the colonization ability of MG-63 cells on the surface, which was more important than other surface parameters (roughness, wettability). Some chemical surface treatments also show the advantages of improving the surface properties of titanium alloys. Khodaei et al. [58] used an H_2O_2 solution to treat the surface of a titanium dental implant, and it was found that the optimum treatment time was approximately 6 h in H_2O_2 solution at 80 °C, making it more suitable for dental implantation. Zhao et al. [59] investigated the doped thermochromic VO_2 film on a Ti surface via rapid thermal annealing (RTA), and a lower transition temperature of 44.9 °C and an extremely narrow hysteresis width of 2.36 °C were formed due to the smaller and more uniform surface particles. Figure 8 demonstrates the microstructures of VO_2 films with different Ti contents. With low Ti content, the scattered rod-shaped particles are formed (Figure 8a,b). After Ti doping, the size of the massive particles becomes significantly smaller, and the distribution of particles becomes uniform (Figure 8c,d). Figure 8e,f show that the thickness of the V film is 102 nm and that of the VO$_2$ film is 148 nm after RTA. Zhao et al. [60] utilized alkali treatments to adjust the microstructures of microarc-oxidized coatings of a Ti2448 titanium alloy, and the coatings showed excellent apatite induction properties when the concentration was less than 15 mol/L. Luo et al. [61] studied the adsorption properties of SO$_2$ gas on N-, Ti-, and N-Ti-doped graphene coatings using the density functional theory, and the N-Ti graphene coating was the most optimally adsorbent because of its low adsorption energy (−2.836 eV) and remarkable charge transfer, which is beneficial to the development of gas sensors to detect SO$_2$.

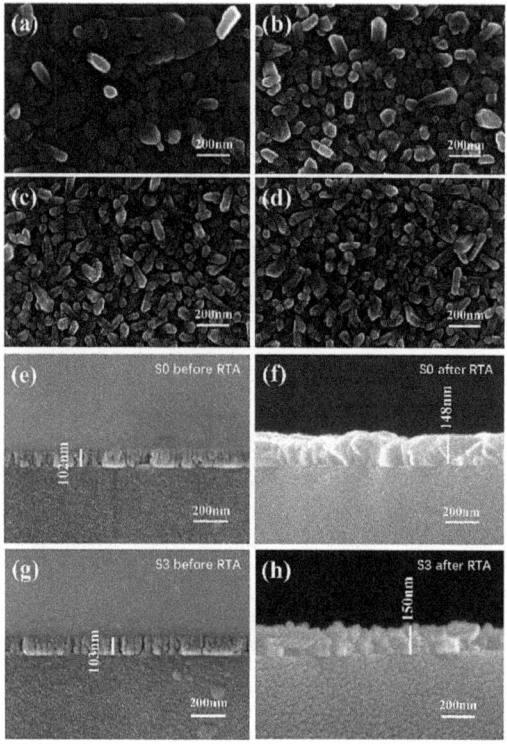

Figure 8. SEM photographs of VO$_2$ films with different Ti contents. (**a–d**) Samples with sputtering times from 0 to 6 min and Ti surface dopant contents from 0% to 0.43%, respectively; (**e–h**) the cross-sections of samples before and after rapid thermal annealing (RTA) [59]. (Reused with permission from ref. [59]. Copyright 2022, Elsevier).

Based on the discussion in this section, much work on electrochemical/chemical surface treatment has been focused on improving the physical and chemical properties of titanium alloy surfaces, as well as the application of titanium alloys as functional materials in corrosion resistance, surface free energy variation, etc. In the field of electrochemical/chemical surface treatment, different surface treatment methods can be adopted to strengthen titanium alloy surfaces, including alkaline degreasing, anodizing, the PEO method, alkali treatments, H$_2$O$_2$ solution treatment, etc., which provides some simple ways to obtain a functional surface.

After electrochemical/chemical surface treatment, some oxides formed on the titanium alloys, which influenced the surface morphology and wettability and provided more nucleation sites for apatite when immersed in the simulated body fluid. The electrochemical/chemical surface treatment was able to improve the surface physicochemical properties;

the hardness of the coatings was improved, but stress-induced cracking was sometimes introduced, which may have reduced the corrosion resistance of the coatings. Consequently, the comprehensive properties of titanium alloys after electrochemical/chemical surface treatment should be considered, including hardness, corrosion resistance, etc. This can provide a theoretical and experimental basis for protecting the surfaces of light metals.

2.5. Laser Surface Modification of Titanium Alloys

Due to the good monochromaticity and high energy of lasers, they have been applied in many fields as special forms of energy. Research works investigating the surface interaction between lasers and titanium have been carried out for many years, especially regarding the influence of lasers on the microstructure and the mechanical properties of titanium alloys, which is still a hot topic of research now. Laser surface nitriding was utilized to modify titanium alloy surfaces for orthopedic implant applications, and Ti-35.5Nb-7.3Zr-5.7Ta (β_{Ti}) was the most appealing choice for joint replacement applications because of the higher mechanical compatibility found in the work of Shirazi et al. [62]. Guo et al. [63] utilized a nanosecond laser to fabricate the hierarchical structures of titanium alloy surfaces, and the micro-/nanostructures of the pitted surfaces showed superhydrophobic properties (as shown in Figure 9). The nanosecond lasers induced micron-sized grooves on titanium surface in the study by Wang et al. [64], as shown in Figure 10. This caused the cells to form a cytoskeleton with a high aspect ratio and limited the migration and spread of the cytoskeleton. The surface fine structure shown in Figures 9 and 10 reveals the advantages of a nanosecond laser; different structures can be obtained via adjusting laser energy and scanning spaces.

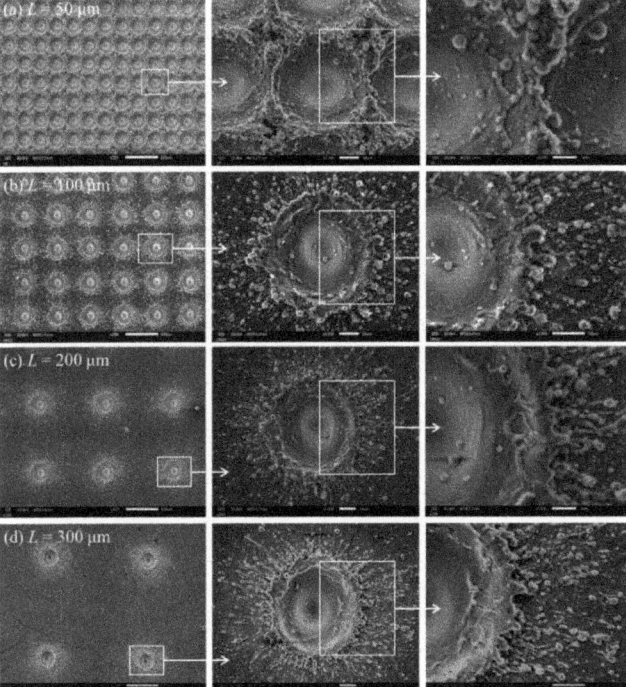

Figure 9. SEM images of the pitted surfaces of N = 10 processed with different scanning spaces. (**a**) L = 50 μm; (**b**) L = 100 μm; (**c**) L = 200 μm; (**d**) L = 300 μm. [63]. (Reused with permission from ref. [63]. Copyright 2021, Elsevier).

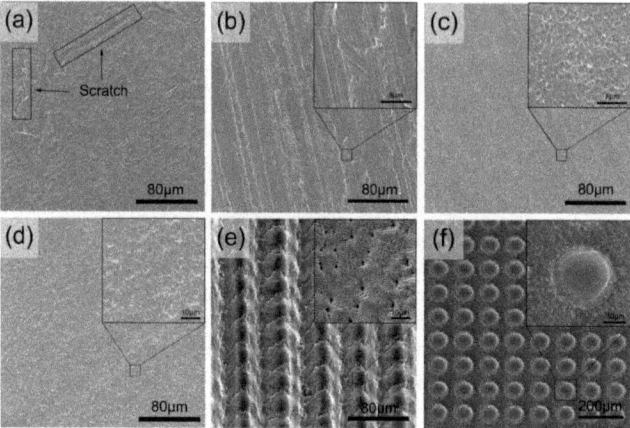

Figure 10. Surface microstructure after treatment. (**a**) Metallographic grinding; (**b**) parallel grinding; (**c**) diamond blasting and acid etching; (**d**) Al_2O_3 blasting and acid etching; (**e**) laser groove array; (**f**) laser pore array [64]. (Reused with permission from ref. [64]. Copyright 2020, Elsevier).

Laser shock peening was adopted in the work of Pan et al. [65] to introduce the refinement surface of Ti-6Al-4V, and the surface texture was transformed from a [0110] fiber orientation to a [1210] orientation. The dynamic recrystallization accompanying the finer grains was ascribed to the massive potential nucleation sites offered by shear bands. Additionally, nitrided layers on a titanium alloy's surface were formed via the laser nitriding method in the work of Yao et al. [66], and the hardness of the nitrided layer was more than 9 GPa. The combined effect of laser texturing and carburizing improved the wettability and specific surface area of a Ti-6Al-4V surface in Dong et al.'s work [67], providing a mechanical self-locking and matching chemical property between the substrate and diamond-like carbon film coatings. Han et al. [68] added Cu in Ti-6Al-4V wires via directed energy deposition, and the formation of Ti_2Cu nano-particles and refined grains improved both the strength and the plasticity of Ti-6Al-4V-8.5Cu. The grain refinement was the main strengthening mechanism after adding Cu. Pan et al. [69] investigated the femtosecond laser-induced surface modification of Ti-6Al-4V. The surface's compressive residual stress reached −746 MPa, and the hardness was improved by 16.6%. The wear mass loss and the average coefficient of friction (COF) after treatment were reduced by 90% and 68.9%.

The results in this section indicate that laser surface treatment is a very popular surface treatment method, especially in the preparation of surface micro/nano structures. Due to the high laser resolution on the order of micrometers, it shows a great advantage for the micro/nano manufacturing on surface fine structures. The surface micro/nano structures are distinct pits with hierarchical structures formed on the surfaces. Besides the laser surface treatment, other high-energy surface treatments also indicate the advantages of improving the surface properties of titanium alloys, such as micro arc oxidation, electron beam melting, etc. Combining the advantages of micro arc oxidation and the electrochemical polymerization of eugenol, the corrosion resistance of a titanium alloy was increased to 81.34% in the work by AlMashhadani et al. [70]. Zhang et al. [71] reviewed the application of electron beam melting on additive manufactured titanium alloys, and the porous structures in titanium alloys improved the biocompatibility further, as shown in Figure 11. The porous structure could provide porosity to encourage bone tissue growth and nutrient flow in order to achieve a better recovery effect. The porous size and structure can be adjusted via changing the electron beam energy and scanning speed.

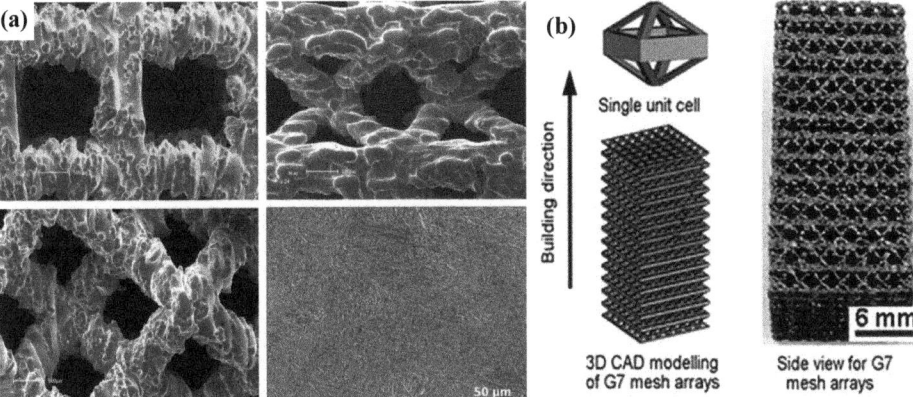

Figure 11. (a) The microstructures of Ti-6Al-4V porous samples with electron beam melting. (b) Porous structure model and side view of the Ti-24Nb-4Zr-8Sn component processed via electron beam melting [71]. (Reused with permission from ref. [71]. Copyright 2017, WILEY-VCH Verlag GmbH & Co. KGaA).

Usually, the laser surface treatment is carried out in combination with other treatment methods, such as grinding, sandblasting and acidizing, nitriding, etc. However, the laser parameters for a possible choice are diverse, including laser energy, scanning speed, powder feeding speed, etc. The selection of these parameters directly affects the performance of the surface coatings; for instance, excessive laser power could introduce tensile residual stresses and cracks on the surface. Therefore, the selection of appropriate laser parameters is particularly important, especially in combination with other processing methods, which are still the focuses of future research and may be resolved via new methods of machine learning or big data processing.

2.6. Surface Mechanical Treatment of Titanium Alloys

In order to improve the mechanical properties of titanium alloy components, especially the fatigue properties, surface mechanical treatment is usually adopted to adjust the surface microstructure and the residual stress distribution, as it allows for excellent surface comprehensive mechanical properties to be obtained. Ren et al. [72] used ultrasonic rolling to modify the surface properties of Ti-6Al-4V, and, according to the simulation results, the best surface finishing was obtained while the static force was 1000 N and the amplitude was 8 μm, as shown in Figure 12. Du et al. [73] investigated the effect of cold rolling on the deformation mechanism of titanium alloys, and, with the increasing deformation level, the crystal orientation of matrix obviously changed. Pu et al. [74] utilized semi-solid stirring-assisted ultrasonic vibration to fabricate TiP/VW94 composites, and the tensile properties of the composites were improved because of the reinforcement of Ti particles. A strong interfacial bonding of the MnTi layer was formed due to the diffusion of Mn atoms. Wang et al. [75] used shot peening and the ultrasonic surface rolling process (USRP) method to improve the fretting fatigue performance of Ti-6Al-4V, which was ascribed to the lower surface roughness compared to that achieved using the shot peening process. In the work by Liu et al. [46], a new ultrasonic nano-crystal surface modification (UNSM) technique was adopted to strengthen the surfaces of titanium alloys. The surface hardness, residual compressive stress, and fatigue-related and tribological properties of the materials were improved.

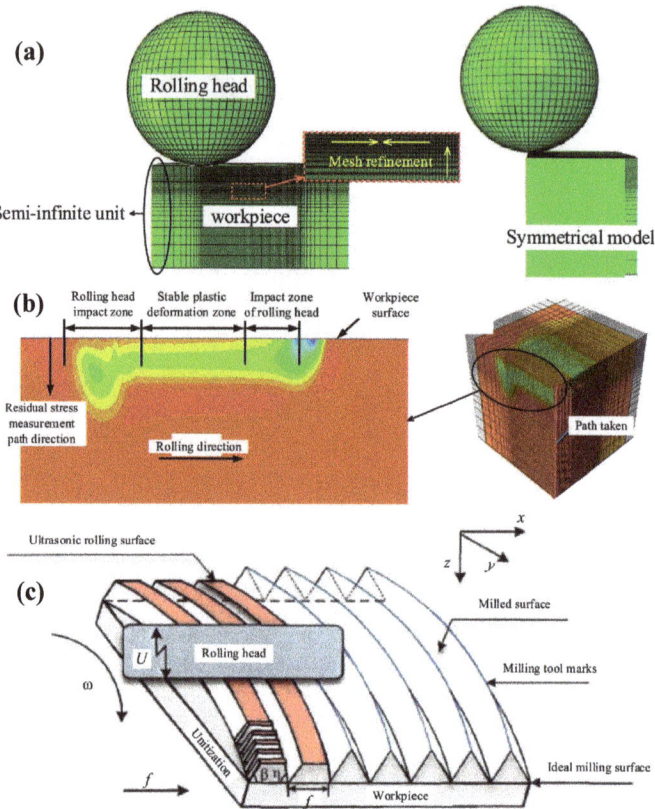

Figure 12. (**a**) Mesh division of a 3D finite element model; (**b**) residual stress distribution (left) and junction path (right) after ultrasonic rolling; the green area indicates the stable residual compressive stress layer near the surface of the processed workpiece; (**c**) schematic diagram of ultrasonic rolling [72]. (Reused with permission from ref. [72]. Copyright 2022, Elsevier).

Similarly, ultrasonic surface rolling treatment was utilized to modify the rolling contact fatigue (RCF) behavior of 17Cr2Ni2MoVNb steel in Zhang et al.'s work [76]. The material, under 1000 N, exhibited a maximum mean RCF life of 3.71×10^6 cycles, which was ascribed to the grain refinement and the residual compressive stress. Klimenov et al. [77] carried out surface modification on VT1-0 (α-phase) and VT6 ($\alpha + \beta$ phase) titanium alloys using ultrasonic treatment. The microhardness was enhanced with the increase in the initial surface roughness, which was ascribed to the surface nanocrystals as 100 nm for VT1-0 and 50 nm for VT6. The roughness of the surface of VT1-0 changed from $Rz = 5$ μm to 100 μm, causing by the successively located ridges and roots of a certain height at a constant pitch, which of VT6 under same modes showed an obviously lower surface roughness as compared to samples of VT1-0. As a result, with the decrease in the plasticity, the metal deforms less, the build-up reduces, vibrations reduce, and the cleanliness level of the treated surface increases. Li et al. [78] optimized the diffusion bonding of dissimilar TC17/TC4 titanium alloys using surface nano-crystallization (SN) treatment. Moreover, the shear strength of the SN-TC17/TC4 bond increased from 275 MPa at 973 K to 634 MPa at 1013 K due to the acceleration of void shrinkage. A schematic illustration of the evolution mechanism is shown in Figure 13. Comparing Figure 13a–c to Figure 13d–f, the surface SN layer on the SN-TC17 side enhances the grain/phase boundary diffusion and volume diffusion processes and the plastic flow effect, which accelerates the void shrinkage of the SN-TC17/TC4 bond and promotes the shear strength of the SN-TC17/TC4 bond.

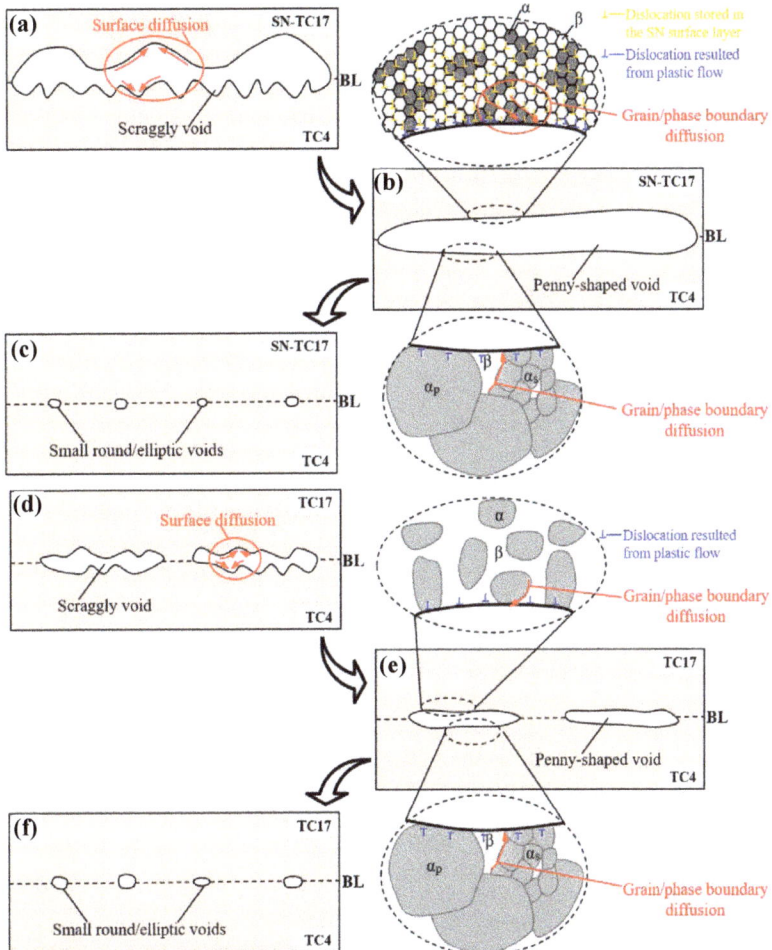

Figure 13. Evolution mechanism of voids in the bonding interface of SN-TC17/TC4 and TC17/TC4 bond: (**a**,**d**) are scraggly voids, (**b**,**e**) are penny-shaped voids, and (**c**,**f**) are round/elliptic voids [78]. (Reused with permission from ref. [78]. Copyright 2022, Elsevier).

The surface mechanical treatment of titanium alloys is usually related to fatigue properties. The relation between them is the microstructure, and the main mechanism is ascribed to the surface's micro-/nanostructure and residual compressive stress. The residual stress distribution was found to be more uniform and the static force and amplitude were increased after surface mechanical treatment. Regarding the mechanical treatment methods, shot peening is the most conventional method, but it also has some shortcomings, such as the high surface roughness and increased surface defects under severe plastic deformation. Therefore, the conventional shot-peening method combined with other treatment methods is a more reasonable choice. According to the discussion in this section, the surface microstructures and defects of titanium alloys can be improved by microstructure optimization and compressive residual stress introduced by surface mechanical treatment, which are ascribed to the surface nanolayer and the compressive residual stress distribution. Surface treatments of titanium alloys have been used in many different fields [79,80]. Each method has its own advantages and characteristics, as well as its own suitable materials and application range. Thus, researchers can choose the appropriate surface treatment method according to the relevant characteristics.

3. Conclusions and Prospects

Regarding the influence of surface treatments on titanium alloys, three aspects should be noted: (1) the effect of surface treatment on the microstructure and mechanical properties of titanium alloys; (2) the influence of surface treatment on the friction and wear properties of titanium alloys; and (3) the improvement of the biocompatibility of titanium alloys after surface treatment. Based on the above three aspects, the progress of research into surface treatment technologies for titanium alloys was reviewed in this work, and many new single methods and more hybrid methods have been introduced.

The single methods of surface treatment on titanium alloys were adopted from the mechanical surface treatments, the electrochemical/chemical surface treatments, etc. The representative single-treatment methods of shot peening, laser shot peening, nitriding, nitrocarburizing, etc., were summarized. Although some new and better results have been obtained via a single surface treatment, limitations still exist. Shot peening is the most conventional method of the mechanical treatment, but high surface roughness and increased surface defects are formed under severe plastic deformation. Nitrocarburizing provides a pathway towards an advanced material with both high strength and good plasticity, but the influence of the hardening layer on surface properties is still a difficult point that needs to be investigated further. Now, the laser surface treatment is a very popular method and some great results have obtained. The laser parameters for a possible choice are diverse, including laser energy, scanning speed, powder feeding speed, etc.; however, excessive laser power could introduce tensile residual stresses and cracks on the surface.

Therefore, the single-surface-treatment method combined with other treatment methods (hybrid surface treatment) is a more reasonable choice. The hybrid surface treatment methods were developed by focusing on combining two or more methods, for example, combining the mechanical surface treatments and the electrochemical/chemical surface treatments; the surface treatment coatings, etc. The representative hybrid surface treatment methods, like the shot peening combined with deposition, the laser boriding combined with composite alloying, etc., were also discussed. Laser surface treatment is carried out combined with other treatment methods, like grinding, sandblasting and acidizing, nitriding, etc. Hybrid surface treatments can bring many advantages, but the demands of its equipment are more complex, and the parameters become so as well; thus, more time is needed in order to find a reasonable process with an optimized set of parameters. With the development of surface treatment technologies for titanium alloys, hybrid surface treatments will be carried out more often in the future.

When a combination of two or more surface modification technologies is used on a titanium alloy, it can lead to better surface functions and achieve the comprehensive performance of "1 + 1 > 2". Moreover, the selection of appropriate parameters is particularly important, especially in combination with other processing methods. These are still research focuses for the future and may be resolved via new methods of machine learning or big data processing.

Author Contributions: B.X.: conceptualization, experiment, analysis, methodology, writing—original draft. K.G.: conceptualization, resources, supervision, writing—review and editing. All authors have read and agreed to the published version of the manuscript.

Funding: This research received no external funding.

Institutional Review Board Statement: Not applicable.

Informed Consent Statement: Not applicable.

Data Availability Statement: The raw/processed data required to reproduce these findings cannot be shared at this time as the data also form part of an ongoing study.

Conflicts of Interest: The authors declare that they have no known competing financial interests or personal relationships that could have appeared to influence the work reported in this paper.

References

1. Lu, W.; Zhang, D.; Zhang, X.; Wu, R.; Sakata, T.; Mori, H. Microstructural characterization of TiB in in situ synthesized titanium matrix composites prepared by common casting technique. *J. Alloys Compd.* **2001**, *327*, 240–247. [CrossRef]
2. Panigrahi, A.; Bönisch, M.; Waitz, T.; Schafler, E.; Calin, M.; Eckert, J.; Skrotzki, W.; Zehetbauer, M. Phase transformations and mechanical properties of biocompatible Ti–16.1Nb processed by severe plastic deformation. *J. Alloys Compd.* **2015**, *628*, 434–441. [CrossRef]
3. Attar, H.; Prashanth, K.G.; Chaubey, A.K.; Calin, M.; Zhang, L.C.; Scudino, S.; Eckert, J. Comparison of wear properties of commercially pure titanium prepared by selective laser melting and casting processes. *Mater. Lett.* **2015**, *142*, 38–41. [CrossRef]
4. Haghighi, S.E.; Lu, H.B.; Jian, G.Y.; Cao, G.H.; Habibi, D.; Zhang, L.C. Effect of α'' martensite on the microstructure and mechanical properties of beta-type Ti–Fe–Ta alloys. *Mater. Des.* **2015**, *76*, 47–54. [CrossRef]
5. Calin, M.; Zhang, L.C.; Eckert, J. Tailoring of microstructure and mechanical properties of a Ti-based bulk metallic glass-forming alloy. *Scr. Mater.* **2007**, *57*, 1101–1104. [CrossRef]
6. De los Rios, E.R.; Walley, A.; Milan, M.T.; Hammersley, G. Fatigue crack initiation and propagation on shot-peened surfaces in A316 stainless steel. *Int. J. Fatigue* **1995**, *17*, 493–499. [CrossRef]
7. Torres, M.A.S.; Voorwald, H.J.C. An evaluation of shot peening, residual stress and stress relaxation on the fatigue life of AISI 4340 steel. *Int. J. Fatigue* **2002**, *24*, 877–886. [CrossRef]
8. Kobayashi, M.; Matsui, T.; Murakami, Y. Mechanism of creation of compressive residual stress by shot peening. *Int. J. Fatigue* **1998**, *20*, 351–357. [CrossRef]
9. Fu, Y.; Loh, N.L.; Batchelor, A.W.; Liu, D.; Zhu, X.; He, J.; Xu, K. Improvement in fretting wear and fatigue resistance of Ti_6Al_4V by application of several surface treatments and coatings. *Surf. Coat. Technol.* **1998**, *106*, 193–197. [CrossRef]
10. Nalla, R.K.; Altenberger, I.; Noster, U.; Liu, G.Y.; Scholtes, B.; Ritchie, R.O. On the influence of mechanical surface treatments—Deep rolling and laser shock peening—On the fatigue behavior of Ti_6Al_4V at ambient and elevated temperatures. *Mater. Sci. Eng. A* **2003**, *355*, 216–230. [CrossRef]
11. Golden, P.J.; Hutson, A.; Sundaram, V.; Arps, J.H. Effect of surface treatments on fretting fatigue of Ti_6Al_4V. *Int. J. Fatigue* **2007**, *29*, 1302–1310. [CrossRef]
12. Liu, C.; Bi, Q.; Matthews, A. Tribological and electrochemical performance of PVD TiN coatings on the femoral head of Ti_6Al_4V artificial hip joints. *Surf. Coat. Technol.* **2003**, *163–164*, 597–604. [CrossRef]
13. Warren, J.; Hsiung, L.M.; Wadley, H.N.G. High temperature deformation behavior of physical vapor deposited Ti_6Al_4V. *Acta Metall. Mater.* **1995**, *43*, 2773–2787. [CrossRef]
14. Perry, S.S.; Ager, J.W., III; Somorjai, G.A.; McClelland, R.J.; Drory, M.D. Interface characterization of chemically vapor deposited diamond on titanium and Ti-6Al-4V. *J. Appl. Phys.* **1993**, *74*, 7542–7550. [CrossRef]
15. Baek, S.H.; Mihec, D.F.; Metson, J.B. The Deposition of Diamond Films by Combustion Assisted CVD on Ti and Ti_6Al_4V. *Chem. Vap. Depos.* **2002**, *8*, 29–34. [CrossRef]
16. Tsipas, S.A.; Vázquez-Alcázar, M.R.; Navas, E.M.R.; Gordo, E. Boride coatings obtained by pack cementation deposited on powder metallurgy and wrought Ti and Ti_6Al_4V. *Surf. Coat. Technol.* **2010**, *205*, 2340–2347. [CrossRef]
17. Çelikkan, H.; Öztürk, M.K.; Aydin, H.; Aksu, M.L. Boriding titanium alloys at lower temperatures using electrochemical methods. *Thin Solid Films* **2007**, *515*, 5348–5352. [CrossRef]
18. Sanders, A.P.; Tikekar, N.; Lee, C.; Chandran, K.S.R. Surface Hardening of Titanium Articles with Titanium Boride Layers and its Effects on Substrate Shape and Surface Texture. *J. Manuf. Sci. Eng.* **2009**, *131*, 031001. [CrossRef]
19. Wang, Z.M.; Ezugwu, E.O. Performance of PVD-Coated Carbide Tools When Machining Ti_6Al_4V©. *Tribol. Trans.* **1997**, *40*, 81–86. [CrossRef]
20. Senthil Selvan, J.; Subramanian, K.; Nath, A.K.; Kumar, H.; Ramachandra, C.; Ravindranathan, S.P. Laser boronising of Ti_6Al_4V as a result of laser alloying with pre-placed BN. *Mater. Sci. Eng. A* **1999**, *260*, 178–187. [CrossRef]
21. Wilson, A.D.; Leyland, A.; Matthews, A. A comparative study of the influence of plasma treatments, PVD coatings and ion implantation on the tribological performance of Ti_6Al_4V. *Surf. Coat. Technol.* **1999**, *114*, 70–80. [CrossRef]
22. Hutchings, R.; Oliver, W.C. A study of the improved wear performance of nitrogen-implanted Ti_6Al_4V. *Wear* **1983**, *92*, 143–153. [CrossRef]
23. Vreeling, J.A.; Ocelík, V.; De Hosson, J.T.M. Ti_6Al_4V strengthened by laser melt injection of WCp particles. *Acta Mater.* **2002**, *50*, 4913–4924. [CrossRef]
24. Man, H.C.; Zhao, N.Q.; Cui, Z.D. Surface morphology of a laser surface nitrided and etched Ti_6Al_4V alloy. *Surf. Coat. Technol.* **2005**, *192*, 341–346. [CrossRef]

25. Kloosterman, A.B.; Kooi, B.J.; De Hosson, J.Th.M. Electron microscopy of reaction layers between SiC and Ti_6Al_4V after laser embedding. *Acta Mater.* **1998**, *46*, 6205–6217. [CrossRef]
26. Lee, H.; Mall, S. Stress relaxation behavior of shot-peened Ti_6Al_4V under fretting fatigue at elevated temperature. *Mater. Sci. Eng. A* **2004**, *366*, 412–420. [CrossRef]
27. John, R.; Buchanan, D.J.; Jha, S.K.; Larsen, J.M. Stability of shot-peen residual stresses in an α+β titanium alloy. *Scr. Mater.* **2009**, *61*, 343–346. [CrossRef]
28. Liu, K.K.; Hill, M.R. The effects of laser peening and shot peening on fretting fatigue in Ti_6Al_4V coupons. *Tribol. Int.* **2009**, *42*, 1250–1262. [CrossRef]
29. Wang, G.; Wang, S.; Yang, X.; Yu, X.; Wen, D.; Chang, Z.; Zhang, M. Fretting wear and mechanical properties of surface-nanostructural titanium alloy bone plate. *Surf. Coat. Technol.* **2021**, *405*, 126512. [CrossRef]
30. Tan, T.; Zhao, Q.; Kuwae, H.; Ueno, T.; Chen, P.; Tsutsumi, Y.; Mizuno, J.; Hanawa, T.; Wakabayashi, N. Surface properties and biocompatibility of sandblasted and acid-etched titanium–zirconium binary alloys with various compositions. *Dent. Mater. J.* **2022**, *41*, 266–272. [CrossRef]
31. Liu, P.; Hao, Y.; Zhao, Y.; Yuan, Z.; Ding, Y.; Cai, K. Surface modification of titanium substrates for enhanced osteogenetic and antibacterial properties. *Colloids Surf. B Biointerfaces* **2017**, *160*, 110–116. [CrossRef] [PubMed]
32. Chae, K.; Jang, W.Y.; Park, K.; Lee, J.; Kim, H.; Lee, K.; Lee, C.K.; Lee, Y.; Lee, S.H.; Seo, J. Antibacterial infection and immune-evasive coating for orthopedic implants. *Sci. Adv.* **2020**, *6*, eabb0025. [CrossRef] [PubMed]
33. Bai, H.; Zhong, L.; Kang, L.; Liu, J.; Zhuang, W.; Lv, Z.; Xu, Y. A review on wear-resistant coating with high hardness and high toughness on the surface of titanium alloy. *J. Alloys Compd.* **2021**, *882*, 160645. [CrossRef]
34. Ren, B.; Wan, Y.; Liu, C.; Wang, H.; Yu, M.; Zhang, X.; Huang, Y. Improved osseointegration of 3D printed Ti_6Al_4V implant with a hierarchical micro/nano surface topography: An in vitro and in vivo study. *Mater. Sci. Eng. C* **2021**, *118*, 111505. [CrossRef] [PubMed]
35. Peng, R.; Zhang, P.; Tian, Z.; Zhu, D.; Chen, C.; Yin, B.; Hua, X. Effect of textured DLC coatings on tribological properties of titanium alloy under grease lubrication. *Mater. Res. Express* **2020**, *7*, 066408. [CrossRef]
36. Todea, M.; Vulpoi, A.; Popa, C.; Berce, P.; Simon, S. Effect of different surface treatments on bioactivity of porous titanium implants. *J. Mater. Sci. Technol.* **2019**, *35*, 418–426. [CrossRef]
37. Zhang, L.-C.; Chen, L.-Y.; Wang, L. Surface Modification of Titanium and Titanium Alloys: Technologies, Developments, and Future Interests. *Adv. Eng. Mater.* **2020**, *22*, 1901258. [CrossRef]
38. Cui, W.; Niu, F.; Tan, Y.; Qin, G. Microstructure and tribocorrosion performance of nanocrystalline TiN graded coating on biomedical titanium alloy. *Trans. Nonferrous Met. Soc. China* **2019**, *29*, 1026–1035. [CrossRef]
39. Takesue, S.; Kikuchi, S.; Misaka, Y.; Morita, T.; Komotori, J. Rapid nitriding mechanism of titanium alloy by gas blow induction heating. *Surf. Coat. Technol.* **2020**, *399*, 126160. [CrossRef]
40. Zhang, Y.S.; Zhao, Y.H.; Zhang, W.; Lu, J.W.; Hu, J.J.; Huo, W.T.; Zhang, P.X. Core-shell structured titanium-nitrogen alloys with high strength, high thermal stability and good plasticity. *Sci. Rep.* **2017**, *7*, 40039. [CrossRef]
41. Li, J.; Pang, X.; Fan, A.; Zhang, H. Friction and wear behaviors of Ti/Cu/N coatings on titanium alloy surface by DC magnetron sputtering. *J. Wuhan Univ. Technol. Mater Sci. Ed.* **2017**, *32*, 140–146. [CrossRef]
42. Takesue, S.; Kikuchi, S.; Akebono, H.; Morita, T.; Komotori, J. Characterization of surface layer formed by gas blow induction heating nitriding at different temperatures and its effect on the fatigue properties of titanium alloy. *Results Mater.* **2020**, *5*, 100071. [CrossRef]
43. Seo, D.M.; Hwang, T.W.; Moon, Y.H. Carbonitriding of Ti_6Al_4V alloy via laser irradiation of pure graphite powder in nitrogen environment. *Surf. Coat. Technol.* **2019**, *363*, 244–254. [CrossRef]
44. Zammit, A.; Attard, M.; Subramaniyan, P.; Levin, S.; Wagner, L.; Cooper, J.; Espitalier, L.; Cassar, G. Enhancing surface integrity of titanium alloy through hybrid surface modification (HSM) treatments. *Mater. Chem. Phys.* **2022**, *279*, 125768. [CrossRef]
45. Zhang, M.-X.; Miao, S.-M.; Shi, Y.-N. A Novel Surface Treatment Technique for Titanium Alloys. *JOM* **2020**, *72*, 4583–4593. [CrossRef]
46. Liu, R.; Yuan, S.; Lin, N.; Zeng, Q.; Wang, Z.; Wu, Y. Application of ultrasonic nanocrystal surface modification (UNSM) technique for surface strengthening of titanium and titanium alloys: A mini review. *J. Mater. Res. Technol.* **2021**, *11*, 351–377. [CrossRef]
47. Baris, N.M.; Golkovsky, M.G.; Tushinsky, L.I. SR-study of coatings obtained by complex electron-beam alloying of titanium alloys and steels. In Proceedings of the Proceedings KORUS 2000. The 4th Korea-Russia International Symposium on Science and Technology, Ulsan, Republic of Korea, 27 June–1 July 2000; Volume 3, pp. 407–410.
48. Casadei, F.; Pileggi, R.; Valle, R.; Matthews, A. Studies on a combined reactive plasma sprayed/arc deposited duplex coating for titanium alloys. *Surf. Coat. Technol.* **2006**, *201*, 1200–1206. [CrossRef]
49. Li, Y.; Zhou, Q.; Liu, M. Effect of novel surface treatment on corrosion behavior and mechanical properties of a titanium alloy. *Baosteel Technol. Res.* **2021**, *15*, 11–19. [CrossRef]
50. Tarbokov, V.A.; Pavlov, S.K.; Remnev, G.E.; Nochovnaya, N.A.; Eshkulov, U.É. Titanium Alloy Surface Complex Modification. *Metallurgist* **2019**, *62*, 1187–1193. [CrossRef]
51. Takesue, S.; Kikuchi, S.; Misaka, Y.; Morita, T.; Komotori, J. Combined Effect of Gas Blow Induction Heating Nitriding and Post-Treatment with Fine Particle Peening on Surface Properties and Wear Resistance of Titanium Alloy. *Mater. Trans.* **2021**, *62*, 1502–1509. [CrossRef]

52. Zammit, A.; Attard, M.; Subramaniyan, P.; Levin, S.; Wagner, L.; Cooper, J.; Espitalier, L.; Cassar, G. Investigations on the adhesion and fatigue characteristics of hybrid surface-treated titanium alloy. *Surf. Coat. Technol.* **2022**, *431*, 128002. [CrossRef]
53. Song, C.; Liu, M.; Deng, Z.-Q.; Niu, S.-P.; Deng, C.-M.; Liao, H.-L. A novel method for in-situ synthesized TiN coatings by plasma spray-physical vapor deposition. *Mater. Lett.* **2018**, *217*, 127–130. [CrossRef]
54. Hou, F.; Gorthy, R.; Mardon, I.; Tang, D.; Goode, C. Low voltage environmentally friendly plasma electrolytic oxidation process for titanium alloys. *Sci. Rep.* **2022**, *12*, 6037. [CrossRef]
55. Kazek-Kęsik, A.; Leśniak, K.; Zhidkov, I.S.; Korotin, D.M.; Kukharenko, A.I.; Cholakh, S.O.; Kalemba-Rec, I.; Suchanek, K.; Kurmaev, E.Z.; Simka, W. Influence of Alkali Treatment on Anodized Titanium Alloys in Wollastonite Suspension. *Metals* **2017**, *7*, 322. [CrossRef]
56. Rudawska, A.; Zaleski, K.; Miturska, I.; Skoczylas, A. Effect of the Application of Different Surface Treatment Methods on the Strength of Titanium Alloy Sheet Adhesive Lap Joints. *Materials* **2019**, *12*, 4173. [CrossRef] [PubMed]
57. Vlcak, P.; Fojt, J.; Koller, J.; Drahokoupil, J.; Smola, V. Surface pre-treatments of Ti-Nb-Zr-Ta beta titanium alloy: The effect of chemical, electrochemical and ion sputter etching on morphology, residual stress, corrosion stability and the MG-63 cell response. *Results Phys.* **2021**, *28*, 104613. [CrossRef]
58. Khodaei, M.; Amini, K.; Valanezhad, A.; Watanabe, I. Surface treatment of titanium dental implant with H_2O_2 solution. *Int. J. Miner. Metall. Mater.* **2020**, *27*, 1281–1286. [CrossRef]
59. Zhao, J.; Chen, D.; Hao, C.; Mi, W.; Zhou, L. The optimization and role of Ti surface doping in thermochromic VO_2 film. *Opt. Mater.* **2022**, *133*, 112960. [CrossRef]
60. Zhao, G.; Xia, L.; Zhong, B.; Wu, S.; Song, L.; Wen, G. Effect of alkali treatments on apatite formation of microarc-oxidized coating on titanium alloy surface. *Trans. Nonferrous Met. Soc. China* **2015**, *25*, 1151–1157. [CrossRef]
61. Luo, Q.; Yin, S.; Sun, X.; Tang, Y.; Feng, Z.; Dai, X. Density functional theory study on the adsorption properties of SO_2 gas on graphene, N, Ti, and N–Ti doped graphene. *Micro Nanostruct.* **2022**, *171*, 207401. [CrossRef]
62. Shirazi, H.A.; Chan, C.-W.; Lee, S. Elastic-plastic properties of titanium and its alloys modified by fibre laser surface nitriding for orthopaedic implant applications. *J. Mech. Behav. Biomed. Mater.* **2021**, *124*, 104802. [CrossRef]
63. Guo, C.; Zhang, M.; Hu, J. Fabrication of hierarchical structures on titanium alloy surfaces by nanosecond laser for wettability modification. *Opt. Laser Technol.* **2022**, *148*, 107728. [CrossRef]
64. Wang, Y.; Yu, Z.; Li, K.; Hu, J. Effects of surface properties of titanium alloys modified by grinding, sandblasting and acidizing and nanosecond laser on cell proliferation and cytoskeleton. *Appl. Surf. Sci.* **2020**, *501*, 144279. [CrossRef]
65. Pan, X.; Wang, X.; Tian, Z.; He, W.; Shi, X.; Chen, P.; Zhou, L. Effect of dynamic recrystallization on texture orientation and grain refinement of Ti6Al4V titanium alloy subjected to laser shock peening. *J. Alloys Compd.* **2021**, *850*, 156672. [CrossRef]
66. Yao, X.; Shi, Y.; Li, T.; Wensheng, L. Forming Characteristics and Analysis of Nitrided Layers During the Laser Nitriding Titanium Alloy. *Rare Met. Mater. Eng.* **2019**, *48*, 4060–4067.
67. Dong, B.; Guo, X.; Zhang, K.; Zhang, Y.; Li, Z.; Wang, W.; Cai, C. Combined effect of laser texturing and carburizing on the bonding strength of DLC coatings deposited on medical titanium alloy. *Surf. Coat. Technol.* **2022**, *429*, 127951. [CrossRef]
68. Han, J.; Zhang, G.; Chen, X.; Cai, Y.; Luo, Z.; Zhang, X.; Su, Y.; Tian, Y. High strength Ti alloy fabricated by directed energy deposition with in-situ Cu alloying. *J. Mater. Process. Technol.* **2022**, *310*, 117759. [CrossRef]
69. Pan, X.; He, W.; Cai, Z.; Wang, X.; Liu, P.; Luo, S.; Zhou, L. Investigations on femtosecond laser-induced surface modification and periodic micropatterning with anti-friction properties on Ti_6Al_4V titanium alloy. *Chin. J. Aeronaut.* **2022**, *35*, 521–537. [CrossRef]
70. Abdulkareem AlMashhadani, H.; Khadom, A.A.; Khadhim, M.M. Effect of Polyeugenol coating on surface treatment of grade 23 titanium alloy by micro arc technique for dental application. *Results Chem.* **2022**, *4*, 100555. [CrossRef]
71. Zhang, L.-C.; Liu, Y.; Li, S.; Hao, Y. Additive Manufacturing of Titanium Alloys by Electron Beam Melting: A Review. *Adv. Eng. Mater.* **2018**, *20*, 1700842. [CrossRef]
72. Ren, Z.; Li, Z.; Zhou, S.; Wang, Y.; Zhang, L.; Zhang, Z. Study on surface properties of Ti_6Al_4V titanium alloy by ultrasonic rolling. *Simul. Model. Pract. Theory* **2022**, *121*, 102643. [CrossRef]
73. Du, J.; He, Q.; Chen, R.; Liu, F.; Zhang, J.; Yang, F.; Zhao, X.; Cui, X.; Cheng, J. Rolling reduction-dependent deformation mechanisms and tensile properties in a β titanium alloy. *J. Mater. Sci. Technol.* **2022**, *104*, 183–193. [CrossRef]
74. Pu, D.; Chen, X.; Ding, Y.; Sun, Y.; Feng, B.; Zheng, K.; Pan, F. Effect of Ti particles size on the microstructure and mechanical properties of TiP/VW94 composites. *Mater. Sci. Eng. A* **2022**, *858*, 144140. [CrossRef]
75. Wang, N.; Zhu, J.; Liu, B.; Zhang, X.; Zhang, J.; Tu, S. Influence of Ultrasonic Surface Rolling Process and Shot Peening on Fretting Fatigue Performance of Ti_6Al_4V. *Chin. J. Mech. Eng.* **2021**, *34*, 90. [CrossRef]
76. Zhang, Y.-L.; Lai, F.-Q.; Qu, S.-G.; Liu, H.-P.; Jia, D.-S.; Du, S.-F. Effect of ultrasonic surface rolling on microstructure and rolling contact fatigue behavior of 17Cr2Ni2MoVNb steel. *Surf. Coat. Technol.* **2019**, *366*, 321–330. [CrossRef]
77. Klimenov, V.A.; Vlasov, V.A.; Borozna, V.Y.; Klopotov, A.A. Ultrasonic Surface Treatment of Titanium Alloys. The Submicrocrystalline State. *IOP Conf. Ser. Mater. Sci. Eng.* **2015**, *91*, 012040. [CrossRef]
78. Li, L.; Sun, L.; Li, M. Diffusion bonding of dissimilar titanium alloys via surface nanocrystallization treatment. *J. Mater. Res. Technol.* **2022**, *17*, 1274–1288. [CrossRef]

79. Urtekin, L.; Aydın, Ş.; Sevim, A.L.I.; GÖK, K.; Uslan, İ. Experimental Determination of Biofilm and Mechanical Properties of Surfaces Obtained by CO_2 Laser Gas-Assisted Nitriding of Ti_6Al_4V Alloy. *Surf. Rev. Lett.* **2022**, *29*, 2250154. [CrossRef]
80. Urtekin, L.; Keleş, Ö. Investigation of Mechanical Properties of TiN-coated Ti_6Al_4V Alloy for Biomedical Applications. *J. Def. Sci.* **2019**, *18*, 91–108.

Disclaimer/Publisher's Note: The statements, opinions and data contained in all publications are solely those of the individual author(s) and contributor(s) and not of MDPI and/or the editor(s). MDPI and/or the editor(s) disclaim responsibility for any injury to people or property resulting from any ideas, methods, instructions or products referred to in the content.

MDPI
St. Alban-Anlage 66
4052 Basel
Switzerland
www.mdpi.com

Coatings Editorial Office
E-mail: coatings@mdpi.com
www.mdpi.com/journal/coatings

Disclaimer/Publisher's Note: The statements, opinions and data contained in all publications are solely those of the individual author(s) and contributor(s) and not of MDPI and/or the editor(s). MDPI and/or the editor(s) disclaim responsibility for any injury to people or property resulting from any ideas, methods, instructions or products referred to in the content.

www.ingramcontent.com/pod-product-compliance
Lightning Source LLC
LaVergne TN
LVHW070416100526
838202LV00014B/1470